Applied Mathematics

Applied Mathematics

Editor: Frank West

NY RESEARCH PRESS
New York

Published by NY Research Press
118-35 Queens Blvd., Suite 400,
Forest Hills, NY 11375, USA
www.nyresearchpress.com

Applied Mathematics
Edited by Frank West

International Standard Book Number: 978-1-63238-558-1 (Hardback)

Cataloging-in-Publication Data

Applied mathematics / edited by Frank West.
p. cm.
Includes bibliographical references and index.
ISBN 978-1-63238-558-1
1. Mathematics. I. West, Frank.
QA36 .A67 2017
510--dc23

Printed in the United States of America.

Contents

Preface

Every book is initially just a concept; it takes months of research and hard work to give it the final shape in which the readers receive it. In its early stages, this book also went through rigorous reviewing. The notable contributions made by experts from across the globe were first molded into patterned chapters and then arranged in a sensibly sequential manner to bring out the best results.

Applied mathematics seeks mathematical solutions to problems that are encountered in the real world. This field has progressed rapidly over the past two decades and its applications are finding their way across multiple areas such as computer science, business, engineering, etc. The various advancements in applied mathematics are glanced at and their applications as well as ramifications are looked at in detail. Topics included in this book encompass the multi-disciplinary aspect of this field and its various applications in numerical and computational modeling. This book will help new researchers by foregrounding their knowledge in this branch. It will prove useful for students and teachers in the fields of game theory, mathematical physics and industrial mathematics.

It has been my immense pleasure to be a part of this project and to contribute my years of learning in such a meaningful form. I would like to take this opportunity to thank all the people who have been associated with the completion of this book at any step.

Editor

A generalised likelihood uncertainty estimation mixed-integer programming model: Application to a water resource distribution network

Godfrey Chagwiza[1*], Brian C. Jones[1], Senelani D. Hove-Musekwa[1] and Sobona Mtisi[2]

*Corresponding author: Godfrey Chagwiza, Department of Applied Mathematics, National University of Science & Technology, P.O. Box AC939, Ascot, Bulawayo, Zimbabwe
E-mail: chagwizag@gmail.com

Reviewing editor: Yong Hong Wu, Curtin University of Technology, Australia

Abstract: A framework for incorporating uncertainty in water distribution models that uses generalised likelihood unbiased estimation (GLUE) and mixed-integer programming (MIP) is proposed and applied to a small water distribution system. Model parameters with uncertainty are first modelled in GLUEWIN and the mean estimates are used instead of single-point estimates. The MIP is solved in GAMS using CPlex solver with single-point estimates and then GLUE-generated estimates. There is a large difference between the results from GLUEMIP and those of the general MIP. It is therefore recommended that GLUEMIP framework be used so as to avoid penalties associated with failure to meet demand and for better planning.

Subjects: Advanced Mathematics; Discrete Mathematics; Mathematics & Statistics; Operational Mathematics; Science

Keywords: water resources; uncertainty modelling; allocation problem

ABOUT THE AUTHORS

Godfrey Chagwiza is a PhD student at National University of Science and Technology, Zimbabwe. His research interests focus on optimisation, mathematical modelling and mathematical programming.

Brian C. Jones is a professor emeritus of Applied Mathematics, National University of Science and Technology, Zimbabwe. His research interests include modelling inflow of material into black holes, as well as outflow under radiation pressure and magnetic forces, implications of spherical/cylindrical symmetries on observational parameters and protean operations research problems.

Senelani D. Hove-Musekwa is an associate professor of Applied Mathematics, National University of Science and Technology, Zimbabwe. Her research interests include application of mathematics to the functioning of the human body, the interaction of the human body with infectious pathogens and the economic epidemiology.

Sobona Mtisi is a senior research officer, Water Policy Programme, Rural Policy and Governance Group. His research interests include socio-economic, political and institutional analyses of water resources governance.

PUBLIC INTEREST STATEMENT

This paper provides an insight of how uncertainty in future water demand, annual capital and operational costs affect water resource distribution systems. Modelling of uncertainty is an important aspect of water distribution networks. Use of underestimated input parameters into the water resource distribution models results in water authorities making wrong decisions on water budgets. If water authorities do not consider the uncertainties that exist in water distribution systems, it is likely that they will face some challenges in future mainly due to shortfalls in funds that are needed for effective and efficient water resource distribution systems. Water authorities have high chances of incurring high penalties associated with failure to meet water demand because the difference between optimum costs obtained using single-point estimates of input parameters mentioned earlier on and that of incorporating uncertainties is much large. It is rather better to prepare water budgets using results of optimisation models that incorporate uncertainties.

1. Introduction

The aim of the paper is to introduce generalised likelihood uncertainty estimation (GLUE) to mixed-integer programming (MIP) to model uncertainty and produce a model for a water distribution network. MIP is the generalisation of linear programming (LP) and sometimes called mixed-integer linear programming (MILP). It has a linear objective function and constraints. MIP has the strength of its expressiveness achieved through mixing LP and integer variables. MIP problems belong to NP hard problems (Garey & Johnson, 1979). MIP is defined as follows

$$Z = C^T x$$

subject to

$$\begin{aligned} & Ax \leq b \\ & l \leq x \leq u \\ & x \in \mathbb{R}^n \\ & x_j \in \mathbb{Z};\ \forall j \in I \end{aligned} \tag{1}$$

where $C \in \mathbb{Q}$ is the cost vector, $A \in \mathbb{Q}^{m \times n}$ is the constraint matrix, $b \in \mathbb{Q}^m$ is the right-hand side, and $l \in (\mathbb{Q} \cup \{-\infty\})^m$ and $u \in (\mathbb{Q} \cup \{\infty\})^m$ are the bounds of the variables which are treated differently from other constraints. Integer variables x_j with the bounds $0 \leq x \leq 1$ are called binary variables.

The advantage of using MIP is that it is powerful when representing decision- making problems under uncertainty. The weakness of MIP is that it lacks the ability to reason about objects, classes of objects and relations. Every combination of discrete solutions is not examined explicitly by MIP, but it examines a subset of possible solutions and uses optimisation theory to prove that no other solution can be better than the one found (Smith & Taskin, 2008). MIP can be used for planning (Vossen, Ball, & Smith, 1999), task allocation under uncertainty (Alighanbari & How, 2005) and reason about uncertainty due to action of agents in a nonzero- game (Sandholm, Gilpin, & Conitzer, 2005).

Gonen and Foote (1981) used MIP to optimise a distribution system and the advantage of using this approach was that they linearised a nonlinear cost function. Some nonlinear problems can be transformed into MIP (Garfinkel & Nemhauser, 1972; Kallrath & Wilson, 1997; Williams, 1993) and this property of MIP makes it a good tool to model water resource allocation.

MIP can be used to schedule irrigation canals (Anwar & Clarke, 2001; de Vries & Anwar, 2004; Suryavanshi & Reddy, 1986). Ramesh, Venugopal, and Karunakaran (2009) used MIP for daily scheduling of laterals from a canal taking into account the constraints of the system and they found MIP as an efficient and simple tool to use. Many researches that focus on optimisation of water distribution concentrate on design of optimised configurations of pipe-interconnected reservoirs (Biscos, Mulholland, Le Lann, Buckley, & Brouckaert, 2003). Research has been focusing on the optimisation of new networks with very little focus on developing techniques for operational optimisation of existing water infrastructure. The main targets of such optimisation are to minimise pumping costs, supply and leak prevention (Cembrano, Wells, Quevedo, Pérez, & Argelaguet, 2000).

A research carried out by Creasey (1988) reviewed the appropriate mathematical techniques in order to solve the problem of operational optimisation of a water network distribution. Creasey concluded that £10 million per year can be saved through integer programming in the UK. Chow (1995) formulated a simulation model into a network for water resources management.

Previous research addressed three types of uncertainties and these are: uncertainty of model parameter values, structural uncertainty and uncertainty originating from error of mathematical relations to describe the physical reality (Freni, Mannina, & Viviani, 2009; Mannina, 2005; Refsgaard, Van der Sluijs, Højberg, & Vanrolleghem, 2007; Willems, 2000). Sen and Higle (1999) presented a tutorial of stochastic linear programming optimisation under uncertainty. They presented two-stage recourse models where some decision variables are used before the outcome of the random variables and some implemented after the outcome of the random variables. This method has a problem of inflexibility due to uncertainty and these were formulated as an extension of deterministic models.

Dye, Stougle, and Tomasgard (2003) introduced a stochastic network optimisation problem with an uncertain demand as an objective function. The researchers concluded that if one uses a limited number of instances, one can develop a heuristic with a constant worst-case performance ratio. Bertsimas and Sim (2003) suggested a robust optimisation approach. This method allowed them to control how conservative their solution was through setting probabilistic bounds on the constraint violations. This method was specifically for discrete optimisation and network flow problems. Sample-path technique was developed and applied to a network design problem to optimise a problem with both the objective and constraints being stochastic, where supply and demand are uncertain (Gurkan, Ozge, & Robinson, 1999).

There are some weaknesses of formulating a MIP to solve water distribution problems. The first weakness is that MIP assumes deterministic parameters, that is, known costs of supply and demand. This is not ideal for water distribution network optimisation where uncertainties are present. These uncertainties can be from unknown actual future demand and cost of building infrastructure, i.e. price of building materials changes. The other problem is that of changing operational costs. Labour and energy costs can change from time to time and also political constraints cannot be predetermined. Heuristics and mathematical programming have the same weakness of deterministic parameters. Due to the presence of uncertainty in water distribution problems, solving a MIP that uses deterministic input of demand and costs is not a better method of modelling real-world water problems. This research seeks to introduce a new way of incorporating uncertainty in a water distribution network model. GLUE will model uncertainty for the parameters and the result will be used to compute MIP instead of using single estimates of the model parameters. The research is motivated by the success of GLUE to solve problems with uncertainty and the strength of MIP when solving combinatorial problems and availability of MIP solvers such as Cplex.

The remainder of the paper is arranged as follows. In Section 2, we develop a model for water distribution systems without incorporating uncertainty. We develop a mathematical model in Section 3 that incorporates uncertainty. In Section 4, we implement the model to a numerical example using the GLUEMIP methodology and compare the results with those of the commercial Cplex solver and in Section 5, conclusions are drawn.

2. Model development

Let us denote the water resource network by $G = [N, E]$ where N is the node (the demand points) and surface water sources. E represents the set of arcs (actual and potential connections i.e. pipes and canals). Our objective is to minimise the total operational and installation costs of the water distribution network.

Define the model variables as

a_{ijt} - the flow along arc (i,j) at time period t,

v_{ijt} - additional capacity along arc (i,j) at time period l,

x_{it} - indicate if an additional supply is installed at node i and it is binary at time period t,

y_{it} - percentage supply capacity to be used at node i at time period t.

Define the model parameters as

d_{it} - demand at each demand node i at time period t,

k^*_{it} - potential plus current supply at each supply node i at time period t,

u_{ijt} - current pipe capacity at time period t,

u^*_{ijt} - current plus additional pipe capacity at time period t,

e^d_{it} - annual capital cost of installing maximum capacity at each supply node (cost of building new water plants which depends on whether or not a new plant has been built at node $i(x_i)$) at time period t,

c^d_{it} - annual operating costs of supply costs (operating, maintenance and environmental costs) which depend on the amount of water supplied from node i expressed as percentage of nodes maximum capacity y_i at time period t,

e^u_{ijt} - annual capital cost of adding capacity to pipelines (cost of building additional capacity) which depends on how much new capacity has been added (v_{ij}) between node i and j to accommodate the new required capacity at time period t,

c^u_{ijt} - annual operating cost of pipelines (includes unit operating, maintenance and environmental costs) which depends on the flow between node i and j (a_{ijt}) at time period t.

2.1. Objective function

$$\min \sum_{(ij)\in E,\ t\in T} (e^u_{ijt})v_{ijt} + c^u_{ijt}a_{ijt} + \sum_{(i)\in N,\ t\in T} e^d_{it}x_{it} + c^d_{it}y_{it} \tag{1}$$

The first part of the objective function represents cost of water transmission pipes and the second represents cost of water supply nodes (surface and ground supply points only).

2.2. Flow balance constraint

This enforces that the flow of water in less flow out should be equal to the water supply or demand at that node. It can be presented as

$$\sum_{j:(i,\,j)\in E,\ t\in T} a_{ijt} - \sum_{j:(j,\,i)\in E,\ t\in T} a_{jit} = d_{it} + k^*_{it}y_{it},\ i\in N,\ t\in T \tag{2}$$

2.3. Capacity on the arcs constraint

We have a constraint on the capacity of each water source. Flow along arc i,j should be less than or equal to current pipe capacity added to current plus additional pipe capacity times its additional capacity needed along arc i,j at time period t. We can present it as

$$a_{ijt} \le u_{ijt} + u^*_{ijt}v_{ijt},\ (i,j)\in E, t\in T \tag{3}$$

2.4. Supply capacity—supply installed constraint

Percentage supply capacity to be used at node i should be less than or equal to additional supply installed at node i at time period t and is shown by Section 4 as

$$y_{it} \le x_{it},\ i\in N,\ t\in T \tag{4}$$

2.5. Sign restriction constraints

$$v_{ijt} \in \{0,1\},\quad \{i,j\}\in E,\ t\in T \tag{5}$$

$$x_{it} \in \{0,1\},\quad i\in N,\ t\in T \tag{6}$$

$$0 \le y_{it} \le 1,\quad i\in N,\ t\in T \tag{7}$$

The resultant MIP is a combination of Equations 1–7 and is

$$\min Z = \sum_{(i,j)\in E,\ t\in T} (e^{u}_{ijt})v_{ijt} + c^{u}_{ijt}a_{ijt} + \sum_{i\in N,\ t\in T} e^{d}_{it}x_{it} + c^{d}_{it}y_{it} \tag{8}$$

subject to

$$\sum_{j:(i,j)\in E,\ t\in T} a_{ijt} - \sum_{j:(j,i)\in E,\ t\in T} a_{jit} = d_{it} + k^{*}_{it}y_{it};\ \forall i \in N,\ \forall t \in T \tag{9}$$

$$a_{ijt} \le u_{ijt} + u^{*}_{ijt}v_{ijt},\ \forall(i,j) \in E,\ \forall t \in T \tag{10}$$

$$y_{it} \le x_{it},\ \forall i \in N,\ \forall t \in T \tag{11}$$

$$v_{ijt} \in \{0,1\},\ \forall\{i,j\} \in E,\ \forall t \in T \tag{12}$$

$$x_{it} \in \{0,1\},\ \forall i \in N,\ \forall t \in T \tag{13}$$

$$0 \le y_{it} \le 1,\ \forall i \in N,\ \forall t \in T \tag{14}$$

3. Uncertainty

Instead of using single-point estimates of model parameters, it is however better to use GLUE for it generates a probability distribution for each parameter. GLUE adopts the idea of equifinality of models, parameters and variables and rejects the idea of model optimisation in favour of retaining a set of behavioural or acceptable models (Beven & Binley, 1992). Equations 8–14 form a MIP which assumes deterministic parameters. In this water distribution problem, uncertain future demand of water is one of the MIP parameters. Uncertainty exists as to actual future demand of water, future costs of infrastructure and operational costs.

3.1. The GLUE methodology

Equifinality was introduced long back by von Bertalanffy (1968) to mean that a system can reach the final stage from different initial conditions. In this case, equifinality exists in water resource distribution systems as to water demand, operational and capital costs. The GLUE methodology assumes that all model parameter sets have equal likelihood of being accepted. Firstly, we identify parameters that most affect output. In this research, future water demand, operational and infrastructure costs were identified. Generation of a high number of parameter sets is done by incorporating prior knowledge about the distribution of the parameters. Sensitivity analysis of the identified parameters is carried out for each of the sets and the results are compared to the observed data of the annual maximum peak of each of the parameters (see Cameron, Beven, & Naden, 2000).

We assess the performance of each trial through calculating the likelihood as in Equation 36. All the parameter sets and corresponding models that reach a minimum threshold are retained. The cumulative distribution function (cdf) is obtained using the retained solutions. The median of the output distribution and the corresponding uncertainty are obtained from the cdf. The median estimate after a considerable number of runs at a specific confidence level is then used as parameter input into the MIP. In this research, only 100 parameter sets are averaged to produce a single decision. These several decisions are significant as they assist in finding a better estimate. These decisions are different from the conventional decision of using observed data and can lead to an improvement in the resultant model output. It has been found out that water resource distribution system models are behavioural, that is, they fit model parameter values that are widely dispersed in the parameter space. Problems of uncertainty and limitations on parameter value predictions of water resource distribution systems such as those mentioned earlier can be reduced or solved through GLUE methodology. In a study of Bunea and Bedford (2002) it was shown that a model that captures uncertainty has substantial impact when optimising costs. To ensure accurate model predictions, proper estimation of model parameters is vital (Makowski, Wallach, & Tremblay, 2002).

The GLUE methodology has the ability of searching sets of parameter values that would give reliable simulations of model inputs. This solves the problem of searching for an optimum parameter set (Candela, Noto, & Aronica, 2005) and it is an ideal scenario in water resource distribution systems with uncertainty. Assumption of distributions of parameters is the main weakness of the GLUE methodology, but the problem can be solved by adopting formal distributions and this improves the method (Romanowicz & Beven, 1998). Notably, GLUE differs from the Monte Carlo simulation in that the GLUE-generated parameter distributions account for covariance implicitly, whereas the latter does not.

3.2. Modelling population growth rate

It is important to consider the population growth rate of the area under the study to deal with uncertain future water demand. Growth rate can be modelled as

$$Q_t = \frac{R_t - R_{t-1}}{R_{t-1}}$$

where Q_t is the growth rate and R_t is the population size at time period t. Instead of using the mean growth rate estimate, we can apply second-order dynamic linear model (2nd DLM) to the data to calculate the growth rate as a normal distribution with a mean and variance. West and Harrison (1997) introduced DLMs to Bayesian statistics and found that 2nd DLMs are useful for describing time series trends and sufficient for short-term forecasting. The resultant 2nd DLM is

$$Q_t = \eta_{1,t} + \epsilon_t,\ \epsilon_t \sim N(0, \sigma_\epsilon^2) \tag{15}$$

$$\eta_{1,t} = \eta_{1,t-1} + \eta_{2,t} + \psi_{1,t} \tag{16}$$

$$\eta_{2,t} = \eta_{2,t-1} + \psi_{2,t} \tag{17}$$

$$\psi_t = (\psi_{1,t}, \psi_{2,t})' \sim N(0, \sigma_\psi^2) \tag{18}$$

where $\eta_{(.),t}$ is the mean growth rate at time t.

Per capita water demand for each person in a year can be estimated in order to find future water demand. Water consumption data can then be fitted to a normal distribution. After forecasting the future population and water demand, the demand can be allocated to demand points in the model Equations 29–35. Population distribution is unlikely to be the same in different parts of any region, as some areas will gain while others decline in population. Therefore, water demand is not going to be exactly proportional to the population of each part of the region. It is important to generate variability in the problem. We can predict the population by multiplying the current population at time t by the growth rate Q_t and add the output to the population at t. The demand per day is presented by Equation 19, where 50 is the basic amount of litres of water per day per person according the World Health Organisation standards (Madden & Carmichael, 2007). The value of d_{it} is the one that is used to construct the posterior distribution in GLUEWIN which results in $d_{\lambda it}$, where

$$d_{it} = 50(Q_t \times R_t + R_t) \tag{19}$$

The Markov chain Monte Carlo (MCMC) allows us to design a Markov chain so that the stationary distribution of the chain will be exactly the distribution that one is interested in sampling from and which converges to a posterior probability distribution (Gelman, Carlin, Stren, & Rubin, 1995). MCMC method in combination with GLUE improves effectiveness and efficiency of GLUE (Blasone et al., 2008). Metropolis sampling method of MCMC distribution can be formulated creating a Markov chain that produces a sequence of values and these reflect the samples from the target distribution.

Let $P(\theta)$, $-\infty < \theta < \infty$, be our sample space. A Markov chain is created as $\theta^{(1)}, \theta^{(2)}, \ldots, \theta^{(t)}, \cdots$ where $\theta^{(t)}$ is the Markov chain at time t. We initialise $\theta^{(1)}$ the first state to some value. We generate θ^{*}, the candidate point which is conditional to the previous sampler using the proposed distribution $f(\theta|\theta^{(t-1)})$. The probability of accepting the proposal is given as

$$m_{\lambda t} = \frac{\left(1,\ P(\theta^{*})\right)}{P(\theta^{(t-1)})} \tag{20}$$

We generate a uniform deviate μ and if $\mu \leq m_{\lambda t}$, we accept the proposal and the following state will be $\theta^{(t)} = \theta^{*}$. If $\mu > m_{\lambda t}$, we reject the proposal and the next state will be $\theta^{(t)} = \theta^{(t-1)}$. The value of $m_{\lambda t}$ is used in Equation 22.

3.3. Modelling costs

Costs of adding new water sources, pipelines and additional operational costs result in uncertainty. Desalination, softening and pumping costs of water from new water sources are difficult to predict, therefore requiring uncertainty capturing forecasting methods. The annualised capital investment of a desalination plant can be obtained by multiplying the annuity factor to the investment cost as

$$e_{ijt}^{(.)} = \frac{i}{1-\left(\frac{1}{1+i}\right)^{n}} \tag{21}$$

where i is the interest rate and n is the lifetime of the new source. The value of $e_{ijt}^{(.)}$ is the one that is used to generate $e_{\lambda ijt}^{(.)}$ in GLUE which is used in Equation 29.

Inclusion of uncertainty in the model Equations 8–14 results in

$$\min Z = \sum_{(i,j)\in E,\ t\in T} e_{ijt}^{u} v_{ijt} + c_{ijt}^{u} a_{ijt} + \sum_{i\in N,\ t\in T} e_{it}^{d} x_{it} + c_{it}^{d} y_{it} + \sum_{t\in T} m_{\lambda t} h_{\lambda t} \tag{22}$$

subject to

$$\sum_{j:(i,j)\in E,\ t\in T} a_{ijt} - \sum_{j:(j,i)\in E,\ t\in T} a_{jit} = d_{it} + k_{it}^{*} y_{it};\ \forall i \in N,\ \forall t \in T \tag{23}$$

$$a_{ijt} \leq u_{ijt} + u_{ijt}^{*} v_{ijt},\ \forall (i,j) \in E,\ \forall t \in T \tag{24}$$

$$y_{it} \leq x_{it},\ \forall i \in N,\ \forall t \in T \tag{25}$$

$$v_{ijt} \in \{0,1\},\ \forall \{i,j\} \in E,\ \forall t \in T \tag{26}$$

$$x_{it} \in \{0,1\},\ \forall i \in N,\ \forall t \in T \tag{27}$$

$$0 \leq y_{it} \leq 1,\ \forall i \in N,\ \forall t \in T \tag{28}$$

where λ is an index denoting presence of uncertainty, $m_{\lambda t}$ is the probability at time period t and $h_{\lambda t}$ is the expected cost under uncertainty which is found by Equations 29–35.

$$h_{\lambda t} = \min Z = \sum_{(i,j)\in E,\ t\in T} (e_{\lambda ijt}^{u}) v_{ijt} + c_{\lambda ijt}^{u} a_{ijt} + \sum_{i\in N,\ t\in T} e_{\lambda it}^{d} x_{it} + c_{\lambda it}^{d} y_{it} \tag{29}$$

subject to

$$\sum_{j:(i,j)\in E,\ t\in T} a_{ijt} - \sum_{j:(j,i)\in E,\ t\in T} a_{jit} = d_{\lambda it} + k_{it}^{*} y_{it};\forall i \in N,\ \forall t \in T \tag{30}$$

$$a_{ijt} \leq u_{ijt} + u_{ijt}^{*} v_{ijt},\ \forall (i,j) \in E,\ \forall t \in T \tag{31}$$

$$y_{it} \leq x_{it},\ \forall i \in N,\ \forall t \in T \tag{32}$$

$$v_{ijt} \in \{0,1\},\ \forall \{i,j\} \in E,\ \forall t \in T \tag{33}$$

$$x_{it} \in \{0,1\},\ \forall i \in N,\ \forall t \in T \tag{34}$$

$$0 \leq y_{it} \leq 1,\ \forall i \in N,\ \forall t \in T \tag{35}$$

4. GLUEMIP methodology application examples

4.1. Example of a network that uses synthetic data

The first step is to estimate parameters that result in uncertainty with GLUE and later use the results from GLUE to solve the MIP Equations 8–14. Desalination and softening of water costs were analysed from three water sources. Means and variances were calculated after implementing MCMC method to forecast desalination and softening costs to generate more data. Normality tests were done to see if the parameters are normally distributed. Based on the results, a multivariate normal distribution was used for the desalination and softening costs. The annualised costs of desalination and softening were calculated by Equation 21 at 10% interest for the next 10 years for each new water source as predicted in the Bulawayo City Council (BCC) master plan (BCC, 2012). Figure 1 presents the flow chart of how GLUEMIP methodology is implemented. Water demand, annual operational costs and capital costs are input parameters that are considered to be influenced by uncertainty. We input the other parameters other than the ones mentioned in the MIP first, without using GLUE to find the mean estimates.

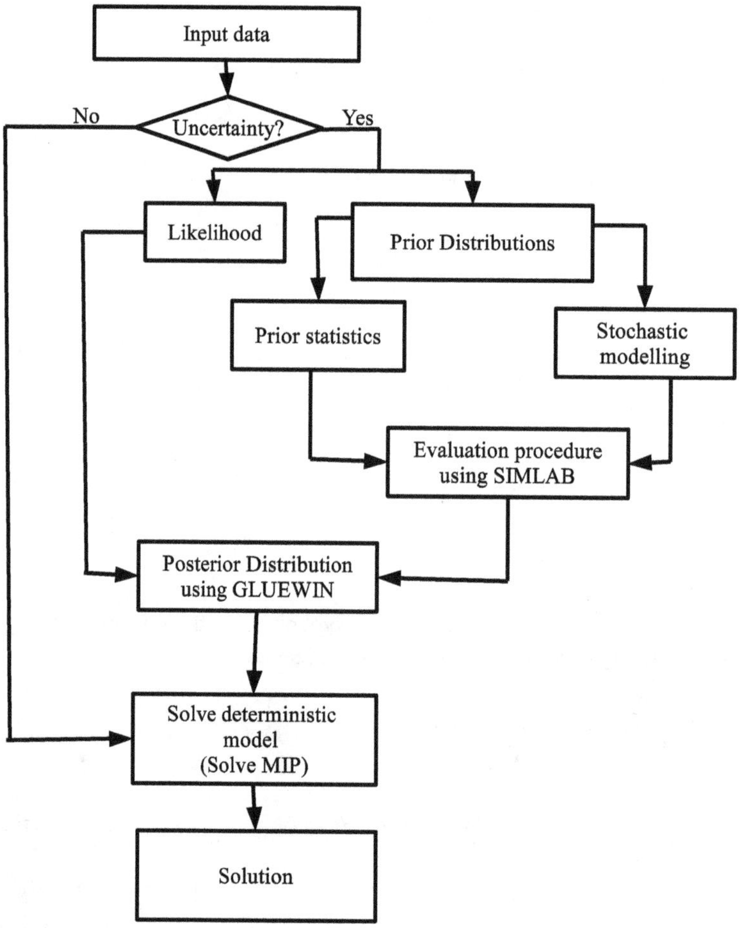

Figure 1. GLUEMIP flow chart.

Table 1. Uncertainty parameters

Parameter	Resultant uncertainty parameter	unit
d_{it}	$d_{\lambda it}$	Ω.
e^u_{ijt}	$e^u_{\lambda ijt}$	US$
e^d_{ijt}	$e^d_{\lambda ijt}$	US$
c^u_{ijt}	$c^u_{\lambda ijt}$	US$
c^d_{ijt}	$c^d_{\lambda ijt}$	US$

MATLAB was used to generate N multivariate normal realisation of parameters in Table 1. Demands at node level were fitted to a uniform Dirichlet distribution which allows "string-cutting" problem formulation after using per capita demand of an individual obtained using MCMC and annual population growth forecast obtained using Equations 15–18 as $b_i = \psi_i^* N_i^* \beta_i^*$ where ψ_i^* is the sample ith component from the water allocation distribution vector, N_i^* the sample from posterior predictive distribution of the population and β_i^* sample from posterior predictive distribution of the per capita consumption. Three runs of GLUE were done to ensure that smooth posterior distributions are obtained. The likelihood value, $L(\theta|R)$ was computed from generated observation, R, with two replicates for each variable, for each N and for θ_i, the generated parameter vector as

$$L(\theta|R) = \prod_{j=1}^{M} \frac{1}{\sqrt{2\pi\sigma_R^2}} \exp\left(-\frac{\left[R_j - P_j(\theta_i)\right]^2}{2\sigma_R^2}\right), \; i = 1, 2, \ldots, N \tag{36}$$

where θ_i is the ith parameter, R_j the jth observation of R, $P_j(\theta_i)$ the jth type of the model output of the set θ_i, σ_R^2 the variance of model errors and M is the number of observation replicates. Mean demand is generated using Equation 19.

A sample of 12,413 households of different socio-economic groups was extracted from BCC (2006) master plan (see Table 2). Parameter sets of 100,000 were generated for each of the components, demand, d_{it}, annualised capital cost, $e^{(.)}_{ijt}$ and annualised operational cost, $c^{(.)}_{ijt}$, in SIMLAB 2.2.1 software. SIMLAB is a software designed to solve problems using Monte Carlo-based uncertainty and sensitivity analysis (Saltelli, 2003). SIMLAB generates samples suitable for uncertainty and sensitivity analysis of the model components. The FAST sampling method was used which was introduced by (Saltelli, Tarantola, & Chan, 1999). The predicted mean values over 1,000 runs, probability, cumulative density distributions and variance were used to characterise prediction uncertainty with 90% confidence interval in GLUEWIN (see Figures 2–6). An explanation of the GLUEWIN methodology, implementation, data requirements and analysis can be found in Ratto and Saltelli (2001). Figures 2–6 show the mean values of the five parameters and these values are the ones that are passed on to the MIP.

GLUE estimates, $d_{\lambda it}$, $c^u_{\lambda ijt}$, $c^d_{\lambda ijt}$, $e^u_{\lambda ijt}$ and $e^d_{\lambda ijt}$, are then used in the main MIP. The MIP is solved in GAMS using CPlex solver. The same MIP is solved using single-point estimates of d_{it}, $e^{(.)}_{ijt}$ and $c^{(.)}_{ijt}$ in

Table 2. Household water consumption data

Residential area	No. of housing units	Average household size	Average household consumption (l/d)
Selbourne Park	830	6	200
Suburbs	698	6	200
Romney Park	457	7	150
Mahatshula	2,091	8	100
Sizinda	2,082	13	100
Emganwini	6,255	11	100

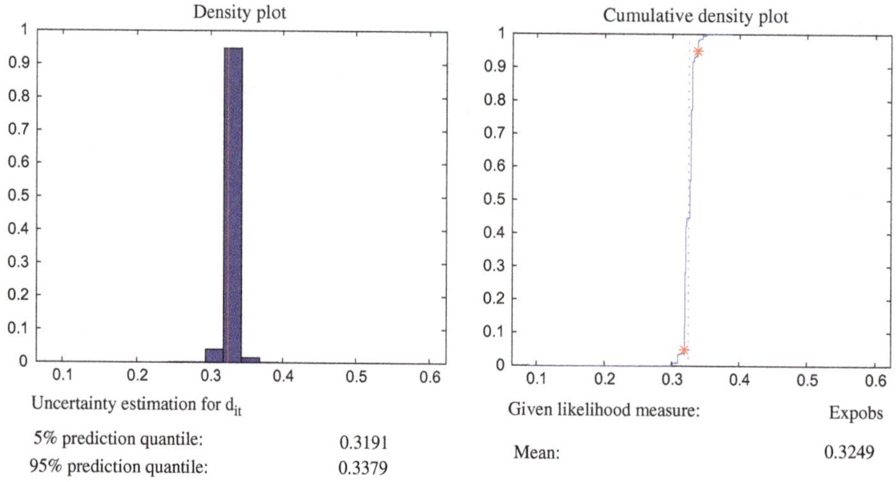

Figure 2. Average daily demand in 10, 000 m^3 per day.

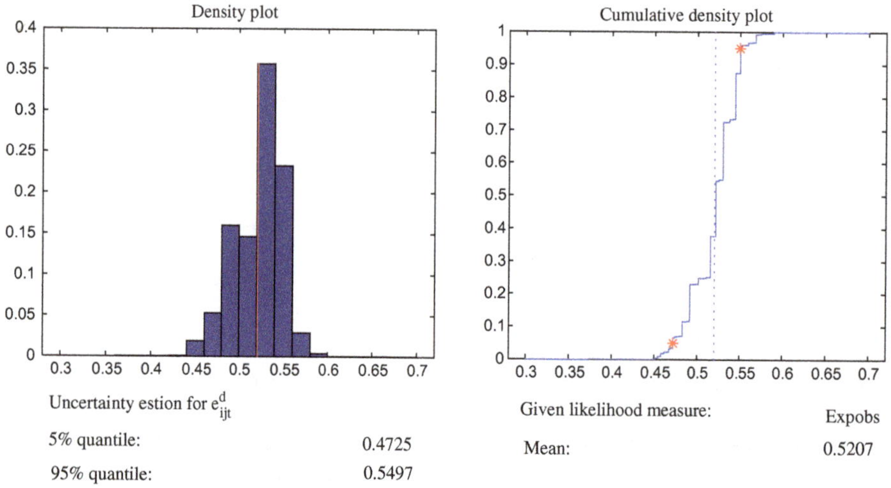

Figure 3. Average annual capital costs of building new water plants.

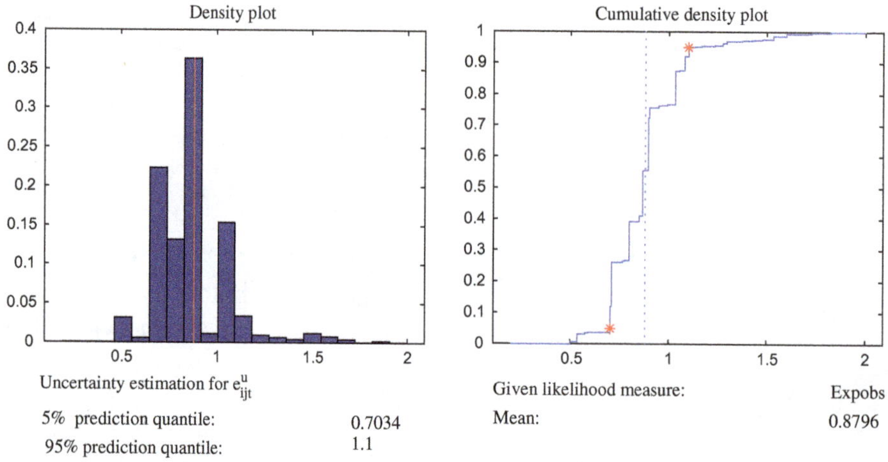

Figure 4. Average annual capital costs of adding capacity to pipelines.

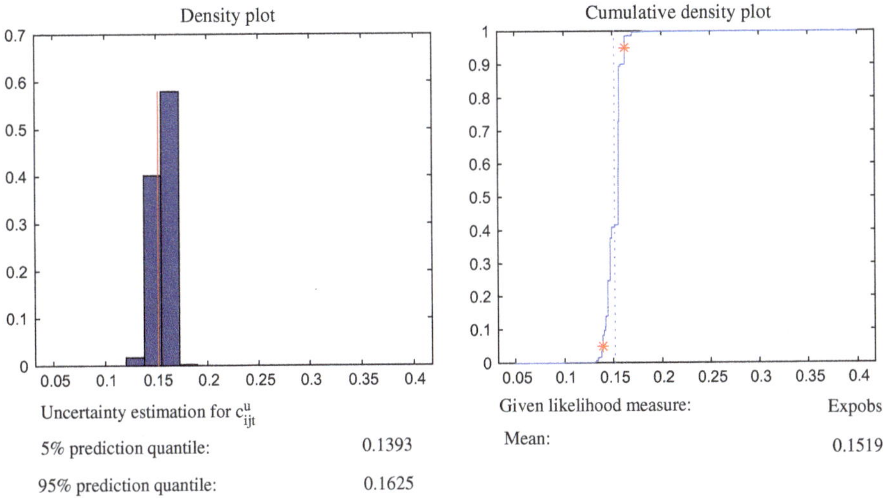

Figure 5. Average annual pipelines operating costs.

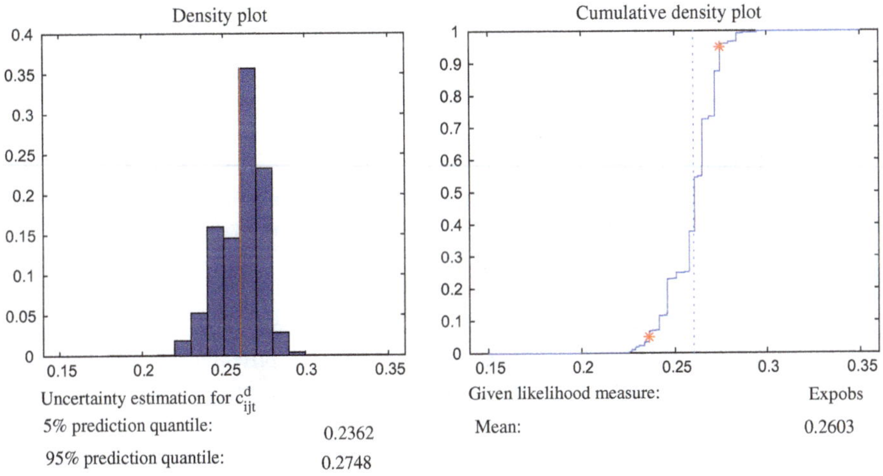

Figure 6. Average annual supply operating costs.

GAMS using CPlex solver, see Table 3 for results and parameter values that were adjusted through GLUE.

The results, in our numerical example, show a great difference in the total cost of water distribution using the two frameworks, that is, general MIP and GLUEMIP. The difference is a result of using underestimated parameter values. Usually, demand, annual capital and operational costs are underestimated if a model that does not incorporate uncertainty is used. The results are supported by that of Bunea and Bedford (2002) and Lal, Obeysekera, and van Zee (1997). The optimum solution is highly sensitive to

Table 3. Summary of input parameters and model results

Single-point estimate		GLUE mean estimate	
Parameter	**Value**	**Parameter**	**Value**
d_{it} (m³/ day)	293,084.72	$d_{\lambda it}$	324,900. 56
c^{u}_{ijt} (US$/m³)	0.11	$c^{u}_{\lambda ijt}$	0.15
c^{d}_{ijt} (US$/m³)	0.18	$c^{d}_{\lambda ijt}$	0.26
e^{u}_{ijt} (US$/m³)	0.79	$e^{u}_{\lambda ijt}$	0.88
e^{d}_{ijt} (US$/m³)	0.45	$e^{d}_{\lambda ijt}$	0.52
Optimum solution (US$/ day)	178,000	Optimum solution (US$/ day)	201,000

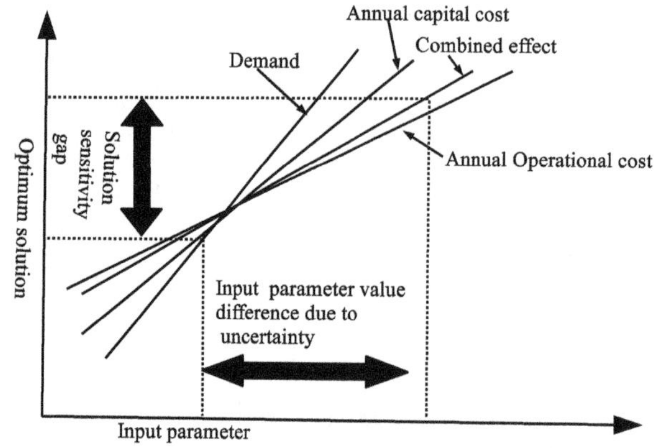

Figure 7. Sensitivity of optimum solution to water demand, annual capital and operational costs.

changes in the input parameter values (see Figure 7). The other reason for the large difference, besides underestimation of input parameters, is due to the fact that the GLUEMIP uses future annual capital and operational costs, while the general MIP uses present costs. The difference is large to the extent that if water authorities do not consider uncertainties in water distribution, it can lead to serious operational challenges. The water authorities will face challenges arising from penalties for failing to meet demand.

Second-order dynamic linear model seems to be an efficient tool for forecasting population growth as evidenced by a large difference between point estimate demand and the demand forecast obtained using predicted population growth, which is the most influential factor of predicting demand (Polebitski & Palmer, 2010). The drawback of using the second-order DLM, coupled with normal distribution of a water consumption series, to predict future water demand, which is subject to marked changes in the regime, is that it cannot represent the time series throughout longer periods. An analysis of the second-order DLM is done for 15 years to come up with the appropriate time series. The time series is limited to 10 years and after these years, the shift in water demand seems to be exaggerated. The observation of the performance of the second-order DLM, in our results, is supported by observations of Yelland and Lee (2003), who observed the DLM's results as compared to other forecasting techniques.

GLUEMIP proves to be a better method for forecasting future water distribution costs rather than using single-point estimates. In our results, the water authority may underestimate the costs by $23,000 and this high value slack is capable of halting the whole water resource distribution system for at least four days. GLUEMIP seems to be a significant method of finding optimal solutions of

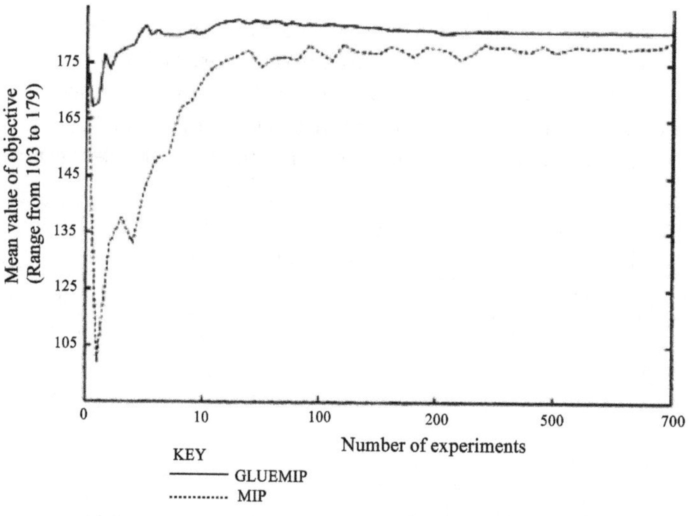

Figure 8. Experimental analysis of the GLUEMIP and MIP.

water resources management problems because it captures the important aspects in water resources management and planning. Water authorities plan for future water resource distribution aiming at meeting future water demand. The GLUEMIP methodology makes use of the important aspects of meeting future demand, that is, population growth and expansion of the water distribution networks by adjusting annual capital and operational costs. Figure 8 illustrates and compares the experiments that were carried out using different input values of water demand, annual capital and operational costs to both the GLUEMIP and general MIP. It is shown that the GLUEMIP methodology appears to be useful in optimising water resource distribution systems with uncertainties and its results are useful for better planning purposes.

4.2. Example of looped network using real data
This example uses real value data of water consumption, operational and maintenance costs. The network has 50 nodes and a total population of 1200 water consumers. The daily demand is $21,500\text{m}^3$. The daily operational and maintenance costs are $\$900.00$ and $\$575.00$, respectively. The procedure of solving this problem is carried out the same way as is in Section 4.1. The results show that there is a large difference between the results obtained by MIP and that of GLUEMIP. The optimum costs, using MIP, is $\$72,060.00$ per year, while it costs $\$75,700.00$/ year for the same water network using GLUEMIP. Again, the water authority will underestimate the costs by $\$3640.00$. This proves the superiority of the GLUEMIP methodology over the MIP.

5. Conclusion
GLUEMIP is a new way of incorporating uncertainty into the MIP. It seems a better way to model a water distribution network that is characterised with uncertainties arising from population growth and future operational costs. Incorporating new water sources in a water distribution network requires estimating uncertainties in pipe costs, desalination, softening and operational costs. Probability of the solution, component probabilities and the probabilities of unique sets of components can be offered through GLUEMIP output. In general, models incorporating uncertainties are preferred to deterministic models when modelling water distribution networks. The GLUEMIP results appear to be remarkably conditioned on the assumptions made in this study. Lastly, the results confirm the impact of uncertainty when optimising costs.

Funding
The authors received no direct funding for this research.

Author details
Godfrey Chagwiza[1]
E-mail: chagwizag@gmail.com
ORCID ID: http://orcid.org/0000-0002-6343-7125
Brian C. Jones[1]
E-mail: amabooksbyo@gmail.com
Senelani D. Hove-Musekwa[1]
E-mail: sdhmusekwa@gmail.com
ORCID ID: http://orcid.org/0000-0002-2194-6343
Sobona Mtisi[2]
E-mail: sobonamtisi@yahoo.co.uk
ORCID ID: http://orcid.org/0000-0002-1294-5779
[1] Department of Applied Mathematics, National University of Science & Technology, P.O. Box AC939, Ascot, Bulawayo, Zimbabwe.
[2] Overseas Development Institute, London, SE1 7JD, UK.

References
Alighanbari, M., & How, J. P. (2005). Cooperative task assignment of unmanned aerial vehicles in adversarial environments. In *Proceedings Institute of Electrical and Electronics Engineers (IEEE) American Control Conference* (pp. 4661–4667). Portland, OR.
Anwar, A. A., & Clarke, D. (2001). Irrigation scheduling using mixed-integer linear programming. *Journal of Irrigation and Drainage Engineering, 127*, 63–69.
Bertsimas, D., & Sim, M. (2003). Robust discrete optimization and network flows. *Mathematical Programming, Series B, 98*, 49–71.
Beven, K. J., & Binley, A. M. (1992). The future of distributed models: Model calibration and uncertainty prediction. *Hydrological Processes, 6*, 279–298.
Biscos, C., Mulholland, M., Le Lann, M. V., Buckley, C. A., & Brouckaert, C. J. (2003). Optimal operation of water distribution networks by predictive control using MINLP. *Water SA, 29*, 393–404.
Blasone, R. S., Vrugt, J. A., Madsen, H., Rosbjerg, D., Robinson, B. A., & Zyvoloski, G. A. (2008). Generalized Likelihood Uncertainty Estimation (GLUE) using adaptive Markov chain Monte Carlo sampling. *Advances in Water Resources, 31*, 630–648.
Bulawayo City Council. (2006). *Water report for the period January–December 2006. Annual Water Report*. Bulawayo, Zimbabwe.

Bulawayo City Council. (2012). *Water report for the period January–December 2012. Annual Water Report*. Bulawayo, Zimbabwe.

Bunea, C., & Bedford, T. (2002). The effect of model uncertainty on maintenance optimization. *IEEE Transactions on Realiability, 51*, 486–493.

Cameron, D. S., Beven, K. J., & Naden, P. (2000). Flood frequency estimation by continuous simulation under climate change (with uncertainty). *Hydrology and Earth System Science, 4*, 393–405.

Candela, A., Noto, L., & Aronica, G. (2005). Influence of surface roughness in hydrological response of semiarid catchments. *Journal of Hydrology, 313*, 119–131.

Cembrano, G., Wells, G., Quevedo, J., Pérez, R., & Argelaguet, R. (2000). Optimal control of a water distribution network in a supervisory control system. *Control Engineering Practice, 8*, 1177–1188.

Chow, N. H. (1995). A study on optimal management of water resources in Kaoping area Taiwan(I). In *Project Completion Representation to National Science Council*. Taipei, Taiwan, R.O.C.

Creasey, J. D. (1988). Pump scheduling in water supply: More than a mathematical problem. In B. Coulbeck & C. H. Orr (Eds.), *Computer application in water supply. Systems optimization and control*. (Vol. 2, pp. 279–289). Taunton: Research Studies Press Ltd.

de Vries, T. T., & Anwar, A. A. (2004). Irrigation scheduling 1: Integer programming approach. *Journal of Irrigation and Drainage Engineering, 130*, 9–16.

Dye, S., Stougle, L., & Tomasgard, A. (2003). The stochastic single resource service-provision problem. *Naval Research Logistics, 50*, 869–887.

Freni, G., Mannina, G., & Viviani, G. (2009). Assessment of data availability influence on integrated urban drainage modelling uncertainty. *Environment Model Software, 24*, 1171–1181.

Garey, M. R., & Johnson, D. S. (1979). *Computers and intractability: A guide to the theory of NP-completeness*. New York, NY: WH Freeman and Company.

Garfinkel, R. S., & Nemhauser, G. L. (1972). *Integer programming* (2nd ed.). New York, NY: John Wiley.

Gelman, A., Carlin, J. B., Stren, H. S., & Rubin, D. B. (1995). *Bayesian data analysis*. New York, NY: Chapmann and Hall.

Gonen, T., & Foote, B. L. (1981). Distribution system planning using mixed-integer programming. *Generation, Transmission and Distribution, IEEE Conference, 128*, 70–79.

Gurkan, G., Ozge, A. Y., & Robinson, S. M. (1999). Solving stochastic optimization problems with stochastic constraints: An application in network design. In P. A. Farrington, H. B. Nembhard, D. T. Sturrock, & G. W. Evans (Eds.), *Proceedings of the 1999 Winter Simulation Conference* (pp. 471–478). Washington, DC: Omnipress.

Kallrath, J., & Wilson, J. M. (1997). *Business optimisation using mathematical programming*. London: Macmillan.

Lal, W., Obeysekera, J., & van Zee, R. (1997). Sensitivity and uncertainty analysis of a regional simulation model for the natural system in South Florida. In *Managing water: Coping with scarcity and abundance*. San Francisco, CA.

Madden, C., & Carmichael, A. (2007). *Every last drop*. Sydney: Random House Australia.

Makowski, D., Wallach, D., & Tremblay, M. (2002). Using a bayesian approach to parameter estimation: Comparison of the glue and mcmc methods. *Agronomie, 22*, 191–203.

Mannina, G. (2005). *Integrated urban drainage modelling with uncertainty for stormwater pollution management* (PhD thesis). Universita di Catania, Italy.

Polebitski, A. & Palmer, R. (2010). Seasonal residential water demand forecasting for census tracts. *Journal of Water Resources Planning and Management, 136*, 27–36.

Ramesh, B. R., Venugopal, K., & Karunakaran, K. (2009). Zero-one programming model for daily operation scheduling for irrigation canal. *Journal of Agricultural Science, 1*, 13–20.

Ratto, M., & Saltelli, A. (2001). Model assessment in integrated procedures for environmental impact evaluation: Software prototypes. IMPACT project, deliverable 18, Joint Research Centre of European Commission, Institute for the Protection and Security of the Citizen. Ispra.

Refsgaard, J. P., Van der Sluijs, J. P., Højberg, A. L., & Vanrolleghem, P. A. (2007). Uncertainty in the environmental modelling process–A framework and guidance. *Environment Model Software, 22*, 1543–1556.

Romanowicz, R. & Beven, K. J. (1998). Dynamic real-time prediction of flood inundation probabilities. *Hydrology Science Journal, 43*, 181–196.

Saltelli, A. (2003). SIMLAB 2.2 manual, simulation environment for uncertainty and sensitivity analysis. SIMLAB 2.2 Manual, JRC/ POLIS ScaRL, SIMLAB.

Saltelli, A., Tarantola, S., & Chan, K. P. S. (1999). A quantitative model-independent method for global sensitivity analysis of model output. *Technometrics, 41*, 39–56.

Sandholm, T., Gilpin, A., & Conitzer, V. (2005). Mixed-integer programming methods for finding Nash equilibria. *American Association for Artificial Intelligence (AAAI)-05, National Conference on Artifcial Intelligence* (pp. 495–501). Pennsylvania: Pittsburgh.

Sen, S. & Higle, J. L. (1999). An introductory tutorial on stochastic linear programming models. *Interfaces, 29*, 33–61.

Smith, J. C., & Taskin, Z. C. (2008). A tutorial guide to mixed-integer programming models and solution techniques. In G. J. Lim & E. K. Lee (Eds.), *Optimization in Medicine and Biology*. Auerbach Publications: Taylor and Francis.

Suryavanshi, A. R. & Reddy, J. M. (1986). Optimal operation schedule of irrigation distribution systems. *Agricultural Water Management, 11*, 23–30.

von Bertalanffy, L. (1968). *General system theory*. New York, NY: Braziller.

Vossen, T., Ball, M., & Smith, R. H. (1999). On the use of integer programming models in AI planning. In *Sixth International Joint Conference on Artificial Intelligence (IJCAI)-99* (pp. 304–309). San Francisco, CA: Morgan Kaufmann.

West, M., & Harrison, J. (1997). *Bayesian forecasting and dynamic models. Springer series in statistics* (2nd ed.). New York, NY: Springer-Verlag.

Willems, P. (2000). Quantification and relative comparison of different types of uncertainties in sewer water quality modelling. *Water Research, 42*, 3539–3551.

Williams, H. P. (1993). *Model building in mathematical programming* (3rd ed.). Chichester: John Wiley.

Yelland, P. M., & Lee, E. (2003). *Forecasting product sales with dynamic linear mixture models*. SMLI TR-2003-122. Sun Microsystems Inc., Mountain View, CA.

2

Numerical approximations of Sturm–Liouville eigenvalues using Chebyshev polynomial expansions method

Uğur Yücel[1]*

*Corresponding author: Uğur Yücel, Faculty of Arts and Sciences, Department of Mathematics, Pamukkale University, Denizli 20070, Turkey
E-mail: uyucel@pau.edu.tr
Reviewing editor: Kok Lay Teo, Curtin University, Australia

Abstract: In this paper, an efficient technique based on the Chebyshev polynomial expansions for computing the eigenvalues of second- and fourth-order Sturm–Liouville boundary value problems is proposed. This technique reduces the given Sturm–Liouville problem to an integral equation. The resulting integral equation is then transformed into the eigenvalue equation by calculating the integrals at the grid points using the Chebyshev expansions. Thus, the required eigenvalues of the given problem are obtained by solving this eigenvalue equation. The excellent performance of this scheme is illustrated through some numerical examples, and comparison with other methods is presented.

Subjects: Applied Mathematics; Mathematics & Statistics; Science

Keywords: Boundary value problems; Chebyshev expansions; eigenvalues; Schrödinger equation; Sturm–Liouville problems

1. Introduction

Sturm–Liouville problems (SLPs) arise throughout engineering mathematics. They arise directly as eigenvalue problems in one space dimension. For example, they describe the vibrational modes of various systems, such as the vibrations of a string or the energy eigenfunctions of a quantum mechanical oscillator, in which case the eigenvalues correspond to the resonant frequencies of vibration or energy levels. They also commonly arise from linear PDEs in several space dimensions when the equations are separable in some coordinate system, such as cylindrical or spherical coordinates.

Mathematicians have studied SLPs for over 200 years. It has a highly developed theory and remains an active area of interest. Therefore, in this work, we consider numerical solutions of both second- and fourth-order SLPs described below.

ABOUT THE AUTHOR

Uğur Yücel is a professor in the Department of Mathematics at Pamukkale University, Denizli, Turkey. He received his PhD degree in Applied Mathematics from Lehigh University, Bethlehem, PA, USA. His research interests are focused on labyrinth seals in turbomachinery and numerical solutions of ordinary and partial differential equations.

PUBLIC INTEREST STATEMENT

"Sturm–Liouville problems" are famous boundary value problems that naturally arise when solving certain partial differential equation problems using a "separation of variables" method. Mathematicians have studied these problems for over 200 years. They have a highly developed theory and remain an active area of interest. In this work, a numerical approach is developed to obtain approximate eigenvalues of these problems.

A regular second-order SLP is a second-order linear, homogeneous, ordinary differential equation of the form

$$-\left(p(x)y'\right)' + q(x)y = \lambda w(x)y, \quad a < x < b, \tag{1}$$

together with the seperated (each with one boundary point) homogeneous boundary conditions

$$\alpha_1 y(a) + \alpha_2 y'(a) = 0, \quad \alpha_1^2 + \alpha_2^2 \neq 0$$

$$\beta_1 y(b) + \beta_2 y'(b) = 0, \quad \beta_1^2 + \beta_2^2 \neq 0 \tag{2}$$

where α_i and β_i, $i = 1, 2$, are the real constants. In Equation 1, λ is the unknown eigenvalue pertinent to the differential operator defined by the left-hand side of Equation 1, the interval (a, b) is finite, the functions p, p', q, and w are in $L^1(a, b)$ with $p(x) > 0$ and $w(x) > 0$ for $x \in (a, b)$. This is distinguished from the case when p or w vanishes at some point in the interval $[a, b]$ or when the interval is of infinite length, in which case the problem is called a singular SLP. For a regular second-order SLP, it is found that all eigenvalues are real and non-negative. In addition, there are infinitely many eigenvalues that are discretely distributed and hence do not fill out any interval. More information on the mathematical theory of second-order SLPs may be found in Amrein, Hinz, and Pearson (2005) and Pryce (1993).

A non-singular fourth-order SLP consists of a fourth-order linear ODE of the form

$$\left(P(x)y''\right)'' - \left(S(x)y'\right)' + Q(x)y = \lambda W(x)y, \quad a < x < b \tag{3}$$

together with seperated, self-adjoint boundary conditions specified at both ends of the domain (a, b), two boundary conditions at the end $x = a$, and another two boundary conditions at the end $x = b$. Basically, there are three types of boundary conditions and their combinations commonly used with Equation 3 in applications. These boundary conditions are given as $y = 0,\ y' = 0$ for clamped end, $y = 0,\ y'' = 0$ for hinged (or simply supported) end, and $y'' = 0,\ y'' = 0$ for free end. The technical conditions for the problem to be non-singular are: the interval (a, b) is finite; the functions $P(x)$, $S(x)$, $Q(x)$, $W(x)$, and $1/P(x)$ are in $L^1(a, b)$; and the essential infima of $P(x)$ and $W(x)$ are both positive. Under these assumptions, it is well known that the eigenvalues are bounded from below. They can be ordered as $\lambda_0 \leq \lambda_1 \leq \lambda_2 \leq \cdots \leq \lambda_k \leq \cdots$, where $\lambda_k \to \infty$ as $k \to \infty$, and where each eigenvalue has multiplicity at most 2. More information on the mathematical theory of fourth-order SLPs may be found in Greenberg (1991) and Greenberg and Marletta (1995).

In most cases, it is not possible to obtain the eigenvalues of the above problems analytically. However, there are various approximate methods as, for example, the weighted residual methods, the variational methods, and finite difference and finite elements methods. Extensive literature reviews for the numerical approximations of second-order and fourth-order SLPs can be found in Yücel (2006) and Yücel and Boubaker (2012), respectively. In order to obtain more efficient numerical results, several ways have been devised in the last years. Recently, El-gamel and Abd El-hady (2013) discussed and compared two useful schemes for second-order Sturm–Liouville eigenvalue problems: differential quadrature method (DQM) and collocation method with sinc functions. Saleh Taher, Malek, and Momeni-Masuleh (2013) proposed a new technique based on Chebyshev differentiation matrix for computing eigenvalues of a general class of regular fourth-order SLPs. Yuan, Ye, Xiao, Kennedy, and Williams (2014) introduced the exact dynamic stiffness vibration method for solving regular second- and fourth-order SLPs. Most recently, Amodio and Settanni (2015) discussed the solution of regular and singular SLPs by means of high-order finite difference schemes. They described a method to define a discrete problem and its numerical solution by means of linear algebra techniques.

In the literature, to the best of the author's knowledge, there is no study on the Chebyshev polynomial expansions (El-gendi's method) (El-gendi, 1969) applied for SLPs. On the other hand, this method is an efficient discretization technique for obtaining accurate numerical solutions of differential, integral, and integro-differential equations. El-gendi's method is based on expanding the unknown function in terms of Chebyshev polynomials. Based on the discrete ortogonality relationships of these polynomials, several methods for solving linear and non-linear ODEs (Fox, 1962) and integral differential equations (Elliott, 1963) were proposed at about the same time. They were found to have considerable advantage over the finite difference methods. Since then, these methods have become standard. They are now part of the larger family of spectral methods (Boyd, 2000). They consist in expanding the unknown function in a series of Chebyshev polynomials, trancating this series, and then substituting the approximation in the actual equation, and finally determining equations for the coefficients. However, El-gendi (1969) described a new method for the numerical solution of differential, integral, and integro-differential equations by computing directly the values of the functions rather than the Chebyshev coefficients. These two approaches are equivalent in the sense that if the function values at some grid points are known, the Chebyshev coefficents for the function can be directly computed. The method has been extended to solve initial value problems in time-dependent quantum field theory and second-order boundary value problems in fluid (Mihaila & Mihaila, 2002). Recently, it has been applied to solve the generalized Kuramoto–Sivashinsky equation (Khater & Temsah, 2008) and convection–diffusion equation (Temsah, 2009).

Our goal in this paper is to extend Chebyshev polynomial expansions, known as El-gendi's method (El-gendi, 1969), to deal with the second- and fourth-order SLPs given above. To achieve this goal, we first consider the numerical approximations of second-order SLP (1) with Drichlet boundary conditions

$$y(a) = 0, \quad y(b) = 0 \tag{4}$$

Then, we consider numerical approximations of a simple form of Equation 3 with some specified end conditions to demonstrate directly the use of El-gendi's method on the fourth-order problems. In addition, we choose the fourth-order problems (3) to be squares of the second-order problems (1) for numerical illustrations. This is because if we have a second-order problem with Drichlet boundary conditions given above and the differential operator,

$$\ell y = -\left(p(x)y'\right)' + q(x)y \tag{5}$$

then the corresponding fourth-order problem (3) has the differential operator

$$\ell^2 y = \left(P(x)y''\right)'' - \left(S(x)y'\right)' + Q(x)y \tag{6}$$

where

$$P = p^2, \quad S = 2pq - pp'', \quad Q = q^2 - pq'' - p'q' \tag{7}$$

The boundary conditions, in this case, are

$$\begin{aligned} y(a) = 0, &\quad p'(a)y'(a) + p(a)y''(a) = 0 \\ y(b) = 0, &\quad p'(b)y'(b) + p(b)y''(b) = 0 \end{aligned} \tag{8}$$

The eigenvalues of the fourth-order problem are then simply the squares of the eigenvalues of the second-order problem which means that they are all non-negative (Greenberg & Marletta, 1995).

This paper is organized as follows. In Section 2, we first review the Chebyshev polynomial expansions method (El-gendy's method), then we give some new formulas for the matrix approximation of integrals which are necessary to deal with higher order problems. The method is then applied to second-order and a simple form of the fourth-order SLPs in Section 3. Numerical examples are discussed in Section 4, and some conclusions are drawn in Section 5.

2. The method of Chebyshev polynomial expansions

In this section, we describe the method of Chebyshev expansion by following closely the procedures outlined in El-gendi (1969). It is assumed that a function $y(x)$, which is continuous and of bounded variation in the interval [-1, 1], can be approximated by

$$y(x) = \sum_{i=0}^{N}{}'' a_i T_i(x) \tag{9}$$

where the summation symbol with double prime indicates that the terms with suffixes $i = 0$ and $i = N$ are to be halved, $T_i(x)$ is the Chebyshev polynomials of the first kind of degree $i \leq N$, and the coefficients a_i are defined by

$$a_i = \frac{2}{N} \sum_{j=0}^{N}{}'' y(x_j) T_i(x_j), \quad i = 0, 1, \dots, N \tag{10}$$

In the equation above, the grid points x_j, the so-called Chebyshev points, are given by

$$x_j = \cos\left(\frac{j\pi}{N}\right), \quad j = 0, 1, \dots, N \tag{11}$$

The approximate formulae (Equation 9) is exact at $x = x_j$ given by Equation 11.

It is known by the Chebyshev trancation theorem (Boyd, 2000) that the error in approximating $y(x)$ by the sum of its first N terms is bounded by the sum of the absolute values of all the neglected coefficients. A detailed discussion of this theorem and the convergence theory for Chebyshev polynomials can be found in Boyd (2000).

Using the above approximation for the aforementioned function $y(x)$, the integral

$$\int_{-1}^{x} y(t)\,dt \quad , \quad -1 \leq x \leq 1 \tag{12}$$

at the grid points

$$x_k = -\cos\left(\frac{k\pi}{N}\right), \quad k = 0, 1, \dots, N \tag{13}$$

can be approximated as

$$\int_{-1}^{x} y(t)dt = \sum_{k=0}^{N}{}'' y(x_k) \frac{2}{N} \sum_{i=0}^{N}{}'' T_i(x_k) \int_{-1}^{x} T_i(t)dt \tag{14}$$

We can now write this equation in matrix form as

$$\left[\int_{-1}^{x} y(t)\,dt\right] = \mathbf{D}\,\mathbf{y} \tag{15}$$

where $\mathbf{D}$ is a square matrix of order $N + 1$ and the actual values of the elements of this matrix $\left(d_{ij},\ i = 0, 1, \dots, N,\ j = 0, 1, \dots, N\right)$ can be easily calculated using the right-hand side of Equation 14. This can be achieved using modern computing tools like Maple and Matlab. For $N = 4$, the elements of the matrix $\mathbf{D}$, obtained by Maple, are given below.

$$\mathbf{D} = \begin{bmatrix} 0 & 0 & 0 & 0 & 0 \\ 0.11940355 & 0.19006343 & -0.02426406 & 0.01328673 & -0.00559644 \\ 0.03333333 & 0.62022005 & 0.40000000 & -0.08688672 & 0.03333333 \\ 0.07226310 & 0.52004659 & 0.82426406 & 0.34326990 & -0.05273689 \\ 0.06666666 & 0.53333333 & 0.80000000 & 0.53333333 & 0.06666666 \end{bmatrix}$$

The equation $\sum_{j=0}^{N} d_{ij} = 1 - \cos\left(\frac{i\pi}{N}\right)$ for i = 0, 1, 2, ..., N can be used for checking purposes. The elements of the column vector **y** in Equation 15 are given by

$$y_i = y\left(-\cos\left(\frac{i\pi}{N}\right)\right), \quad i = 0, 1, \dots, N \tag{16}$$

It can be seen that the approximation of the integral (12) based on the Chebyshev expansion (9) can be obtained without computing the Chebyshev coefficients (Equation 10). In other words, the right-hand side of Equation 15 gives the approximate values of the integral (12) at the points (13), rather than the Chebyshev coefficients of the integral (10). However, the two approaches are equivalent in the sense that if we know the values of the integral at x_k given by Equation 13, its Chebyshev coefficients can be directly computed by using a formula similar to Equation 10. The main advantage of using this approach is that for a certain value of N, the elements of matrix **D** can be eveluated once and for all.

The following approximations which are necessary for practical applications to be discussed later can be deduced from relation (Equation 15).

$$\int_{-1}^{1} y(t)\, dt = \tilde{\mathbf{d}}\, \mathbf{y} \tag{17}$$

where $\tilde{\mathbf{d}}$ is an (N + 1) row vector whose elements are the last row of matrix **D**.

$$\left[\int_{-1}^{x}\int_{-1}^{x'} y(t)\, dt\, dx'\right] = \mathbf{E}\,\mathbf{y} \tag{18}$$

where **E** = $\mathbf{D}^2$.

$$\int_{-1}^{1}\int_{-1}^{x} y(t)\, dt\, dx = \tilde{\mathbf{e}}\,\mathbf{y} \tag{19}$$

where $\tilde{\mathbf{e}}$ is an (N + 1) row vector whose elements are the last row of matrix **E**.

$$\left[\int_{-1}^{x}\int_{-1}^{x'} \phi(t) y(t)\, dt\, dx'\right] = \hat{\mathbf{E}}\,\mathbf{y} \tag{20}$$

where $\hat{\mathbf{E}}$ is an (N + 1) × (N + 1) matrix whose elements are given by $\hat{e}_{ij} = \phi_j e_{ij}$, i, j = 0, 1, 2, ..., N, $\phi_j = \phi\left(t_j\right)$, and e_{ij} the elements of matrix **E**.

$$\left[\int_{-1}^{1}\int_{-1}^{x'} \phi(t) y(t)\, dt\, dx'\right] = \tilde{\mathbf{s}}\,\mathbf{y} \tag{21}$$

where $\tilde{\mathbf{s}}$ is an (N + 1) row vector whose elements are the last row of matrix $\mathbf{D}\hat{\mathbf{D}}$; $\hat{\mathbf{D}}$ is an (N + 1) × (N + 1) matrix whose elements are given by $\hat{d}_{ij} = \phi_j d_{ij}$.

$$\left[\int_{-1}^{x}\int_{-1}^{x'}\int_{-1}^{x''}\int_{-1}^{x'''} y(t)\,dt\,dx'''dx''\,dx'\right] = \mathbf{H}\,\mathbf{y} \tag{22}$$

where $\mathbf{H} = \mathbf{D}^4$.

$$\int_{-1}^{1}\int_{-1}^{x}\int_{-1}^{x'}\int_{-1}^{x''} y(t)\,dt\,dx''dx'dx = \tilde{\mathbf{h}}\,\mathbf{y} \tag{23}$$

where $\tilde{\mathbf{h}}$ is an $(N+1)$ row vector whose elements are the last row of matrix $\mathbf{H}$.

Standard Chebyshev polynomials of the first kind $T_r(x)$ are valid over the interval [−1, 1]. However, this interval can be normalized using the change of variable $x \to (x+1)/2$. In this case, the Chebyshev expansion of the integral will be in terms of the shifted Chebyshev polynomials of the first kind defined by

$$T_r^*(x) = T_r(2x-1), \quad 0 \le x \le 1 \tag{24}$$

which are valid over the interval [0, 1]. All the properties of $T_r^*(x)$ can be deduced from those of $T_r(2x-1)$ (Fox & Parker, 1968). Therefore, when the range is [0, 1], we have the approximation

$$\left[\int_{0}^{x} y(t)\,dt\right] = \bar{\mathbf{D}}\,\mathbf{y} \tag{25}$$

where $\bar{\mathbf{D}} = \frac{1}{2}\mathbf{D}$ and the right-hand side defines the integral at the grid points

$$x_k = \frac{1}{2}\left[1-\cos\left(\frac{k\pi}{N}\right)\right], \quad k = 0,1,\ldots,N \tag{26}$$

The elements of the column $\mathbf{y}$ in Equation 25 are also defined at these points. Hence, for practical applications, the operators from Equations 18 to 23 can be similarly defined by replacing the matrix $\mathbf{D}$ by $\bar{\mathbf{D}}$ and the points Equation 13 by Equation 26.

As can be seen from the above formulas, the range of x in the computational domain should either be [−1, 1] or [0, 1]. For practical applications, the physical domain is neither [−1, 1] nor [0, 1], but rather $[a, b]$, then for this case, we can perform a coordinate transformation. This can be easily done by remembering that any finite range, $a \le t \le b$, can be transformed to the basic range $-1 \le x \le 1$ and/or the range $0 \le x \le 1$ with the change of variables

$$t = \frac{1}{2}(b-a)\,x + \frac{1}{2}(b+a), \quad t = (b-a)\,x + a \tag{27}$$

3. Applications to SLPs

In this section, we consider the numerical approximations of second- and fourth-order Sturm–Liouville eigenvalues using the matrix representations of the integrals obtained above.

3.1. Second-order problem

The general SLP (1) can be easily reduced to the so-called Liouville normal form (or equivalently the Schrödinger equation)

$$-y''(t) + \tilde{q}(t)y(t) = \lambda\,y(t)\,, \quad \tilde{a} < t < \tilde{b} \tag{28}$$

where now all the properties of the original equation can be associated with the properties of the coefficient $\tilde{q}(\bullet)$ on the transformed interval $(\tilde{a}, \tilde{b})$. In this form, the properties of the original differential equation may be easier to consider for both analytical and numerical purposes. The endpoint classification at a, b of Equation 1 is invariant under the Liouville transformation and is identical with the endpoint classification at $\tilde{a}, \tilde{b}$. Therefore, we consider the above differential equation with the boundary conditions (4).

We first perform a coordinate transformation of the first form of Equation 27. Then, Equation 28 becomes

$$-y''(x) + \hat{q}(x)y(x) = \hat{\lambda}\, y(x)\,, \quad -1 < x < 1 \tag{29}$$

where $\hat{q}(x) = \left((\tilde{b} - \tilde{a})/2\right)^2 \tilde{q}(x)$ and $\hat{\lambda} = \left((\tilde{b} - \tilde{a})/2\right)^2 \lambda$. The boundary conditions, in this case, are

$$y(-1) = 0, \quad y(1) = 0 \tag{30}$$

It should be noted that translating the equation from the t domain to the x domain leaves the eigenvalues unaltered.

To apply the method developed in the previous section, we first integrate Equation 29 from the lower limit -1 to an arbitrary value of $x \in [-1, 1]$ to obtain

$$-y'(x) + y'(-1) + \int_{-1}^{x} \hat{q}(t)y(t)\,dt = \hat{\lambda} \int_{-1}^{x} y(t)\,dt \tag{31}$$

Similarly, integrating the resulting Equation 31 once more yields

$$-y(x) + (x+1)y'(-1) + \int_{-1}^{x}\int_{-1}^{t} \hat{q}(t')y(t')dt'dt \;=\; \hat{\lambda}\int_{-1}^{x}\int_{-1}^{t} y(t')\,dt'dt \tag{32}$$

where we have made use of the boundary condition $y(-1) = 0$. In this equation, $y'(-1)$ is an unknown that has to be determined by introducing boundary conditions (Equation 30). To do this, we first specialize Equation 32 for $x = 1$, together with the boundary conditions given by Equation 30. Then, we obtain

$$y'(-1) = -\frac{1}{2}\int_{-1}^{1}\int_{-1}^{t} \hat{q}(t')y(t')dt'dt + \frac{\hat{\lambda}}{2}\int_{-1}^{1}\int_{-1}^{t} y(t')\,dt'dt \tag{33}$$

We can now substitute the so obtained $y'(-1)$ into Equation 32 to obtain the final integrated form

$$\begin{aligned} &-y(x) - \frac{(x+1)}{2}\int_{-1}^{1}\int_{-1}^{t} \hat{q}(t')y(t')dt'dt + \int_{-1}^{x}\int_{-1}^{t} \hat{q}(t')y(t')dt'dt \\ &= \hat{\lambda}\left[\int_{-1}^{x}\int_{-1}^{t} y(t')\,dt'dt - \frac{(x+1)}{2}\int_{-1}^{1}\int_{-1}^{t} y(t')dt'dt\right] \end{aligned} \tag{34}$$

Using the techniques developed in the previous section to calculate integrals, this integral equation can now be transformed into the following eigenvalue equation:

$$\mathbf{A\,y} = \hat{\lambda}\,\mathbf{B\,y} \tag{35}$$

where **A** and **B** are $(N+1) \times (N+1)$ matrices whose entries are given by

$$a_{ij} = -\delta_{ij} - u_i\tilde{s}_j + \hat{e}_{ij}, \quad b_{ij} = e_{ij} - u_i\tilde{e}_j, \quad i,j = 0, 1, 2, \ldots, N \tag{36}$$

In the above equation, δ_{ij} is the Kronecker delta, $u_i = (x_i + 1) / 2$ the elements of an $(N + 1)$ column vector $\mathbf{u}$, $\tilde{s}_j$ the elements of $(N + 1)$ row vector $\tilde{\mathbf{s}}$ given by Equation 21 with $\phi = \hat{q}$, $\hat{e}_{ij}$ the elements of $(N + 1) \times (N + 1)$ matrix $\hat{\mathbf{E}}$ given by Equation 20 with $\phi = \hat{q}$, e_{ij} the elements of $(N + 1) \times (N + 1)$ matrix $\mathbf{E}$, and $\tilde{e}_j$ the elements of $(N + 1)$ row vector $\tilde{\mathbf{e}}$ given by Equation 19. It should be noted that proper definition of Equation 36 can be easily obtained when the transformed range is [0, 1]. It is important to note that the first and last columns of **B** have no effect (since multuplied by zero). Hence, the eigenvalue Equation 35 can be reduced to a $(N - 1) \times (N - 1)$ system by eliminating the first and last rows and columns of matrices **A** and **B**.

Equation 35 is a generalized eigenvalue problem. When **B** is non-singular, this equation is mathematically equivalent to $\left(\mathbf{B}^{-1}\mathbf{A}\right)\mathbf{y} = \hat{\lambda}\mathbf{y}$, and when **A** is non-singular, it is equivalent to $\left(\mathbf{A}^{-1}\mathbf{B}\right)\mathbf{y} = (1 / \hat{\lambda})\ \mathbf{y}$. Thus, in theory, if one of the matrices **A** or **B** is known to be non-singular, the problem could be reduced to a standard eigenvalue problem. However, for this reduction to be satisfactory from the point of view of numerical stability, it is necessary not only that **B** (or **A**) should be non-singular, but that it should be well-conditioned with respect to inversion. There are many underlying computational routines that can be used for solving these problems. Once we have the computed $\hat{\lambda}$ values, they can be used to obtain the eigenvalues of the original problem (28) using $\lambda = \hat{\lambda} / \left((\tilde{b} - \tilde{a})/2\right)^2$.

3.2. A simple case of the fourth-order problem

Although the numerical results of Equation 3 with some specified end conditions, obtained using the solutions of the second-order problem, will be given in the next section, we consider here a simple case of Equation 3 to demonstrate directly the use of El-gendi's method on the fourth-order problems. Therefore, we put $P(x) = W(x) = 1$, $S(x) = Q(x) = 0$, $a = 0$, and $b = 1$ in Equation 3. Then, we obtain the following simple equation, the steady-state Euler-Bernoulli equation for the deflection $y(x)$ of a vibrating beam of length 1:

$$y^{(4)}(x) = \lambda y(x), \quad 0 < x < 1 \tag{37}$$

As mentioned earlier, there are three types of boundary conditions and their combinations commonly used with this equation in applications. In this work, the formulations are given only for one of them. Other types of boundary conditions can be handled similarly. We assume that the clamped end is at $x = 0$ and the simply supported end is at $x = 1$,

$$y(0) = y'(0) = 0, \quad y(1) = y''(1) = 0 \tag{38}$$

A coordinate transformation is not necessary for this case since the physical domain is already [0, 1]. By successive integration of Equation 37 from the lower limit 0 to an arbitrary value of $x \in [0, 1]$, we obtain

$$y'''(x) - y'''(0) = \lambda \int_0^x y(t)\,dt \tag{39}$$

$$y''(x) - y''(0) - x y'''(0) = \lambda \int_0^x \int_0^{x'} y(t)\,dt dx' \tag{40}$$

$$y'(x) - xy''(0) - \frac{x^2}{2} y'''(0) = \lambda \int_0^x \int_0^{x'} \int_0^{x''} y(t)\,dt dx'' dx' \tag{41}$$

$$y(x) - \frac{x^2}{2} y''(0) - \frac{x^3}{6} y'''(0) = \lambda \int_0^x \int_0^{x'} \int_0^{x''} \int_0^{x'''} y(t)\,dt dx''' dx'' dx' \tag{42}$$

As can be seen from Equation 42, there are two unknowns, $y''(0)$ and $y'''(0)$, to be determined by introducing boundary conditions. Specializing Equations 40 and 42 for $x = 1$, together with the boundary conditions given by Equation 38, and then solving the resulting system of equations for the two unknowns, we obtain the final integrated form of Equation 42 as

$$y(x) = \lambda \left[\int_0^x \int_0^{x'} \int_0^{x''} \int_0^{x'''} y(t)\,dt dx''' dx'' dx' + \left(\frac{x^3 - 3x^2}{2}\right) \int_0^1 \int_0^{x'} \int_0^{x''} \int_0^{x'''} y(t)\,dt dx''' dx'' dx' + \left(\frac{x^2 - x^3}{4}\right) \int_0^1 \int_0^x y(t)\,dt dx \right] \tag{43}$$

Using the techniques developed in Section 2 to calculate integrals, this integral equation can be transformed into the following generalized eigenvalue equation:

$$\mathbf{A\,y} = \lambda\,\mathbf{B\,y} \tag{44}$$

where **A** and **B** are $(N + 1) \times (N + 1)$ matrices whose entries are given by

$$a_{ij} = \delta_{ij}, \quad b_{ij} = h_{ij} + v_i \tilde{h}_j + z_i \tilde{e}_j, \quad i,j = 0, 1, 2, \ldots, N \tag{45}$$

Here, $v_i = \left(x_i^3 - 3x_i^2\right)/\,2$ are the elements of an $(N + 1)$ column vector $\mathbf{v}$, $z_i = \left(x_i^2 - x_i^3\right)/\,4$ the elements of an $(N + 1)$ column vector $\mathbf{z}$, h_{ij} the elements of $(N + 1) \times (N + 1)$ matrix **H** given by Equation 22, $\tilde{h}_j$ the elements of $(N + 1)$ row vector $\tilde{\mathbf{h}}$ given by Equation 23, and $\tilde{e}_j$ the elements of $(N + 1)$ row vector $\tilde{\mathbf{e}}$ given by Equation 19. For the same reason given in Section 3.1, the eigenvalue equation (44) can be reduced to a $(N - 1) \times (N - 1)$ system. A computational routine can now be used to solve Equation 44 for obtaining eigenvalues of the problem.

4. Numerical results

We present here the results obtained by applying our algorithms to the numerical solution of some second- and fourth-order SLPs in practice. First two problems which have exact solutions are chosen to demonstrate the efficiency and accuracy of the method developed in this work. We will use the relative error ε_k which is defined as

$$\varepsilon_k = \left| \frac{\lambda_k^{exact} - \lambda_k^{approx}}{\lambda_k^{exact}} \right|, \quad k = 1, 2, 3, \ldots, \tag{46}$$

to measere the performance of the method. In the last problem, we choose the fourth-order problem (3) to be square of the second-order problem (1) as discussed in Section 1. Numerical calculations are carried out using Maple and Matlab.

Problem 1. Our first example for the second-order SLP is given by

$$-y''(x) = \lambda y(x), \quad y(0) = y(1) = 0 \tag{47}$$

which arises when solving the one-dimensional wave equation $u_{tt} - u_{xx} = 0$ (the standard model for an oscillating string of length 1 with fixed endpoints) using a "seperation of variables" method. The exact eigenvalues of this problem are $\lambda_k = k^2\pi^2$, $k = 1, 2, 3, \ldots$.

Table 1 lists the relative errors of the obtained results with different number of the Chebyshev points N. It is observed from this table that in order to have good approximations to the first kth eigenvalues, at least $2k$ Chebyshev points have to be used. It is also observed that the computed values for the lower eigenvalues have a better accuracy than those for the higher eigenvalues. As the number of grid points further increased to above $2k$, the accuracy of the obtained results especially for the higher eigenvalues can be further improved as shown in Table 1.

Table 1. Relative errors of calculated results for Problem 1

k	λ_k^{exact}	ε_k		
		N = 12	N = 24	N = 36
1	9.8696044010893	6.209399e−14	1.799826e−16	3.599652e−16
2	39.4784176043574	5.488263e−11	5.399477e−16	8.999129e−16
3	88.8264396098042	4.030874e−08	3.199690e−16	4.799535e−16
4	157.9136704174297	7.326063e−06	7.199303e−16	0.000000e+00
5	246.7401100272339	4.333980e−04	2.465041e−14	2.303777e−16
6	355.3057584392169	7.291857e−03	1.841422e−13	3.199690e−16
7	483.6106156533785	5.254565e−02	6.026152e−11	4.348967e−15
8	631.6546816697189	2.216018e−01	3.401451e−10	5.399477e−16
9	799.4379564882380	7.214591e−01	3.163688e−08	1.564293e−15
10	986.9604401089358	2.324092e+00	8.785227e−07	3.340477e−15
11	1194.2221325318123	1.102225e+01	2.189959e−05	1.180448e−13
12	1421.2230337568676		3.179505e−04	1.791827e−13
13	1667.9631437841015		2.639745e−03	2.428031e−11
14	1934.4424626135142		1.345588e−02	3.340653e−11
15	2220.6609902451055		4.634721e−02	8.812858e−10
16	2526.6187266788756		1.202738e−01	7.011934e−08
17	2852.3156719148246		2.596393e−01	1.380718e−06
18	3197.7518259529520		5.050217e−01	1.788821e−05
19	3562.9271887932582		9.438169e−01	1.660706e−04
20	3947.8417604357433		1.799175e+00	1.081240e−03

Problem 2. As a second example, we consider the following fourth-order problem

$$y^{(4)}(x) = \lambda y(x), \quad y(0) = y'(0) = 0, \quad y(1) = y''(1) = 0 \tag{48}$$

which corresponds to the case already discussed in Section 3.2. The exact eigenvalues of problem (48) can be found by solving $\tanh\left(\lambda^{1/4}\right) - \tan\left(\lambda^{1/4}\right) = 0$.

In Table 2, the relative errors of the first 10 eigenvalues of this problem for several values of the number of Chebyshev points are given. It can be seen that the method developed in this work is very efficient for finding the eigenvalues of the given fourth-order problem.

Problem 3. In this example, we consider the Coffey–Evans equation (Greenberg and Marletta, 1995)

$$-y'' + \left(\mu^2 \sin^2 2x - 2\mu \cos 2x\right) y = \lambda y, \quad y(-\pi/2) = y(\pi/2) = 0 \tag{49}$$

which arises in a model of the coupling between dipoles in polarizable liquids such as used in liquid crystal displays (Pryce, 1986). The parameter μ, typically in the range 0–50, measures the strength of coupling.

The lowest eigenvalue λ_0 of this problem is close to zero. Figure 1 shows the computed eigenvalues (λ_1–λ_{19}) of this problem for several values of the parameter μ, $\mu = 10$, $\mu = 20$, and $\mu = 30$. In Table 3, we compare our results with the available results of Pryce (1986) for λ_k where $k \leq 4$. It can be observed that our results are in good agreement with those obtained by Pryce (1986). Pryce's paper was concerned with Prüfer shooting methods for finding eigenvalues and corresponding eigenfunctions of the classical SLP (1) with appropriate boundary conditions given by Equation 2. Such methods use a transformation

Table 2. Relative errors of calculated results for Problem 2

k	λ_k^{exact}	ε_k		
		N = 12	N = 24	N = 36
1	237.72106753	1.954193e−12	4.782363e−16	2.391181e−16
2	2496.48743786	1.838806e−09	1.092929e−15	3.096633e−15
3	10867.58221698	4.789974e−08	4.519194e−15	2.343286e−15
4	31780.09645408	4.814241e−05	2.289470e−16	5.036834e−15
5	74000.84934916	1.933174e−03	1.822901e−13	1.533833e−14
6	148634.47728577	2.572511e−02	4.578776e−12	1.253172e−14
7	269123.43482664	1.626128e−01	2.255631e−10	5.169238e−14
8	451247.99471928	6.953546e−01	3.474120e−09	1.870393e−13
9	713126.24789600	2.832007e+00	1.438088e−07	3.013529e−13
10	1075214.10347396	1.683336e+01	4.069691e−06	8.705049e−14

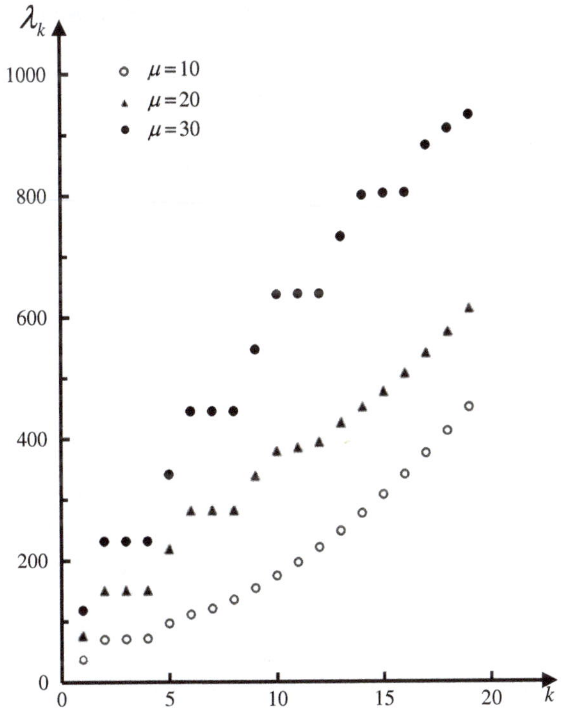

Figure 1. Eigenvalues of Coffey–Evans equation.

to phase amplitude variables (θ, r) in the (y', y)-plane (a Prüfer transformation) and solve the resulting equations by a shooting method using an ODE initial value code. Pryce studied the error control of such methods with specific reference to two published codes based on this technique.

The corresponding fourth-order problem for the Equation 49 can be obtained by using Equations 5–9:

$$y^{(4)} - \left[2\left(\mu^2 \sin^2 2x - 2\mu \cos 2x\right) y'\right]' + \left[\left(\mu^2 \sin^2 2x - 2\mu \cos 2x\right)^2 - 8\left(\mu^2 \cos^2 2x - \mu^2 \sin^2 2x + \mu \cos 2x\right)\right] y = \lambda y \quad (50)$$

with the boundary conditions

Table 3. Comparison of calculated results of the present work for Problem 3 with the results of Pryce (1986)

Index k	$\mu = 10$		$\mu = 20$		$\mu = 30$	
	Pryce (1986)	Present work	Pryce (1986)	Present work	Pryce (1986)	Present work
0	−0.00000066	0.00000005	−0.00000287	−0.00000015	−0.00000005	−0.00006416
1	37.80590010	37.80590023	77.91619396	77.91620152	117.9463064	117.94863349
2	69.79528236	69.79528143	151.4627743	151.46276528	231.6649293	231.65447078
3	70.54751205	70.54750974	151.4633026	151.46322366	231.6649342	231.66492931
4	71.40519810	71.40525148	151.4634228	151.46365429	231.6649044	231.66492931

Table 4. Comparison of calculated results of the present work with the results obtained in Greenberg and Marletta (1995)

Index k	$\mu = 10$		$\mu = 30$	
	Greenberg and Marletta (1995)	Present work	Greenberg and Marletta (1995)	Present work
2	4872.19471	4871.38131	53668.6347	53668.6395
3	4976.95126	4976.95113	53668.6435	53668.6417
4	5098.70998	5098.70994	53668.6435	53668.6279

$$y(-\pi / 2) = y''(-\pi / 2) = 0, \quad y(\pi / 2) = y''(\pi / 2) = 0 \tag{51}$$

As mentioned earlier in Section 1, the eigenvalues of this fourth-order problem are simply the squares of the eigenvalues of the second-order problem (49). The results are shown in Table 4 where the comparisons are made with the available data obtained by Greenberg and Marletta (1995) for $\mu = 10$ and $\mu = 30$. Greenberg and Marletta (1995) developed a shooting method to approximate the eigenvalues and eigenfunctions of the fourth-order problem (3). The problem was first reduced to a Hamiltonian system. They developed an efficient approach to solve this system based on the approximation of the coefficients in the differential equation and a suitable zero-counting algorithm. The zeros which they counted were not zeros of a solution of the fourth-order equation, but nullities (rank deficiencies) of a certain 2×2 matrix formed from solutions of the fourth-order equation.

5. Conclusions

In this paper, we present a method to obtain good approximations to the eigenvalues of second- and fourth-order SLPs defined on finite domains based on a spectral method known as El-gendi's method. This method provides a robust algorithm of computing the integral of a non-singular function defined on a finite domain. Therefore, the method presented in this work converts the given SLP into an integral equation. The resulting integral equation is then transformed into the eigenvalue equation by calculating the integrals at the grid points using the Chebyshev expansions. The required eigenvalues are then obtained by solving this eigenvalue equation. The method is quite general and has some special advantages which were discussed in detail in Boyd (2000). The advantages of El-gendi's method over finite difference methods were also discussed in Boyd (2000) and so will not be repeated here.

Through test examples which have exact solutions, it was found that in order to obtain accurate numerical results for the first kth eigenvalues, at least $2k$ Chebyshev points have to be used. Comparison with other published works in the literature showed that the method produces highly accurate results for the eigenvalues of the SLPs considered in this work. It may be concluded that the presented method is very powerful and efficient in finding the approximate solutions of SLPs arising in science and engineering.

Acknowledgments
The author would like to thank the editor and two anonymous referees for their valuable comments and suggestions to substantially improve the paper.

Funding
This work was supported by the Council of Higher Education of Turkey [Article 39 of Law No. 2547].

Author details
Uğur Yücel[1]
E-mail: uyucel@pau.edu.tr
[1] Faculty of Arts and Sciences, Department of Mathematics, Pamukkale University, Denizli 20070, Turkey.

References

Amodio, P., & Settanni, G. (2015). Reprint of variable-step finite difference schemes for the solution of Sturm–Liouville problems. *Communications in Nonlinear Science and Numerical Simulation, 21*, 12–21. http://dx.doi.org/10.1016/j.cnsns.2014.10.014

Amrein, W. O., Hinz, A. M., & Pearson, D. B. (2005). *Sturm-Liouville theory*. Basel: Birkhäuser Verlag. http://dx.doi.org/10.1007/3-7643-7359-8

Boyd, J. P. (2000). *Chebyshev and Fourier spectral methods*. New York, NY: Dover Publications.

El-gamel, M., & Abd El-hady, M. (2013). Two very accurate and efficient methods for computing eigenvalues of Sturm–Liouville problems. *Applied Mathematical Modelling, 37*, 5039–5046. http://dx.doi.org/10.1016/j.apm.2012.10.019

El-gendi, S. E. (1969). Chebyshev solution of differential, integral and integro-differential equations. *The Computer Journal, 12*, 282–287. http://dx.doi.org/10.1093/comjnl/12.3.282

Elliott, D. (1963). A Chebyshev series method for the numerical solution of Fredholm integral equations. *The Computer Journal, 6*, 102–112. http://dx.doi.org/10.1093/comjnl/6.1.102

Fox, L. (1962). Chebyshev methods for ordinary differential equations. *The Computer Journal, 4*, 318–331. http://dx.doi.org/10.1093/comjnl/4.4.318

Fox, L., & Parker, I. B. (1968). *Chebyshev polynomials in numerical analysis*. London: Oxford University Press.

Greenberg, L. (1991). An oscillation method for fourth-order, selfadjoint, two-point boundary value problems with nonlinear eigenvalues. *SIAM Journal on Mathematical Analysis, 22*, 1021–1042. http://dx.doi.org/10.1137/0522067

Greenberg, L., & Marletta, M. (1995). Oscillation theory and numerical solution of fourth order Sturm–Liouville problems. *IMA Journal of Numerical Analysis, 15*, 319–356. http://dx.doi.org/10.1093/imanum/15.3.319

Khater, A. H., & Temsah, R. S. (2008). Numerical solutions of the generalized Kuramoto–Sivashinsky equation by Chebyshev spectral collocation methods. *Computers & Mathematics with Applications, 56*, 1465–1472. http://dx.doi.org/10.1016/j.camwa.2008.03.013

Mihaila, B., & Mihaila, I. (2002). Numerical approximations using Chebyshev polynomial expansions: El-gendi's method revisited. *Journal of Physics A: Mathematical and General, 35*, 731–746. http://dx.doi.org/10.1088/0305-4470/35/3/317

Pryce, J. D. (1986). Error control of phase-function shooting methods for Sturm–Liouville problems. *IMA Journal of Numerical Analysis, 6*, 103–123. http://dx.doi.org/10.1093/imanum/6.1.103

Pryce, J. D. (1993). *Numerical solution of Sturm–Liouville problems*. Oxford: Oxford University Press.

Saleh Taher, A. H., Malek, A., & Momeni-Masuleh, S. H. (2013). Chebyshev differentiation matrices for efficient computation of the eigenvalues of fourth-order Sturm–Liouville problems. *Applied Mathematical Modelling, 37*, 4634–4642. http://dx.doi.org/10.1016/j.apm.2012.09.062

Temsah, R. S. (2009). Numerical solutions for convection-diffusion equation using El-Gendi method. *Communications in Nonlinear Science and Numerical Simulation, 14*, 760–769. http://dx.doi.org/10.1016/j.cnsns.2007.11.004

Yuan, S., Ye, K., Xiao, C., Kennedy, D., & Williams, F. W. (2014). Solution of regular second- and fourth-order Sturm–Liouville problems by exact dynamic stiffness method analogy. *Journal of Engineering Mathematics, 86*, 157–173. http://dx.doi.org/10.1007/s10665-013-9646-5

Yücel, U. (2006). Approximations of Sturm–Liouville eigenvalues using differential quadrature (DQ) method. *Journal of Computational and Applied Mathematics, 192*, 310–319. http://dx.doi.org/10.1016/j.cam.2005.05.008

Yücel, U., & Boubaker, K. (2012). Differential quadrature method (DQM) and Boubaker polynomials expansion scheme (BPES) for efficient computation of the eigenvalues of fourth-order Sturm–Liouville problems. *Applied Mathematical Modelling, 36*, 158–167. http://dx.doi.org/10.1016/j.apm.2011.05.030

3

Approximating positive solutions of nonlinear first order ordinary quadratic differential equations

Bapurao C. Dhage[1*] and Shyam B. Dhage[1]

*Corresponding author: Bapurao C. Dhage, Kasubai, Gurukul Colony, Ahmedpur, 413 515 Maharashtra, India
E-mail: bcdhage@gmail.com

Reviewing editor: Kok Lay Teo, Curtin University, Australia

Abstract: In this paper, the authors prove the existence as well as approximations of the positive solutions for an initial value problem of first-order ordinary nonlinear quadratic differential equations. An algorithm for the solutions is developed and it is shown that the sequence of successive approximations converges monotonically to the positive solution of related quadratic differential equations under some suitable mixed hybrid conditions. We base our results on the Dhage iteration method embodied in a recent hybrid fixed-point theorem of Dhage (2014) in partially ordered normed linear spaces. An example is also provided to illustrate the abstract theory developed in the paper.

Subjects: Advanced Mathematics; Analysis - Mathematics; Differential Equations; Mathematics & Statistics; Operator Theory; Science

Keywords: quadratic differential equation; initial value problem; Dhage iteration method; approximate positive solution

AMS subject classifications: 34A12; 34A38

1. Introduction

Given a closed and bounded interval $J = [t_0, t_0 + a]$, of the real line $\mathbb{R}$ for some $t_0, a \in \mathbb{R}$ with $t_0 \geq 0, a > 0$, consider the initial value problem (in short IVP) of first-order ordinary nonlinear quadratic differential equation, (in short HDE)

$$\left.\begin{array}{l} \frac{d}{dt}\left[\frac{x(t)}{f(t,x(t))}\right] + \lambda \left[\frac{x(t)}{f(t,x(t))}\right] = g(t, x(t)), \ t \in J, \\ x(t_0) = x_0 \in \mathbb{R}, \end{array}\right\} \tag{1.1}$$

ABOUT THE AUTHORS

The key research project of the authors of the paper is to prove existence and find the algorithms for different nonlinear equations that arise in mathematical analysis and allied areas of mathematics via newly developed Dhage iteration method. The quadratic differential equations form an important class in the theory of differential equations. In the present paper, it is shown that the new method is also applicable to such type of nonlinear quadratic differential equations for proving the existence as well as approximations of the solutions under mixed monotonic and geometric conditions.

PUBLIC INTEREST STATEMENT

It is known that many of the natural, physical, biological, and social processes or phenomena are governed by mathematical models of nonlinear differential equations. So if a person is engaged in the study of such complex universal phenomena and not having the knowledge of sophisticated nonlinear analysis of this paper, then one may convinced the use of the results of this paper, in particular when one comes across a certain dynamic process which is based on a mathematical model of quadratic differential equations. In such situations, the application of the results of the present paper yields numerical concrete solutions under some suitable natural conditions thereby which it is possible to improve the situation for better desired goals.

for $\lambda \in \mathbb{R}$, $\lambda > 0$, where $f : J \times \mathbb{R} \to \mathbb{R} \setminus \{0\}$ and $g : J \times \mathbb{R} \to \mathbb{R}$ are continuous functions.

By a *solution* of the QDE (1.1), we mean a function $x \in C^1(J, \mathbb{R})$ that satisfies

(i) $t \mapsto \dfrac{x}{f(t,x)}$ is a continuously differentiable function for each $x \in \mathbb{R}$, and

(ii) x satisfies the equations in (1.1) on J,where $C(J, \mathbb{R})$ is the space of continuously differentiable real-valued functions defined on J.

The QDE (1.1) with $\lambda = 0$ is well known in the literature and is a hybrid differential equation with a quadratic perturbation of second type. Such differential equations can be tackled with the use of hybrid fixed-point theory (cf. Dhage 1999; 2013; 2014a). The special cases of QDE (1.1) have been discussed at length for existence as well as other aspects of the solutions under some strong Lipschitz and compactness-type conditions which do not yield any algorithm to determine the numerical solutions. See Dhage and Regan (2000), Dhage and Lakshmikantham (2010) and the references therein. Very recently, the study of approximation of the solutions for the hybrid differential equations is initiated in Dhage, Dhage, and Ntouyas (2014) via hybrid fixed-point theory. Therefore, it is of interest and new to discuss the approximations of solutions for the QDE (1.1) along the similar lines. This is the main motivation of the present paper and it is proved that the existence of the solutions may be proved via an algorithm based on successive approximations under weaker partial continuity and partial compactness-type conditions.

2. Auxiliary results

Unless otherwise mentioned, throughout this paper that follows, let E denotes a partially ordered real-normed linear space with an order relation $\preceq$ and the norm $\|\cdot\|$. It is known that E is *regular* if $\{x_n\}_{n\in\mathbb{N}}$ is a nondecreasing (resp. nonincreasing) sequence in E such that $x_n \to x^*$ as $n \to \infty$, then $x_n \preceq x^*$ (resp. $x_n \succeq x^*$) for all $n \in \mathbb{N}$. Clearly, the partially ordered Banach space, $C(J, \mathbb{R})$ is regular and the conditions guaranteeing the regularity of any partially ordered normed linear space E may be found in Nieto and Lopez (2005) and Heikkilä and Lakshmikantham (1994) and the references therein.

We need the following definitions in the sequel.

Definition 2.1 A mapping $\mathcal{T} : E \to E$ is called *isotone* or *nondecreasing* if it preserves the order relation $\preceq$, that is if $x \preceq y$ implies $\mathcal{T}x \preceq \mathcal{T}y$ for all $x, y \in E$.

Definition 2.2 (Dhage, 2010) A mapping $\mathcal{T} : E \to E$ is called *partially continuous* at a point $a \in E$ if for $\epsilon > 0$ there exists a $\delta > 0$ such that $\|\mathcal{T}x - \mathcal{T}a\| < \epsilon$ whenever x is comparable to a and $\|x - a\| < \delta$. $\mathcal{T}$ called partially continuous on E if it is partially continuous at every point of it. It is clear that if $\mathcal{T}$ is partially continuous on E, then it is continuous on every chain C contained in E.

Definition 2.3 A mapping $\mathcal{T} : E \to E$ is called *partially bounded* if $T(C)$ is bounded for every chain C in E. $\mathcal{T}$ is called *uniformly partially bounded* if all chains $\mathcal{T}(C)$ in E are bounded by a unique constant. $\mathcal{T}$ is called *bounded* if $T(E)$ is a bounded subset of E.

Definition 2.4 A mapping $\mathcal{T} : E \to E$ is called *partially compact* if $\mathcal{T}(C)$ is a relatively compact subset of E for all totally ordered sets or chains C in E. $\mathcal{T}$ is called *uniformly partially compact* if $\mathcal{T}(C)$ is a uniformly partially bounded and partially compact on E. $\mathcal{T}$ is called *partially totally bounded* if for any totally ordered and bounded subset C of E, $\mathcal{T}(C)$ is a relatively compact subset of E. If $\mathcal{T}$ is partially continuous and partially totally bounded, then it is called *partially completely continuous* on E.

Definition 2.5 (Dhage, 2009) The order relation $\preceq$ and the metric d on a nonempty set E are said to be *compatible* if $\{x_n\}_{n\in\mathbb{N}}$ is a monotone, that is, monotone nondecreasing or monotone nonincreasing sequence in E and if a subsequence $\{x_{n_k}\}_{n\in\mathbb{N}}$ of $\{x_n\}_{n\in\mathbb{N}}$ converges to x^* implies that the whole sequence $\{x_n\}_{n\in\mathbb{N}}$ converges to x^*. Similarly, given a partially ordered normed linear space $(E, \preceq, \|\cdot\|)$, the order relation $\preceq$ and the norm $\|\cdot\|$ are said to be compatible if $\preceq$ and the metric d defined through the norm $\|\cdot\|$ are compatible.

Clearly, the set $\mathbb{R}$ of real numbers with usual order relation $\leq$ and the norm defined by the absolute value function $|\cdot|$ has this property. Similarly, the finite-dimensional Euclidean space $\mathbb{R}^n$ with usual componentwise order relation and the standard norm possesses the compatibility property.

Definition 2.6 (Dhage, 2010) An upper semi-continuous and nondecreasing function $\psi : \mathbb{R}_+ \to \mathbb{R}_+$ is called a $\mathcal{D}$-function, provided $\psi(0) = 0$. Let $(E, \preceq, \|\cdot\|)$ be a partially ordered normed linear space. A mapping $\mathcal{T} : E \to E$ is called *partially nonlinear $\mathcal{D}$-Lipschitz* if there exists a $\mathcal{D}$-function $\psi : \mathbb{R}_+ \to \mathbb{R}_+$ such that

$$\|\mathcal{T}x - \mathcal{T}y\| \leq \psi(\|x - y\|) \tag{2.1}$$

for all comparable elements $x, y \in E$. If $\psi(r) = kr, k > 0$, then $\mathcal{T}$ is called a partially Lipschitz with a Lipschitz constant k.

Let $(E, \preceq, \|\cdot\|)$ be a partially ordered normed linear algebra. Denote

$$E^+ = \{x \in E \mid x \succeq \theta, \text{ where } \theta \text{ is the zero element of } E\}$$

and

$$\mathcal{K} = \{E^+ \subset E \mid uv \in E^+ \text{ for all } u, v \in E^+\}. \tag{2.2}$$

The elements of the set $\mathcal{K}$ are called the positive vectors in E. The following lemma follows immediately from the definition of the set $\mathcal{K}$, which is oftentimes used in the hybrid fixed-point theory of Banach algebras and applications to nonlinear differential and integral equations.

LEMMA 2.1 *(Dhage, 1999) If $u_1, u_2, v_1, v_2 \in \mathcal{K}$ are such that $u_1 \preceq v_1$ and $u_2 \preceq v_2$, then $u_1u_2 \preceq v_1v_2$.*

Definition 2.7 An operator $\mathcal{T} : E \to E$ is said to be positive if the range $R(\mathcal{T})$ of $\mathcal{T}$ is such that $R(\mathcal{T}) \subseteq \mathcal{K}$.

The Dhage iteration principle or method (in short DIP or DIM) developed in Dhage (2010; 2013; 2014a) may be rephrased as "monotonic convergence of the sequence of successive approximations to the solutions of a nonlinear equation beginning with a lower or an upper solution of the equation as its initial or first approximation" and which forms a useful tool in the subject of existence theory of nonlinear analysis. The Dhage iteration method is different from other iterations methods and embodied in the following applicable hybrid fixed-point theorem of Dhage (2014b), which is the key tool for our work contained in the present paper. A few other hybrid fixed-point theorems containing the Dhage iteration principle appear in Dhage (2010; 2013; 2014a; 2014b).

THEOREM 2.1 *Let $(E, \preceq, \|\cdot\|)$ be a regular partially ordered complete normed linear algebra such that the order relation $\preceq$ and the norm $\|\cdot\|$ in E are compatible in every compact chain of E. Let $\mathcal{A}, \mathcal{B} : E \to \mathcal{K}$ be two nondecreasing operators such that*

(a) $\mathcal{A}$ is partially bounded and partially nonlinear $\mathcal{D}$-Lipschitz with $\mathcal{D}$-function $\psi_{\mathcal{A}}$,
(b) $\mathcal{B}$ is partially continuous and uniformly partially compact,
(c) $M\psi_{\mathcal{A}}(r) < r, r > 0$, where $M = \sup\{\|\mathcal{B}(C)\| : C \in \mathcal{P}_{ch}(E)\}$, and
(d) there exists an element $x_0 \in X$ such that $x_0 \preceq \mathcal{A}x_0\, \mathcal{B}x_0$ or $x_0 \succeq \mathcal{A}x_0\, \mathcal{B}x_0$.

Then the operator equation

$$\mathcal{A}x\,\mathcal{B}x = x \tag{2.3}$$

has a positive solution x^* in E and the sequence $\{x_n\}$ of successive iterations defined by $x_{n+1} = \mathcal{A}x_n\,\mathcal{B}x_n$, $n = 0, 1, \ldots$; converges monotonically to x^*.

Remark 2.1 The compatibility of the order relation $\preceq$ and the norm $\|\cdot\|$ in every compact chain of E is held if every partially compact subset S of E possesses the compatibility property with respect to $\preceq$ and $\|\cdot\|$. This simple fact is used to prove the desired characterization of the positive solution of the QDE (1.1) defined on J.

3. Main results

The QDE (1.1) is considered in the function space $C(J, \mathbb{R})$ of continuous real-valued functions defined on J. We define a norm $\|\cdot\|$ and the order relation $\leq$ in $C(J, \mathbb{R})$ by

$$\|x\| = \sup_{t \in J} |x(t)| \tag{3.1}$$

and

$$x \leq y \iff x(t) \leq y(t) \tag{3.2}$$

for all $t \in J$, respectively. Clearly, $C(J, \mathbb{R})$ is a Banach algebra with respect to above supremum norm and is also partially ordered w.r.t. the above partially order relation $\leq$. It is known that the partially ordered Banach algebra $C(J, \mathbb{R})$ has some nice properties w.r.t. the above order relation in it. The following lemma follows by an application of Arzelá–Ascoli theorem.

LEMMA 3.1 *Let $\left(C(J, \mathbb{R}), \leq, \|\cdot\|\right)$ be a partially ordered Banach space with the norm $\|\cdot\|$ and the order relation $\leq$ defined by (3.1) and (3.2), respectively. Then, $\|\cdot\|$ and $\leq$ are compatible in every partially compact subset of $C(J, \mathbb{R})$.*

Proof The proof of the lemma is given in Dhage and Dhage (in press). Since it is not well known, we give the details of proof for the sake of completeness. Let S be a partially compact subset of $C(J, \mathbb{R})$ and let $\{x_n\}$ be a monotone nondecreasing sequence of points in S. Then, we have

$$x_1(t) \leq x_2(t) \leq \cdots \leq x_n(t) \leq \cdots \tag{ND}$$

for each $t \in J$.

Suppose that a subsequence $\{x_{n_k}\}$ of $\{x_n\}$ is convergent and converges to a point x in S. Then the subsequence $\{x_{n_k}(t)\}$ of the monotone real sequence $\{x_n(t)\}$ is convergent. By monotone characterization, the whole sequence $\{x_n(t)\}$ is convergent and converges to a point $x(t)$ in $\mathbb{R}$ for each $t \in J$. This shows that the sequence $\{x_n(t)\}$ converges pointwise in S. To show the convergence is uniform, it is enough to show that the sequence $\{x_n(t)\}$ is equicontinuous. Since S is partially compact, every chain or totally ordered set and consequently $\{x_n\}$ is an equicontinuous sequence by Arzelá–Ascoli theorem. Hence $\{x_n\}$ is convergent and converges uniformly to x. As a result, $\|\cdot\|$ and $\leq$ are compatible in S. This completes the proof.

We need the following definition in what follows.

Definition 3.1 A function $u \in C^1(J, \mathbb{R})$ is said to be a lower solution of the QDE (1.1) if the function $t \mapsto \dfrac{u(t)}{f(t, u(t))}$ is continuously differentiable and satisfies

$$\left.\begin{aligned}\frac{d}{dt}\left[\frac{u(t)}{f(t,u(t))}\right] + \lambda\left[\frac{u(t)}{f(t,u(t))}\right] &\le g(t,u(t)),\\ u(t_0) &\le x_0\end{aligned}\right\}$$

for all $t \in J$. Similarly, a function $v \in C^1(J,\mathbb{R})$ is said to be an upper solution of the QDE (1.1) if it satisfies the above property and inequalities with reverse sign.

We consider the following set of assumptions in what follows:

(A_0) The map $x \mapsto \dfrac{x}{f(t,x)}$ is injection for each $t \in J$.
(A_1) f defines a function $f{:}J \times \mathbb{R} \to \mathbb{R}_+$.
(A_2) There exists a constant $M_f > 0$ such that $0 < f(t,x) \le M_f$ for all $t \in J$ and $x \in \mathbb{R}$.
(A_3) There exists a $\mathcal{D}$-function ϕ, such that

$$0 \le f(t,x) - f(t,y) \le \phi(x-y),$$

for all $t \in J$ and $x, y \in \mathbb{R}$, $x \ge y$.
(B_1) g defines a function $g{:}J \times \mathbb{R} \to \mathbb{R}_+$.
(B_2) There exists a constant $M_g > 0$ such that $g(t,x) \le M_g$ for all $t \in J$ and $x \in \mathbb{R}$.
(B_3) $g(t,x)$ is nondecreasing in x for all $t \in J$.
(B_4) The QDE (1.1) has a lower solution $u \in C^1(J,\mathbb{R})$.

Remark 3.1 Notice that Hypothesis (A_0) holds in particular if the function $x \mapsto \frac{x}{f(t,x)}$ is increasing in $\mathbb{R}$ for each $t \in J$.

LEMMA 3.2 *Suppose that hypothesis (A_0) holds. Then a function $x \in C(J,\mathbb{R})$ is a solution of the QDE (1.1), if and only if it is a solution of the nonlinear quadratic integral equation (in short QIE),*

$$x(t) = [f(t,x(t))]\left(\frac{ce^{-\lambda t}}{f(t_0,x_0)} + \int_{t_0}^{t} e^{-\lambda(t-s)} g(s,x(s))\,ds\right) \tag{3.3}$$

for all $t \in J$, where $c = x_0 e^{\lambda t_0}$.

THEOREM 3.1 *Assume that hypotheses (A_0)-(A_3) and (B_1)-(B_4) hold. Furthermore, assume that*

$$\left(\left|\frac{x_0}{f(t_0,x_0)}\right| + M_g a\right)\phi(r) < r,\ \ r < 0, \tag{3.4}$$

then the QDE (1.1) has a positive solution x^* defined on J and the sequence $\{x_n\}_{n=1}^{\infty}$ of successive approximations defined by

$$x_{n+1}(t) = [f(t,x_n(t))]\left(\frac{ce^{-\lambda t}}{f(t_0,x_0)} + \int_{t_0}^{t} e^{-\lambda(t-s)} g(s,x_n(s))\,ds\right) \tag{3.5}$$

for $t \in \mathbb{R}$, where $x_1 = u$, converges monotonically to x^*.

Proof Set $E = C(J,\mathbb{R})$ Then, by Lemma 3.1, every compact chain in E possesses the compatibility property with respect to the norm $\|\cdot\|$ and the order relation $\le$ in E.

Define two operators $\mathcal{A}$ and $\mathcal{B}$ on E by

$$\mathcal{A}x(t) = f(t, x(t)), \quad t \in J, \tag{3.6}$$

and

$$\mathcal{B}x(t) = \frac{ce^{-\lambda t}}{f(t_0, x_0)} + \int_{t_0}^{t} e^{-\lambda(t-s)} g(s, x(s))\, ds, \quad t \in J. \tag{3.7}$$

From the continuity of the integral, it follows that $\mathcal{A}$ and $\mathcal{B}$ define the maps $\mathcal{A}, \mathcal{B} : E \to E$. Now by Lemma 3.2, the QDE (1.1) is equivalent to the operator equation

$$\mathcal{A}x(t)\mathcal{B}x(t) = x(t), \quad t \in J. \tag{3.8}$$

We shall show that the operators $\mathcal{A}$ and $\mathcal{B}$ satisfy all the conditions of Theorem 2.1. This is achieved in the series of following steps.

Step I: $\mathcal{A}$and $\mathcal{B}$ are nondecreasing on E.

Let $x, y \in E$ be such that $x \geq y$. Then by hypothesis (A_3), we obtain

$$\mathcal{A}x(t) = f(t, x(t)) \geq f(t, y(t)) = \mathcal{A}y(t)$$

for all $t \in J$. This shows that $\mathcal{A}$ is nondecreasing operator on E into E. Similarly using hypothesis (B_3), it is shown that the operator $\mathcal{B}$ is also nondecreasing on E into itself. Thus, $\mathcal{A}$ and $\mathcal{B}$ are nondecreasing positive operators on E into itself.

Step II: $\mathcal{A}$ is partially bounded and partially $\mathcal{D}$-Lipschitz on E.

Let $x \in E$ be arbitrary. Then by (A_2),

$$|\mathcal{A}x(t)| \leq |f(t, x(t))| \leq M_f$$

for all $t \in J$. Taking supremum over t, we obtain $\|\mathcal{A}x\| \leq M_f$ and so, $\mathcal{A}$ is bounded. This further implies that $\mathcal{A}$ is partially bounded on E.

Next, let $x, y \in E$ be such that $x \geq y$. Then,

$$|\mathcal{A}x(t) - \mathcal{A}y(t)| = |f(t, x(t)) - f(t, y(t))| \leq \phi(|x(t) - y(t)|) \leq \phi(\|x - y\|)$$

for all $t \in J$. Taking supremum over t, we obtain $\|\mathcal{A}x - \mathcal{A}y\| \leq \phi(\|x - y\|)$ for all $x, y \in E$, $x \geq y$. Hence, $\mathcal{A}$ is a partial nonlinear $\mathcal{D}$-LIpschitz on E which further implies that $\mathcal{A}$ is a partially continuous on E.

Step III: *$\mathcal{B}$ is partially continuous on E.*

Let $\{x_n\}_{n \in \mathbb{N}}$ be a sequence in a chain C of E such that $x_n \to x$ for all $n \in \mathbb{N}$. Then, by dominated convergence theorem, we have

$$\begin{aligned}
\lim_{n\to\infty} \mathcal{B}x_n(t) &= \lim_{n\to\infty} \frac{ce^{-\lambda t}}{f(t_0, x_0)} + \lim_{n\to\infty} \int_{t_0}^{t} e^{-\lambda(t-s)} g(s, x_n(s))\, ds \\
&= \frac{ce^{-\lambda t}}{f(t_0, x_0)} + \int_{t_0}^{t} e^{-\lambda(t-s)} \left[\lim_{n\to\infty} g(s, x_n(s))\right] ds \\
&= \frac{ce^{-\lambda t}}{f(t_0, x_0)} + \int_{t_0}^{t} e^{-\lambda(t-s)} g(s, x(s))\, ds \\
&= \mathcal{B}x(t)
\end{aligned}$$

for all $t \in J$. This shows that $\mathcal{B}x_n$ converges monotonically to $\mathcal{B}x$ pointwise on J.

Next, we will show that $\{\mathcal{B}x_n\}_{n\in\mathbb{N}}$ is an equicontinuous sequence of functions in E. Let $t_1, t_2 \in J$ with $t_1 < t_2$. Then

$$\begin{aligned}
|\mathcal{B}x_n(t_2) - \mathcal{B}x_n(t_1)| &\le \left| \frac{ce^{-\lambda t_1}}{f(t_0,x_0)} - \frac{ce^{-\lambda t_2}}{f(t_0,x_0)} \right| \\
&\quad + \left| \int_{t_0}^{t_1} e^{-\lambda(t_1-s)} g(s,x_n(s))\,ds - \int_{t_0}^{t_1} e^{-\lambda(t_2-s)} g(s,x_n(s))\,ds \right| \\
&\quad + \left| \int_{t_0}^{t_1} e^{-\lambda(t_2-s)} g(s,x_n(s))\,ds - \int_{t_0}^{t_2} e^{-\lambda(t_2-s)} g(s,x_n(s))\,ds \right| \\
&\le \left| \frac{ce^{-\lambda t_1}}{f(t_0,x_0)} - \frac{ce^{-\lambda t_2}}{f(t_0,x_0)} \right| + \left| \int_{t_0}^{t_1} \left| e^{-\lambda(t_1-s)} - e^{-\lambda(t_2-s)} \right| |g(s,x_n(s))|\,ds \right| \\
&\quad + \left| \int_{t_2}^{t_1} |g(s,x_n(s))\,ds| \right| \\
&\le \left| \frac{ce^{-\lambda t_1}}{f(t_0,x_0)} - \frac{ce^{-\lambda t_2}}{f(t_0,x_0)} \right| + M_g \int_{t_0}^{t_0+a} \left| e^{-\lambda(t_1-s)} - e^{-\lambda(t_2-s)} \right| ds \\
&\quad + M_g\,|t_1 - t_2| \\
&\to 0 \quad as \quad t_2 - t_1 \to 0
\end{aligned}$$

uniformly for all $n \in \mathbb{N}$. This shows that the convergence $\mathcal{B}x_n \to \mathcal{B}x$ is uniform and hence $\mathcal{B}$ is partially continuous on E.

Step IV: $\mathcal{B}$ is uniformly partially compact operator on E.

Let C be an arbitrary chain in E. We show that $\mathcal{B}(C)$ is a uniformly bounded and equicontinuous set in E. First, we show that $\mathcal{B}(C)$ is uniformly bounded. Let $y \in \mathcal{B}(C)$ be any element. Then there is an element $x \in C$, such that $y = \mathcal{B}x$. Now, by hypothesis (B_2),

$$\begin{aligned}
|y(t)| &\le \left| \frac{ce^{-\lambda t}}{f(t_0,x_0)} + \int_{t_0}^{t} e^{-\lambda(t-s)} g(s,x(s))\,ds \right| \\
&\le \left| \frac{ce^{-\lambda t}}{f(t_0,x_0)} \right| + \left| \int_{t_0}^{t} e^{-\lambda(t-s)} g(s,x(s))\,ds \right| \\
&\le \left| \frac{x_0}{f(t_0,x_0)} \right| + \int_{t_0}^{t_0+a} |g(s,x(s))|\,ds \\
&\le \left| \frac{x_0}{f(t_0,x_0)} \right| + M_g\,a = M
\end{aligned}$$

for all $t \in J$. Taking supremum over t, we obtain $\|y\| = \|\mathcal{B}x\| \le M$ for all $y \in \mathcal{B}(C)$. Hence, $\mathcal{B}(C)$ is a uniformly bounded subset of E. Moreover, $\|\mathcal{B}(C)\| \le M$ for all chains C in E. Hence, $\mathcal{B}$ is a uniformly partially bounded operator on E.

Next, we will show that $\mathcal{B}(C)$ is an equicontinuous set in E. Let $t_1, t_2 \in J$ with $t_1 < t_2$. Then, for any $y \in \mathcal{B}(C)$, one has

$$
\begin{aligned}
|y(t_2) - y(t_1)| &= |\mathcal{B}x(t_2) - \mathcal{B}x(t_1)| \\
&\leq \left| \frac{ce^{-\lambda t_1}}{f(t_0, x_0)} - \frac{ce^{-\lambda t_2}}{f(t_0, x_0)} \right| \\
&\quad + \left| \int_{t_0}^{t_1} e^{-\lambda(t_1 - s)} g(s, x(s))\, ds - \int_{t_0}^{t_1} e^{-\lambda(t_2 - s)} g(s, x(s))\, ds \right| \\
&\quad + \left| \int_{t_0}^{t_1} e^{-\lambda(t_2 - s)} g(s, x(s))\, ds - \int_{t_0}^{t_2} e^{-\lambda(t_2 - s)} g(s, x(s))\, ds \right| \\
&\leq \left| \frac{ce^{-\lambda t_1}}{f(t_0, x_0)} - \frac{ce^{-\lambda t_2}}{f(t_0, x_0)} \right| + \left| \int_{t_0}^{t_1} \left| e^{-\lambda(t_1 - s)} - e^{-\lambda(t_2 - s)} \right| |g(s, x(s))|\, ds \right| \\
&\quad + \left| \int_{t_2}^{t_1} |g(s, x(s))\, ds| \right| \\
&\leq \left| \frac{ce^{-\lambda t_1}}{f(t_0, x_0)} - \frac{ce^{-\lambda t_2}}{f(t_0, x_0)} \right| + M_g \int_{t_0}^{t_0 + a} \left| e^{-\lambda(t_1 - s)} - e^{-\lambda(t_2 - s)} \right| ds \\
&\quad + M_g |t_1 - t_2| \\
&\to 0 \quad \text{as} \quad t_2 - t_1 \to 0
\end{aligned}
$$

uniformly for all $y \in \mathcal{B}(C)$. Hence $\mathcal{B}(C)$ is an equicontinuous subset of E. Now, $\mathcal{B}(C)$ is a uniformly bounded and equicontinuous set of functions in E, so it is compact. Consequently, $\mathcal{B}$ is a uniformly partially compact operator on E into itself.

Step V: u satisfies the operator inequality $u \leq \mathcal{A}u\, \mathcal{B}u$.

By hypothesis (B_4), the QDE (1.1) has a lower solution u defined on J. Then, we have

$$
\left.
\begin{aligned}
\frac{d}{dt}\left[\frac{u(t)}{f(t, u(t))}\right] + \lambda \left[\frac{u(t)}{f(t, u(t))}\right] &\leq g(t, u(t)), \\
u(t_0) &\leq x_0
\end{aligned}
\right\}
\tag{3.9}
$$

for all $t \in J$. Multiplying the above inequality (3.9) by the integrating factor $e^{\lambda t}$, we obtain

$$
\left(e^{\lambda t} \frac{u(t)}{f(t, u(t))} \right)' \leq e^{\lambda t} g(t, u(t)) \tag{3.10}
$$

for all $t \in J$. A direct integration of (3.10) from t_0 to t yields

$$
u(t) \leq [f(t, u(t))] \left(\frac{ce^{-\lambda t}}{f(t_0, x_0)} + \int_{t_0}^{t} e^{-\lambda(t - s)} g(s, u(s))\, ds \right) \tag{3.11}
$$

for all $t \in J$. From definitions of the operators $\mathcal{A}$ and $\mathcal{B}$, it follows that $u(t) \leq \mathcal{A}u(t)\, \mathcal{B}u(t)$, for all $t \in J$. Hence $u \leq \mathcal{A}u\, \mathcal{B}u$.

Step VI: $\mathcal{D}$-function ϕ satisfies the growth condition $M\psi_{\mathcal{A}}(r) < r, r > 0$.

Finally, the $\mathcal{D}$-function ϕ of the operator $\mathcal{A}$ satisfies the inequality given in hypothesis (d) of Theorem 2.1. Now from the estimate given in Step IV, it follows that

$$
M\psi_{\mathcal{A}}(r) \leq \left(\left| \frac{x_0}{f(t_0, x_0)} \right| + M_g\, a \right) \phi(r) < r
$$

for all $r > 0$.

Thus, $\mathcal{A}$ and $\mathcal{B}$ satisfy all the conditions of Theorem 2.1 and we apply it to conclude that the operator equation $\mathcal{A}x\,\mathcal{B}x = x$ has a solution. Consequently the integral Equation 3.3 and the QDE (1.1) has a solution x^* defined on J. Furthermore, the sequence $\{x_n\}_{n=1}^{\infty}$ of successive approximations defined by (3.5) converges monotonically to x^*. This completes the proof.

Remark 3.2 The conclusion of Theorem 3.1 also remains true if we replace the hypothesis (**B**$_4$) with the following:

(B_4') The QDE (1.1) has an upper solution $v \in C^1(J, \mathbb{R})$.

The proof under this new hypothesis is similar to the proof of Theorem 3.1 with appropriate modifications.

Example 3.1 Given a closed and bounded interval $J = [0, 1]$, consider the IVP of QDE,

$$\left.\begin{aligned} &\frac{d}{dt}\left[\frac{x(t)}{f(t,x(t))}\right] = \frac{1}{4}[2 + \tanh x(t)],\ t \in J, \\ &x(0) = 0 \in \mathbb{R} \end{aligned}\right\} \tag{3.12}$$

where the functions $f, g: J \times \mathbb{R} \to \mathbb{R}$ are defined as

$$f(t,x) = \begin{cases} 1, & \text{if } x \le 0, \\ 1 + x, & \text{if } 0 < x < 3, \\ 4, & \text{if } , x \ge 3 \end{cases}$$

and

$$g(t,x) = \frac{1}{4}[2 + \tanh x]$$

Clearly, the functions f and g are continuous on $J \times \mathbb{R}$ into $\mathbb{R}_+$. The function f satisfies the hypothesis (A_3) with $\phi(r) = r$. To see this, we have

$$0 \le f(t,x) - f(t,y) \le x - y$$

for all $x, y \in \mathbb{R}$, $x \ge y$. Therefore, $\phi(r) = r$. Moreover, the function $f(t,x)$ is positive and bounded on $J \times \mathbb{R}$ with bound $M_f = 4$ and so the hypothesis (A_2) is satisfied. Again, since g is positive and bounded on $J \times \mathbb{R}$ by $M_g = \frac{3}{4}$, the hypothesis (B_2) holds. Furthermore, $g(t, x)$ is nondecreasing in x for all $t \in J$, and thus hypothesis (B_3) is satisfied. Also, condition (3.4) of Theorem 3.1 is held. Finally, the QDE (3.12) has a lower solution $u(t) = \frac{t}{4}$ defined on J, thus all hypotheses of Theorem 3.1 are satisfied. Hence, we apply Theorem 3.1 and conclude that the QDE (3.12) has a solution x^* defined on J and the sequence $\{x_n\}_{n=1}^{\infty}$ defined by

$$x_{n+1}(t) = \frac{1}{4}\,[f(t, x_n(t))]\left(\int_0^t [2 + \tan hx_n(s)]\ ds\right) \tag{3.13}$$

for all $t \in J$, where $x_1 = u$, converges monotonically to x^*.

Funding
The authors received no direct funding for this research.

Author details
Bapurao C. Dhage[1]
E-mail: bcdhage@gmail.com
Shyam B. Dhage[1]
E-mail: sbdhage4791@gmail.com
[1] Kasubai, Gurukul Colony, Ahmedpur, 413 515 Maharashtra, India.

References

Dhage, B. C. (1999). Fixed point theorems in ordered Banach algebras and applications. *PanAmerican Mathematical Journal, 9*, 93–102.

Dhage, B. C. (2009). Local asymptotic attractivity for nonlinear quadratic functional integral equations. *Nonlinear Analysis, 70*, 1912–1922.

Dhage, B. C. (2010). Quadratic perturbations of periodic boundary value problems of second order ordinary differential equations. *Differential Equations and Applications, 2*, 465–486.

Dhage, B. C. (2013). Hybrid fixed point theory in partially ordered normed linear spaces and applications to fractional integral equations. *Differential Equations and Applications, 5*, 155–184.

Dhage, B. C. (2014a). Global attractivity results for comparable solutions of nonlinear hybrid fractional integral equations. *Differential Equations and Applications, 6*, 165–186.

Dhage, B. C. (2014b). Partially condensing mappings in partially ordered normed linear spaces and applications to functional integral equations. *Tamkang Journal of Mathematics, 45*, 397–426.

Dhage, B. C., & Dhage, S. B. (in press). Approximating solutions of nonlinear first order ordinary differential equations. *Global Journal of Mathematical Sciences, 3*.

Dhage, B. C., Dhage, S. B., & Ntouyas, S. K. (2014). Approximating solutions of nonlinear hybrid differential equations. *Applied Mathematics Letters, 34*, 76–80.

Dhage, B. C., & Lakshmikantham, V. (2010). Basic results on hybrid differential equations. *Nonlinear Analysis: Hybrid Systems, 4*, 414–424.

Dhage, B. C., & Regan, D. O. (2000). A fixed point theorem in Banach Algebras with applications to functional integral equations. *Functional Differential Equations, 7*, 259–267.

Heikkilä, S., & Lakshmikantham, V. (1994). *Monotone iterative techniques for discontinuous nonlinear differential equations*. New York, NY: Marcel Dekker.

Nieto, J. J., & Rodriguez-Lopez, R. (2005). Contractive mappings theorems in partially ordered sets and applications to ordinary differential equations. *Order, 22*, 223–239.

4

Iterative solvers for the Maxwell–Stefan diffusion equations: Methods and applications in plasma and particle transport

Jürgen Geiser[1*]

*Corresponding author: Jürgen Geiser, The Institute of Theoretical Electrical Engineering, Ruhr University of Bochum, Universitätsstrasse 150, D-44801 Bochum, Germany

E-mail: juergen.geiser@ruhr-uni-bochum.de

Reviewing editor: Yong Hong Wu, Curtin University of Technology, Australia

Abstract: In this paper, we are motivated to discuss a model based on a local thermodynamic equilibrium, weakly ionized plasma-mixture model used for medical and technical applications in etching processes. For studying the model, we consider a simplified model based on the Maxwell–Stefan model, which describes multicomponent diffusive fluxes in the gas mixture. The MS model is more adequate to describe complex mixtures without dominating background species. Based on additional conditions to the fluxes, we obtain an irreducible and quasi-positive diffusion matrix. Such problems result in nonlinear diffusion equations, which are more delicate to solve as simpler standard diffusion equations with Fickian's approach. Here, we propose an efficient explicit time-discretization method, which is embedded to a fast iterative solver for the nonlinearities. Such a combination of coupling discretization and solver methods allows to simulate the delicate nonlinear differential equations more effectively. We present the efficiency and accuracy of the iterative solvers for some first ternary component gaseous mixtures and discuss the details of the numerical methods.

Subject: Applied Mathematics; Mathematics Statistics; Science

Keywords: Maxwell–Stefan approach; plasma model; multicomponent mixture; explicit discretization schemes; iterative schemes

AMS subject classifications: 35K25; 35K20; 74S10; 70G65

ABOUT THE AUTHOR

Jürgen Geiser (http://homepage.ruhr-uni-bochum.de/Juergen.Geiser/), researcher and lecturer at the Ruhr-University of Bochum, Germany, has been involved in teaching and research projects and has collaborated with engineering and physicist groups on numerical modeling of technical and physical models. The research activity refers to the mathematical modeling, numerics, and analysis of transport and flow problems in engineering applications, e.g. groundwater modeling and plasma modelling. He is a specialist in multiscale solvers and iterative solvers and most of the topics of the special issue. Moreover, Juergen Geiser is the author of scientific books and editor of various scientific journals, and thus able to manage the editorial activity.

PUBLIC INTEREST STATEMENT

A multicomponent model based on the Stefan–Maxwell approach is presented. Iterative solver approaches are used to solve the nonlinear modelling problem.

1. Introduction

We are motivated to understand the gaseous mixtures of a normal pressure and room temperature plasma. The understanding of normal pressure and room temperature plasma applications is important for applications in medical and technical processes. Since many years, the increasing importance of plasma chemistry based on the multicomponent plasma is a key factor in understanding the gaseous mixture processes (see for low pressure plasma Senega & Brinkmann, 2006 and for atmospheric pressure regimes Tanaka, 2004).

We consider a simplified Maxwell–Stefan diffusion (MSD) equation to model the gaseous mixture of multicomponent plasma. Here, we consider of a macroscopic model, while the limits to apply to a kinetic (microscopic) model are discussed in Boudin, Grec, and Salvarani (2015). While the most classical description of the diffusion goes back to the Fickian's approach (see Fick, 1995), we apply the modern description of the multicomponent diffusion based on the Maxwell–Stefan's approach (see Maxwell, 1867). The novel approach considers a more detailed description of the flux and concentration, which are indeed not only proportionally coupled as in the simplified Fickian's approach. Here, we deal with an inter-species force balance, which allows to model cross-effects, e.g. the so-called reverse diffusion (uphill diffusion in the direction of the gradients).

Such a more detailed modeling results in irreducible and quasi-positive diffusion matrices, which can be reduced by transforming with reductions or with Perron–Frobenius theorems to the solvable partial differential equations (see Bothe, 2011). The obtained system of nonlinear partial differential equations is delicate to solve with standard discretization and solver methods. Therefore, we have taken into account effective linearization methods, e.g. iterative fix-point schemes, to overcome the nonlinearities. Alternative methods exist and are explained in Böttcher (2010) and Spille-Kohoff, Preuß, and Böttcher (2012). Here, they reduce the MSD equation and solve it explicitly, but such methods are restricted to ternary or quaternary systems. Further multicomponent approaches with MSD equations are discussed for only stationary problems, while the solver methods embed the nonlinear MSD approaches into the finite volume discretization schemes (see Peerenboom, van Boxtel, Janssen, & van Dijk, 2014). Such approximations lack with respect to solve nonstationary problems.

The paper is outlined as follows.

In Section 2, we present our mathematical model. A possible reduced model for the further approximations is derived in Section 3. In Section 4, we discuss the underlying numerical schemes. The first numerical results are presented in Section 5. In the contents, that are given in Section 6, we summarize our results.

2. Mathematical model

For the full plasma model, we assume that the neutral particles can be described as the fluid dynamical model, where the elastic collision defines the dynamics and few inelastic collisions are, among other reasons, responsible for the chemical reactions.

To describe the individual mass densities, as well as the global momentum and the global energy as the dynamical conservation quantities of the system, corresponding conservation equations are derived from Boltzmann equations.

The individual character of each species is considered by mass conservation equations and the so-called difference equations.

The extension of the nonmixtured multicomponent transport model (Senega & Brinkmann, 2006) is done with respect to the collision integrals related to the right-hind side sources of the conservation laws.

The conservation laws of the neutral elements are given as

$$\frac{\partial}{\partial t}\rho_s + \frac{\partial}{\partial \boldsymbol{r}} \cdot \rho_s \boldsymbol{u}_s = m_s Q_n^{(s)},$$
$$\frac{\partial}{\partial t}\rho \boldsymbol{u} + \frac{\partial}{\partial \boldsymbol{r}} \cdot \left(\underline{\underline{P}}^* + \rho \boldsymbol{u}\boldsymbol{u}\right) = -Q_m^{(e)},$$
$$\frac{\partial}{\partial t}\mathcal{E}_{tot}^* + \frac{\partial}{\partial \boldsymbol{r}} \cdot \left(\mathcal{E}_{tot}^* \boldsymbol{u} + \boldsymbol{q}^* + \underline{\underline{P}}^* \cdot \boldsymbol{u}\right) = -Q_{\mathcal{E}}^{(e)},$$

where ρ_s : density of species i, $\rho = \sum_{i=1}^{N} \rho_i$, $\boldsymbol{u}$: velocity, and $\mathcal{E}_{tot}^*$: total energy of the neutral particles.

Further, the variable $Q_n^{(s)}$ is the collision term of the mass conservation equation, $Q_m^{(e)}$ is the collision term of the momentum conservation equation, and $Q_{\mathcal{E}}^{(e)}$ is the collision term of the energy conservation equation.

We derive the collision term with respect to the Chapmen–Enskog method (see Chapman & Cowling, 1990) and achieve for the first derivatives the following results:

$$m_s Q_n^{(s)} = -\nabla \cdot (\rho_i \sum_{j=0} \mathbf{V}_i^j), \tag{1}$$

$$Q_m^{(e)} = -\sum_{i=1}^{n_s} \rho_i F_i, \tag{2}$$

$$Q_{\mathcal{E}}^{(e)} = -\sum_{i=1}^{n_s} \rho_i \rho F_i (\mathbf{u} + \sum_{j=0} V_i^{(j)}), \tag{3}$$

where $i = 1, \ldots, n_s$, F_i is an external force per unit mass (see Boltzmann equation); further, the diffusion velocity is given as:

$$\mathbf{V}_i^0 = 0, \tag{4}$$

$$\mathbf{V}_i^1 = -\sum_{j=1}^{N} D_{ij}(d_j + k_{T_j}\frac{\Delta T}{T}), \tag{5}$$

where $\sum_{i=1}^{N} d_i = 0$,

$$d_i = \nabla x_i + x_i \frac{\nabla p}{p} - \frac{\rho_i}{\rho} F_i, \tag{6}$$

$$d_i = d^i - y_i \sum_j d_j^*, \tag{7}$$

where $x_i = \frac{n_s}{n}$ is the molar fraction of species i.

We have an additional constraint based on the mass fraction of each species:

$$\frac{\partial}{\partial t} y_i + \nabla y_i = R_i(y_1, \ldots, y_N), \tag{8}$$

where y_i is the mass fraction of species i and R_i is the net production rate of species i due to the reactions.

Remark 1 The full model problem considers a fully coupled system of conservation laws and Maxwell-Stefan equations. Each equation is coupled such that the gaseous mixture influences the transport equations and vice versa. In the following, we decouple the equations system and consider only the delicate Maxwell-Stefan equations.

3. Simplified model with Maxwell-Stefan diffusion equations

We discuss in the following a multicomponent gaseous mixture with three species (ternary mixture). The model problem is discussed in the experiments of Duncan and Toor (1962).

Here, they studied an ideal gaseous mixture of the following components:

(1) Hydrogen (H_2, first species),

(2) Nitrogen (N_2, second species), and

(3) Carbon dioxide (CO_2, third species).

The Maxwell-Stefan equations are given for the three species as (see also Boudin, Grec, & Salvarani, 2012):

$$\partial_t \xi_i + \nabla \cdot N_i = 0, \ 1 \leq i \leq 3, \tag{9}$$

$$\sum_{j=1}^{3} N_j = 0, \tag{10}$$

$$\frac{\xi_2 N_1 - \xi_1 N_2}{D_{12}} + \frac{\xi_3 N_1 - \xi_1 N_3}{D_{13}} = -\nabla \xi_1, \tag{11}$$

$$\frac{\xi_1 N_2 - \xi_2 N_1}{D_{12}} + \frac{\xi_3 N_2 - \xi_2 N_3}{D_{23}} = -\nabla \xi_2, \tag{12}$$

where the domain is given as $\Omega \in \mathbb{R}^d, d \in N^+$ with $\xi_i \in C^2$.

For such ternary mixture, we can rewrite the three differential Equations (9) and (11 and 12) with the help of the zero-condition (10) into two differential equations, given as:

$$\partial_t \xi_i + \nabla \cdot N_i = 0, \ 1 \leq i \leq 2, \tag{13}$$

$$\frac{1}{D_{13}} N_1 + \alpha N_1 \xi_2 - \alpha N_2 \xi_1 = -\nabla \xi_1, \tag{14}$$

$$\frac{1}{D_{23}} N_2 - \beta N_1 \xi_2 + \beta N_2 \xi_1 = -\nabla \xi_2, \tag{15}$$

where $\alpha = \left(\frac{1}{D_{12}} - \frac{1}{D_{13}}\right)$ and $\beta = \left(\frac{1}{D_{12}} - \frac{1}{D_{23}}\right)$.

Further, we have the relations:

(1) Third mole fraction: $\xi_3 = 1 - \xi_1 - \xi_2$,

(2) Third molar flux: $N_3 = -N_1 - N_2$.

4. Numerical methods

In the following, we discuss the numerical methods which are based on iterative schemes with embedded explicit discretization schemes (see also Geiser, 2015, in press). We apply the following methods:

(1) Iterative scheme in time (global linearization with matrix method),

(2) Iterative scheme in time (local linearization with Richardson's method).

For spatial discretization, we apply finite volume or finite difference methods. The underlying time discretization is based on a first-order explicit Euler method.

4.1. Iterative scheme in time (global linearization with matrix method)

We solve the iterative scheme:

$$\xi_1^{n+1} = \xi_1^n - \Delta t\, D_+ N_1^n, \tag{16}$$

$$\xi_2^{n+1} = \xi_2^n - \Delta t\, D_+ N_2^n, \tag{17}$$

$$\begin{pmatrix} A & B \\ C & D \end{pmatrix} \begin{pmatrix} N_1^{n+1} \\ N_2^{n+1} \end{pmatrix} = \begin{pmatrix} -D_- \xi_1^{n+1} \\ -D_- \xi_2^{n+1} \end{pmatrix}, \tag{18}$$

for $j = 0, \ldots, J$, where $\xi_1^n = (\xi_{1,0}^n, \ldots, \xi_{1,J}^n)^T$, $\xi_2^n = (\xi_{2,0}^n, \ldots, \xi_{2,J}^n)^T$ and $I_J \in \mathbb{R}^{J+1} \times \mathbb{R}^{J+1}$, $N_1^n = (N_{1,0}^n, \ldots, N_{1,J}^n)^T$, $N_2^n = (N_{2,0}^n, \ldots, N_{2,J}^n)^T$ and $I_J \in \mathbb{R}^{J+1} \times \mathbb{R}^{J+1}$, where $n = 0, 1, 2, \ldots, N_{end}$ and N_{end} are the number of time steps, i.d. $N_{end} = T/\Delta t$.

The matrices are given as:

$$A, B, C, D \in \mathbb{R}^{J+1} \times \mathbb{R}^{J+1}, \tag{19}$$

$$A_{j,j} = \frac{1}{D_{13}} + \alpha \xi_{2,j}, \quad j = 0 \ldots, J, \tag{20}$$

$$B_{j,j} = -\alpha \xi_{1,j}, \quad j = 0 \ldots, J, \tag{21}$$

$$C_{j,j} = -\beta \xi_{2,j}, \quad j = 0 \ldots, J, \tag{22}$$

$$D_{j,j} = \frac{1}{D_{23}} + \beta \xi_{1,j}, \quad j = 0 \ldots, J, \tag{23}$$

$$A_{i,j} = B_{i,j} = C_{i,j} = D_{i,j} = 0, \quad i,j = 0 \ldots, J,\ i \neq J, \tag{24}$$

meaning that the diagonal entries given for the scale case in Equation (13) and the outer diagonal entries are zero.

The explicit form with time discretization is given as:

Algorithm 1 *1.) Initialization $n = 0$:*

$$\begin{pmatrix} N_1^0 \\ N_2^0 \end{pmatrix} = \begin{pmatrix} \tilde{A} & \tilde{B} \\ \tilde{C} & \tilde{D} \end{pmatrix} \begin{pmatrix} -D_- \xi_1^0 \\ -D_- \xi_2^0 \end{pmatrix}, \tag{25}$$

where $\xi_1^0 = (\xi_{1,0}^0, \ldots, \xi_{1,J}^0)^T$, $\xi_2^0 = (\xi_{2,0}^0, \ldots, \xi_{2,J}^0)^T$ *and* $\xi_{1,j}^0 = \xi_1^{in}(j\Delta x)$, $\xi_{2,j}^0 = \xi_2^{in}(j\Delta x)$, $j = 0, \ldots, J$ *and given as for the different intialisations, we have:*

1. *Uphill example*

$$\xi_1^{in}(x) = \begin{cases} 0.8 & \text{if } 0 \le x < 0.25, \\ 1.6(0.75 - x) & \text{if } 0.25 \le x < 0.75, \\ 0.0 & \text{if } 0.75 \le x \le 1.0, \end{cases} \tag{26}$$

$$\xi_2^{in}(x) = 0.2, \text{ for all } x \in \Omega = [0,1]. \tag{27}$$

2. *Diffusion example (Asymptotic behavior)*

$$\xi_1^{in}(x) = \begin{cases} 0.8 \text{ if } 0 \le x \in 0.5, \\ 0.0 \text{ else}, \end{cases} \tag{28}$$

$$\xi_2^{in}(x) = 0.2, \text{ for all } x \in \Omega = [0,1]. \tag{29}$$

The inverse matrices are given as:

$$\tilde{A}, \tilde{B}, \tilde{C}, \tilde{D} \in \mathbb{R}^{J+1} \times \mathbb{R}^{J+1}, \tag{30}$$

$$\tilde{A}_{j,j} = \gamma_j \left(\frac{1}{D_{23}} + \beta \xi_{1,j}^0\right), \; j = 0 \ldots, J, \tag{31}$$

$$B_{j,j} = \gamma_j \, \alpha \xi_{1,j}^0, \; j = 0 \ldots, J, \tag{32}$$

$$C_{j,j} = \gamma_j \, \beta \xi_{2,j}^0, \; j = 0 \ldots, J, \tag{33}$$

$$D_{j,j} = \gamma_j \left(\frac{1}{D_{13}} + \alpha \xi_{2,j}^0\right), \; j = 0 \ldots, J, \tag{34}$$

$$\gamma_j = \frac{D_{13} D_{23}}{1 + \alpha D_{13} \xi_{2,j}^0 + \beta D_{23} \xi_{1,j}^0}, \; j = 0 \ldots, J, \tag{35}$$

$$\tilde{A}_{i,j} = \tilde{B}_{i,j} = \tilde{C}_{i,j} = \tilde{D}_{i,j} = 0, \; i,j = 0 \ldots, J, \; i \ne J. \tag{36}$$

Further the values of the first and the last grid points of N are zero, means $N_{1,0}^0 = N_{1,J}^0 = N_{2,0}^0 = N_{2,J}^0 = 0$ *(boundary condition).*

2.) Next time-steps (till $n = N_{end}$):

2.1) Computation of ξ_1^{n+1} and ξ_2^{n+1}

$$\xi_1^{n+1} = \xi_1^n - \Delta t \, D_+ N_1^n, \tag{37}$$

$$\xi_2^{n+1} = \xi_2^n - \Delta t \, D_+ N_2^n. \tag{38}$$

2.2) Computation of N_1^{n+1} and N_2^{n+1}

$$\begin{pmatrix} N_1^{n+1} \\ N_2^{n+1} \end{pmatrix} = \begin{pmatrix} \tilde{A} & \tilde{B} \\ \tilde{C} & \tilde{D} \end{pmatrix} \begin{pmatrix} -D_- \xi_1^{n+1} \\ -D_- \xi_2^{n+1} \end{pmatrix}, \tag{39}$$

where $\xi_1^n = (\xi_{1,0}^n, \ldots, \xi_{1,J}^n)^T$, $\xi_2^n = (\xi_{2,0}^n, \ldots, \xi_{2,J}^n)^T$ *and the inverse matrices are given as:*

$$\tilde{A}, \tilde{B}, \tilde{C}, \tilde{D} \in \mathbb{R}^{J+1} \times \mathbb{R}^{J+1}, \tag{40}$$

$$\tilde{A}_{j,j} = \gamma_j(\frac{1}{D_{23}} + \beta \xi_{1,j}^{n+1}),\ j = 0 \ldots, J, \tag{41}$$

$$B_{j,j} = \gamma_j\ \alpha \xi_{1,j}^{n+1},\ j = 0 \ldots, J, \tag{42}$$

$$C_{j,j} = \gamma_j\ \beta \xi_{2,j}^{n+1},\ j = 0 \ldots, J, \tag{43}$$

$$D_{j,j} = \gamma_j(\frac{1}{D_{13}} + \alpha \xi_{2,j}^{n+1}),\ j = 0 \ldots, J, \tag{44}$$

$$\gamma_j = \frac{D_{13} D_{23}}{1 + \alpha D_{13} \xi_{2,j}^{n+1} + \beta D_{23} \xi_{1,j}^{n+1}},\ j = 0 \ldots, J, \tag{45}$$

$$\tilde{A}_{i,j} = \tilde{B}_{i,j} = \tilde{C}_{i,j} = \tilde{D}_{i,j} = 0,\ i, j = 0 \ldots, J,\ i \neq J. \tag{46}$$

Further the values of the first and the last grid points of N *are zero, means* $N_{1,0}^n = N_{1,J}^n = N_{2,0}^n = N_{2,J}^n = 0$ *(boundary condition).*

3.) Do $n = n + 1$ *and goto 2.) .*

4.2. Iterative scheme in time (local linearization with Richardson's method

We solve the iterative scheme given in the Richardson iterative scheme:

$$\xi_1^{n+1,k} = \xi_1^n - \Delta t\ D_+ N_1^{n+1}, \tag{47}$$

$$\xi_2^{n+1,k} = \xi_2^n - \Delta t\ D_+ N_2^{n+1}, \tag{48}$$

$$\begin{pmatrix} A^{n+1,k-1} & B^{n+1,k-1} \\ C^{n+1,k-1} & D^{n+1,k-1} \end{pmatrix} \begin{pmatrix} N_1^{n+1} \\ N_2^{n+1} \end{pmatrix} = \begin{pmatrix} -D_- \xi_1^{n+1,k-1} \\ -D_- \xi_2^{n+1,k-1} \end{pmatrix}, \tag{49}$$

for $j = 0, \ldots, J$, where $\xi_1^n = (\xi_{1,0}^n, \ldots, \xi_{1,J}^n)^T$, $\xi_2^n = (\xi_{2,0}^n, \ldots, \xi_{2,J}^n)^T$ and $I_J \in \mathbb{R}^{J+1} \times \mathbb{R}^{J+1}$, $N_1^n = (N_{1,0}^n, \ldots, N_{1,J}^n)^T$, $N_2^n = (N_{2,0}^n, \ldots, N_{2,J}^n)^T$ and $I_J \in \mathbb{R}^{J+1} \times \mathbb{R}^{J+1}$, where $n = 0, 1, 2, \ldots, N_{end}$ and N_{end} are the number of time steps, i.d. $N_{end} = T/\Delta t$.

Further, $k = 1, 2, \ldots, K$ is the iteration index where $\xi_1^{n+1,0} = (\xi_{1,0}^n, \ldots, \xi_{1,J}^n)^T$, $\xi_2^{n+1,0} = (\xi_{2,0}^n, \ldots, \xi_{2,J}^n)^T$, and $I_J \in \mathbb{R}^{J+1} \times \mathbb{R}^{J+1}$ is the start solution given with the solution at $t = t^n$.

The matrices are given as:

$$A^{n+1,k-1}, B^{n+1,k-1}, C^{n+1,k-1}, D^{n+1,k-1} \in \mathbb{R}^{J+1} \times \mathbb{R}^{J+1}, \tag{50}$$

$$A_{j,j}^{n+1,k-1} = \frac{1}{D_{13}} + \alpha \xi_{2,j}^{n+1,k-1}, \quad j = 0 \ldots, J, \tag{51}$$

$$B_{j,j}^{n+1,k-1} = -\alpha \xi_{1,j}^{n+1,k-1}, \quad j = 0 \ldots, J, \tag{52}$$

$$C_{j,j}^{n+1,k-1} = -\beta \xi_{2,j}^{n+1,k-1}, \quad j = 0 \ldots, J, \tag{53}$$

$$D_{j,j}^{n+1,k-1} = \frac{1}{D_{23}} + \beta \xi_{1,j}^{n+1,k-1}, \quad j = 0 \ldots, J, \tag{54}$$

$$A_{i,j}^{n+1,i-1} = B_{i,j}^{n+1,i-1} = C_{i,j}^{n+1,i-1} = D_{i,j}^{n+1,i-1} = 0, \quad i,j = 0 \ldots, J,\ i \neq J, \tag{55}$$

meaning that the diagonal entries given for the scale case in Equation (95) and the outer diagonal entries are zero.

The explicit form with time discretization is given as:

Algorithm 2 *1.) Initialization $n = 0$ with an explicit time-step (CFL condition is given):*

$$\begin{pmatrix} N_1^0 \\ N_2^0 \end{pmatrix} = \begin{pmatrix} \tilde{A} & \tilde{B} \\ \tilde{C} & \tilde{D} \end{pmatrix} \begin{pmatrix} -D_- \xi_1^0 \\ -D_- \xi_2^0 \end{pmatrix}, \tag{56}$$

where $\xi_1^0 = (\xi_{1,0}^0, \ldots, \xi_{1,J}^0)^T$,
$\xi_2^0 = (\xi_{2,0}^0, \ldots, \xi_{2,J}^0)^T$ *and* $\xi_{1,j}^0 = \xi_1^{in}(j\Delta x)$, $\xi_{2,j}^0 = \xi_2^{in}(j\Delta x)$, $j = 0, \ldots, J$ *and given as for the different intializations, we have:*

1. *Uphill example*

$$\xi_1^{in}(x) = \begin{cases} 0.8 & \text{if } 0 \le x < 0.25, \\ 1.6(0.75 - x) & \text{if } 0.25 \le x < 0.75, \\ 0.0 & \text{if } 0.75 \le x \le 1.0, \end{cases} \tag{57}$$

$$\xi_2^{in}(x) = 0.2, \text{ for all } x \in \Omega = [0, 1]. \tag{58}$$

2. *Diffusion example (Asymptotic behavior)*

$$\xi_1^{in}(x) = \begin{cases} 0.8 \text{ if } 0 \le x \in 0.5, \\ 0.0 \text{ else}, \end{cases} \tag{59}$$

$$\xi_2^{in}(x) = 0.2, \text{ for all } x \in \Omega = [0, 1]. \tag{60}$$

The inverse matrices are given as:

$$\tilde{A}, \tilde{B}, \tilde{C}, \tilde{D} \in \mathbb{R}^{J+1} \times \mathbb{R}^{J+1}, \tag{61}$$

$$\tilde{A}_{j,j} = \gamma_j \left(\frac{1}{D_{23}} + \beta \xi_{1,j}^0\right), \; j = 0 \ldots, J, \tag{62}$$

$$B_{j,j} = \gamma_j \; \alpha \xi_{1,j}^0, \; j = 0 \ldots, J, \tag{63}$$

$$C_{j,j} = \gamma_j \; \beta \xi_{2,j}^0, \; j = 0 \ldots, J, \tag{64}$$

$$D_{j,j} = \gamma_j \left(\frac{1}{D_{13}} + \alpha \xi_{2,j}^0\right), \; j = 0 \ldots, J, \tag{65}$$

$$\gamma_j = \frac{D_{13} D_{23}}{1 + \alpha D_{13} \xi_{2,j}^0 + \beta D_{23} \xi_{1,j}^0}, \; j = 0 \ldots, J, \tag{66}$$

$$\tilde{A}_{i,j} = \tilde{B}_{i,j} = \tilde{C}_{i,j} = \tilde{D}_{i,j} = 0, \; i, j = 0 \ldots, J, \; i \neq J. \tag{67}$$

Further the values of the first and the last grid points of N are zero, means $N_{1,0}^0 = N_{1,J}^0 = N_{2,0}^0 = N_{2,J}^0 = 0$ *(boundary condition).*

2.) Next timesteps (till $n = N_{end}$) (iterative scheme restricted via the CFL condition based on the previous iterative solutions in the matrices):

2.1) Computation of $\xi_1^{n+1,I}$ *and* $\xi_2^{n+1,I}$

$$\xi_1^{n+1,k} = \xi_1^n - \Delta t \; D_+ N_1^{n+1}, \tag{68}$$

$$\xi_2^{n+1,k} = \xi_2^n - \Delta t \; D_+ N_2^{n+1}. \tag{69}$$

2.2) Computation of $N_1^{n+1,k-1}$ *and* $N_2^{n+1,k-1}$

$$\begin{pmatrix} N_1^{n+1} \\ N_2^{n+1} \end{pmatrix} = \begin{pmatrix} \tilde{A}^{n+1,k-1} & \tilde{B}^{n+1,k-1} \\ \tilde{C}^{n+1,k-1} & \tilde{D}^{n+1,k-1} \end{pmatrix} \begin{pmatrix} -D_{-}\xi_1^{n+1,k-1} \\ -D_{-}\xi_2^{n+1,k-1} \end{pmatrix}, \tag{70}$$

where $\xi_1^n = (\xi_{1,0}^n, \ldots, \xi_{1,J}^n)^T$, $\xi_2^n = (\xi_{2,0}^n, \ldots, \xi_{2,J}^n)^T$ *and the inverse matrices are given as:*

$$\tilde{A}^{n+1,k-1}, \tilde{B}^{n+1,k-1}, \tilde{C}^{n+1,k-1}, \tilde{D}^{n+1,k-1} \in \mathbb{R}^{J+1} \times \mathbb{R}^{J+1}, \tag{71}$$

$$\tilde{A}_{j,j}^{n+1,k-1} = \gamma_j\left(\frac{1}{D_{23}} + \beta\xi_{1,j}^{n+1,k-1}\right),\ j = 0\ldots, J, \tag{72}$$

$$B_{j,j}^{n+1,k-1} = \gamma_j\ \alpha\xi_{1,j}^{n+1,k-1},\ j = 0\ldots, J, \tag{73}$$

$$C_{j,j}^{n+1,k-1} = \gamma_j\ \beta\xi_{2,j}^{n+1,k-1},\ j = 0\ldots, J, \tag{74}$$

$$D_{j,j}^{n+1,k-1} = \gamma_j\left(\frac{1}{D_{13}} + \alpha\xi_{2,j}^{n+1,k-1}\right),\ j = 0\ldots, J, \tag{75}$$

$$\gamma_j = \frac{D_{13}D_{23}}{1 + \alpha D_{13}\xi_{2,j}^{n+1,k-1} + \beta D_{23}\xi_{1,j}^{n+1,k-1}},\ j = 0\ldots, J, \tag{76}$$

$$\tilde{A}_{i,j}^{n+1,k-1} = \tilde{B}_{i,j}^{n+1,k-1} = \tilde{C}_{i,j}^{n+1,k-1} = \tilde{D}_{i,j}^{n+1,k-1} = 0, \tag{77}$$

$$i, j = 0\ldots, J,\ i \neq J.$$

Further the values of the first and the last grid points of N *are zero, means* $N_{1,0}^{n+1} = N_{1,J}^{n+1} = N_{2,0}^{n+1} = N_{2,J}^{n+1} = 0$ *(boundary condition).*

Further $k = 1, 2, \ldots, K$ *is the iteration index with*
where $\xi_1^{n+1,0} = (\xi_{1,0}^n, \ldots, \xi_{1,J}^n)^T$, $\xi_2^{n+1,0} = (\xi_{2,0}^n, \ldots, \xi_{2,J}^n)^T$ *and*
$I_J \in \mathbb{R}^{J+1} \times \mathbb{R}^{J+1}$ *is the start solution given with the solution at* $t = t^n$.

3.) Do $n = n + 1$ *and goto 2.) .*

5. Numerical experiments

In the following, we concentrate on the following three-component system, which is given as:

$$\partial_t \xi_i + \partial_x N_i = 0, \quad 1 \le i \le 3, \tag{78}$$

$$\sum_{j=1}^{3} N_j = 0, \tag{79}$$

$$\frac{\xi_2 N_1 - \xi_1 N_2}{D_{12}} + \frac{\xi_3 N_1 - \xi_1 N_3}{D_{13}} = -\partial_x \xi_1, \tag{80}$$

$$\frac{\xi_1 N_2 - \xi_2 N_1}{D_{12}} + \frac{\xi_3 N_2 - \xi_2 N_3}{D_{23}} = -\partial_x \xi_2, \tag{81}$$

where the domain is given as $\Omega \in \mathbb{R}^d, d \in N^+$ with $\xi_i \in C^2$.

The parameters and the initial and boundary conditions are given as:

(1) $D_{12} = D_{13} = 0.833$ (means $\alpha = 0$) and $D_{23} = 0.168$ (Uphill diffusion, semi-degenerated Duncan and Toor experiment),

(2) $D_{12} = 0.0833, D_{13} = 0.680$ and $D_{23} = 0.168$ (asymptotic behavior, Duncan and Toor experiment (see Duncan & Toor, 1962)),

(3) $J = 140$ (spatial grid points),

(4) The time step restriction for the explicit method is given as: $\Delta t \leq \frac{(\Delta x)^2}{2\max\{D_{12}, D_{13}, D_{23}\}}$,

(5) The spatial domain is $\Omega = [0, 1]$, the time-domain $[0, T] = [0, 1]$,

(6) The initial conditions are:

(i) Uphill example

$$\xi_1^{in}(x) = \begin{cases} 0.8 & \text{if } 0 \leq x < 0.25, \\ 1.6(0.75 - x) & \text{if } 0.25 \leq x < 0.75, \\ 0.0 & \text{if } 0.75 \leq x \leq 1.0, \end{cases} \tag{82}$$

$$\xi_2^{in}(x) = 0.2, \quad \text{for all } x \in \Omega = [0, 1]. \tag{83}$$

(ii) Diffusion example (Asymptotic behavior)

$$\xi_1^{in}(x) = \begin{cases} 0.8 & \text{if } 0 \leq x \in 0.5, \\ 0.0 & \text{else}, \end{cases} \tag{84}$$

$$\xi_2^{in}(x) = 0.2, \quad \text{for all } x \in \Omega = [0, 1]. \tag{85}$$

(7) The boundary conditions are of no-flux type:

$$N_1 = N_2 = N_3 = 0, \text{ on } \partial\Omega \times [0, 1]. \tag{86}$$

We could reduce to a simpler model problem as:

$$\partial_t \xi_i + \partial_x \cdot N_i = 0, \; 1 \leq i \leq 2, \tag{87}$$

$$\frac{1}{D_{13}} N_1 + \alpha N_1 \xi_2 - \alpha N_2 \xi_1 = -\partial_x \xi_1, \tag{88}$$

$$\frac{1}{D_{23}} N_2 - \beta N_1 \xi_2 + \beta N_2 \xi_1 = -\partial_x \xi_2, \tag{89}$$

where $\alpha = \left(\frac{1}{D_{12}} - \frac{1}{D_{13}}\right), \beta = \left(\frac{1}{D_{12}} - \frac{1}{D_{23}}\right)$.

We rewrite Into:

$$\partial_t \xi_1 + \partial_x \cdot N_1 = 0, \tag{90}$$

$$\partial_t \xi_2 + \partial_x \cdot N_2 = 0, \tag{91}$$

$$\begin{pmatrix} \frac{1}{D_{13}} + \alpha \xi_2 & -\alpha \xi_1 \\ -\beta \xi_2 & \frac{1}{D_{23}} + \beta \xi_1 \end{pmatrix} \begin{pmatrix} N_1 \\ N_2 \end{pmatrix} = \begin{pmatrix} -\partial_x \xi_1 \\ -\partial_x \xi_2 \end{pmatrix}, \tag{92}$$

and we have

$$\partial_t \xi_1 + \partial_x \cdot N_1 = 0, \tag{93}$$

$$\partial_t \xi_2 + \partial_x \cdot N_2 = 0, \tag{94}$$

$$\begin{pmatrix} N_1 \\ N_2 \end{pmatrix} = \frac{D_{13}D_{23}}{1 + \alpha D_{13}\xi_2 + \beta D_{23}\xi_1} \begin{pmatrix} \frac{1}{D_{23}} + \beta\xi_1 & \alpha\xi_1 \\ \beta\xi_2 & \frac{1}{D_{13}} + \alpha\xi_2 \end{pmatrix} \begin{pmatrix} -\partial_x \xi_1 \\ -\partial_x \xi_2 \end{pmatrix}. \tag{95}$$

The next step is to apply the semi-discretization of the partial differential operator $\frac{\partial}{\partial x}$.

We apply the first differential operator in Equations (93) and (94) as an forward upwind scheme given as

$$\frac{\partial}{\partial x} = D_+ = \frac{1}{\Delta x} \cdot \begin{pmatrix} -1 & 0 & \dots & & 0 \\ 1 & -1 & 0 & \dots & 0 \\ \vdots & \ddots & \ddots & \ddots & \vdots \\ 0 & & 1 & -1 & 0 \\ 0 & \dots & 0 & 1 & -1 \end{pmatrix} \in \mathbb{R}^{(J+1)\times(J+1)}, \tag{96}$$

and the second differential operator in Equation (95) as an backward upwind scheme given as

$$\frac{\partial}{\partial x} = D_- = \frac{1}{\Delta x} \cdot \begin{pmatrix} -1 & 1 & 0 & \dots & 0 \\ 0 & -1 & 1 & 0 & \dots \\ \vdots & \ddots & \ddots & \ddots & \ddots \\ 0 & \dots & 0 & -1 & 1 \\ 0 & & \dots & 0 & -1 \end{pmatrix} \in \mathbb{R}^{(J+1)\times(J+1)}. \tag{97}$$

5.1. Experiments with the iterative scheme in time (global linearization)

In the first experiments, we test the first iterative scheme (iterative scheme in time (global linearization)).

We test the different experiments (uphill and diffusion examples) and obtain the results as shown in Figure 1 for the uphill example.

Remark 2 In Figure 1, we obtain a typical complex mixture, without dominating background gas. While the concentration ξ_1 increases and ξ_3 decreases, the important concentration ξ_2 decreases and increases around the value 0.2. Such a behavior cannot be produced with simple standard Fickian's approach and here it is important to deal with the MS approach.

The concentration and their fluxes are given in Figure 2.

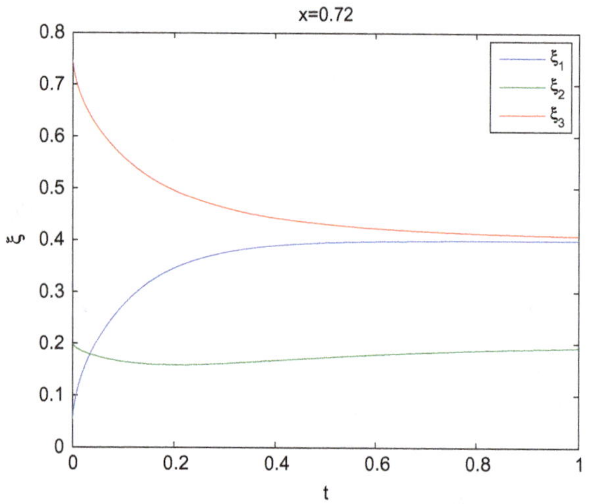

Figure 1. Results of the mole fraction (concentration) ξ_1, ξ_2 and ξ_3.

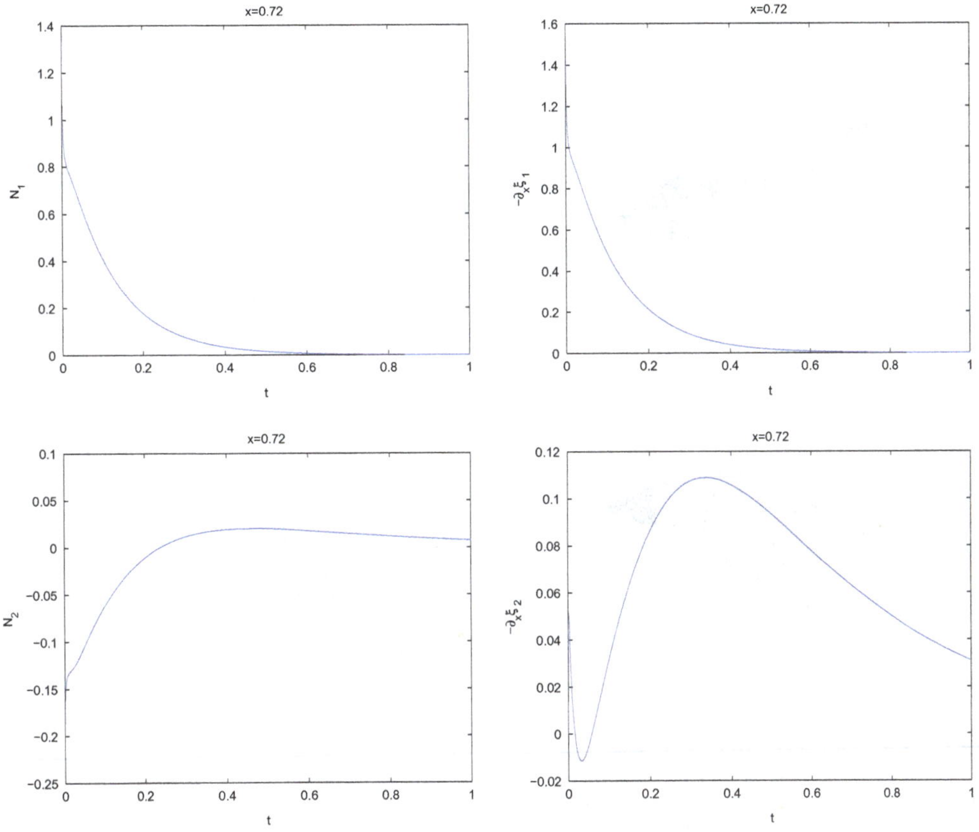

Figure 2. Concentration and flux of iterative scheme in time (global linearization).

Notes: The upper figures present the results of the flux N_1 and concentration gradient$-\partial_x\xi_1$.
The lower figures present the results of flux N_2 and concentration gradient $-\partial_x\xi_2$.

Remark 3 In Figure 2, we obtain a fine resolved behavior of the important second species ξ_2. Here, we see detailed the oscillatory behavior of the flux N_2 and the concentration gradient $-\partial_x\xi_2$ in the time interval $t = [0, 1]$. For the next time interval $t = [1, 2]$, we obtained a stabilized behavior of the complex mixture. Here, it is important to resolve the nonlinearity very accurately for unstable initialization of the complex mixture.

The full plots in time and space of the concentrations and their fluxes are given in Figure 3.

The space–time regions where $-N2\partial_x\xi_2 \geq 0$ for the uphill diffusion and asymptotic diffusion, given in Figure 4.

Remark 4 In Figure 3, we present the delicate unstable behavior of the mixture around time scale $t = [0, 1]$ and the spatial scale $x = [0, 1]$ in a 3D plot with the first and second concentration. The steep gradients of the concentrations have to be resolved with a very fine time step. We also obtain the same delicate results of the initialization process in Figure 4. Here, we present the space–time regions for the uphill and asymptotic diffusions. In both experiments, we see that we have a backflow of the mixture, meaning the second concentration ξ_2 has both decreasing and increasing gradients. For both experiments, we resolve the initial process with fine time steps.

Remark 5 In the first numerical method, we apply global linearization based on the time steps. Meaning, we deal with an explicit time discretization that solves the linearized equation forward. All effects are resolved by taking into account the CFL condition. Therefore, we achieve better results with finer time steps, e.g. $\Delta t_{CFl}/8$, such that the global linearization, via the time step, is important for our first numerical method.

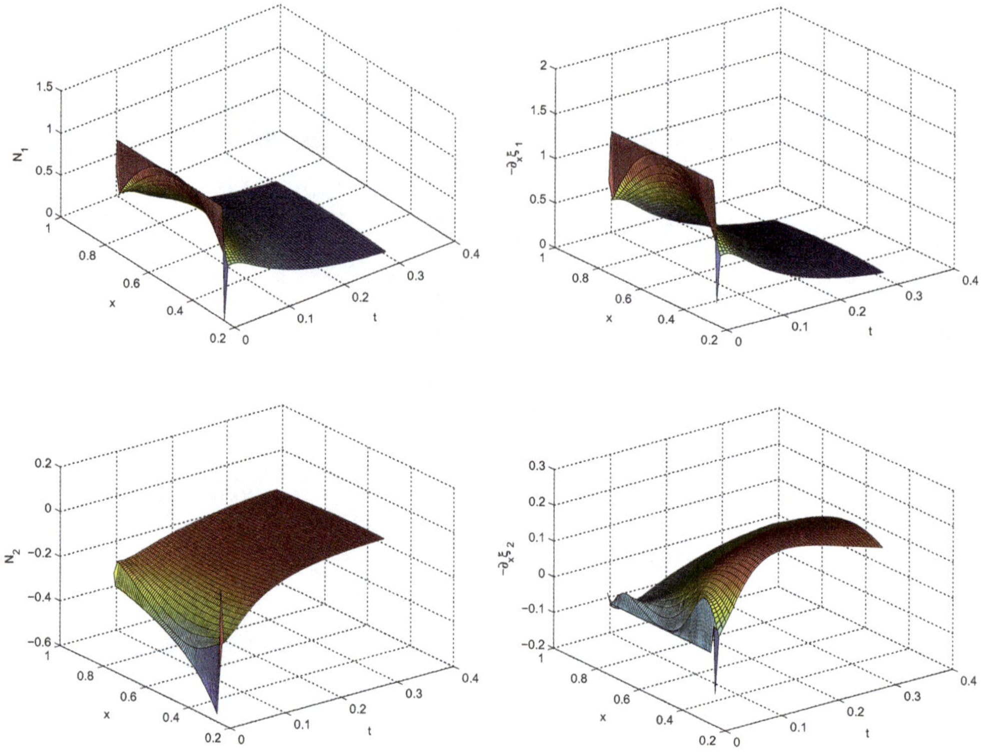

Figure 3. Results of the 3D plots in time and space.

Notes: The upper figures present the results of the flux N_1 and concentration gradient $-\partial_x \xi_1$. The lower figures present the results of flux N_2 and concentration gradient $-\partial_x \xi_2$.

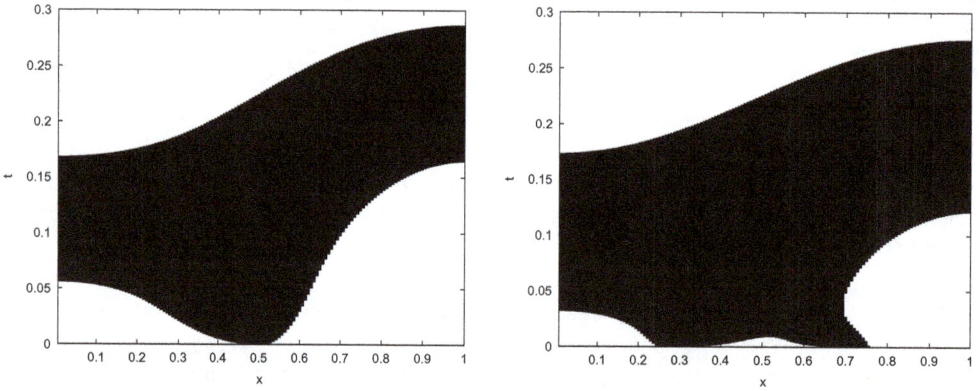

Figure 4. Asymptotic diffusion (left-hand side) and uphill diffusion (right-hand side) in the space–time region.

5.2. Iterative schemes in time (local linearization)

In the next series of experiments, we apply the more refined linearization scheme, meaning the iterative approximation in a single time step. We deal with a backward idea and resolve the nonlinearity with a fix-point method (see Geiser, 2011) in each local time step. Such a modified method allows to enlarge the local time steps while we have an additional fix-point scheme for the nonlinearity (see Kelley, 1995).

We apply the numerical convergence of the schemes with the reference solution of the explicit method by $\Xi_{ref} = (\xi_1, \xi_2)$, where the time step is $\Delta t_{CFL}/8$ for this refined solution, with the error only marginal.

Based on the reference solution, we deal with the following errors:

$$E_{L_{1,\Delta x}}(t) = \int_\Omega |\Xi_{method,J,\Delta x}(x,t) - \Xi_{ref}(x,t)|\, dx \tag{98}$$
$$= \Delta x \sum_{i=1}^{N} |\Xi_{method,J,\Delta x}(x_i,t) - \Xi_{ref}(x_i,t)|,$$

where $method, J$ is the Richardson method (second method, see Section 4.1) with J iterative steps and $\Delta t = \Delta t_{CFL}, \Delta t_{CFL}/2, \Delta t_{CFL}/4$.

Further, $method, expl$ is the explicit method (first method, see Section 4.2) with $\Delta t = \Delta t_{CFL}, \Delta t_{CFL}/2, \Delta t_{CFL}/4$.

We apply different versions of time steps and iterative steps and a reference solution is obtained with a fine time step $\Delta t = \Delta t_{CFL}/4$. We see improvements in Figure 5 and the errors in Figure 6.

Remark 6 In Figure 5, we compare a reference solution (first method) with fine time steps with flexible time step and iterative step solutions (second method). We obtain some accurate solutions with the second method with more larger time steps and more iterative steps. Based on the large time steps, the computational time decreases and additional number of more iterative cycles did not influence the amount of computational work.

Remark 7 In Figure 6, we present the L_1 errors of the second method (local linearization) compared with the reference solution (first method with very fine time steps). Here, we see the benefit of large time steps with $N = 100$ and $K = 800$, meaning we have only 100 time steps and 800 iterative steps, which are not expensive. Therefore, we could gain the same results as with many small time steps $N = 80000$ and only one iterative step $K = 1$, such that the relaxation method benefits with the iterative cycles and we could enlarge the time steps. We obtain the same accurate results for larger time

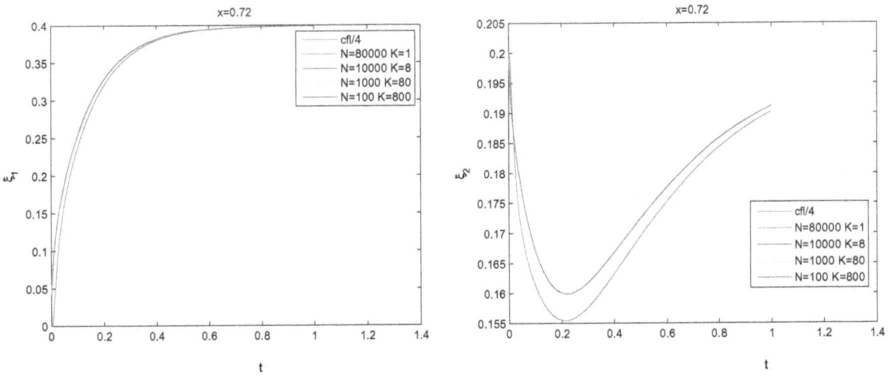

Figure 5. Solutions of different time steps and iterative steps of the Richardson method (left-hand side: concentration ξ_1 and right-hand side: concentration ξ_2).

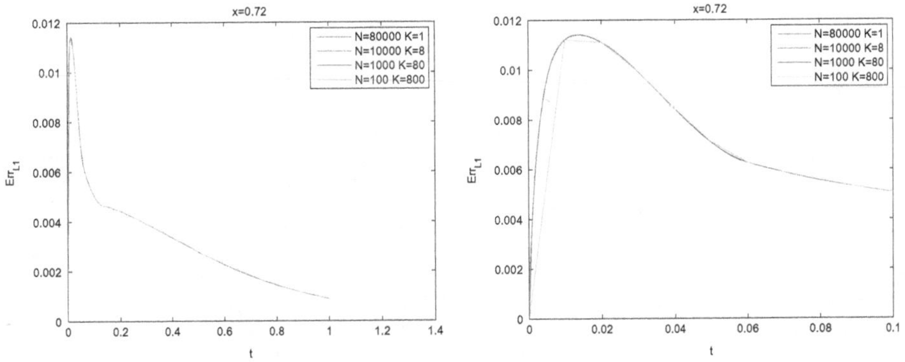

Figure 6. Errors of the different time step and iterative step solutions of the second method (left-hand side: error with the reference solution at the full time interval $t \in [0, 1.4]$ and right-hand side: error with the reference solutions at the initial time interval $t \in [0, 0.1]$).

steps, instead of a computational intensive fine resolution with small time steps. Such a novel treatment with a local linearization and additional iterative cycles reduces the computational amount of the novel second scheme.

Remark 8 The second method applies a linear linearization based on the iterative approaches in each single time step. The spatial and time discretizations are embedded into the iterative solver method. We have the benefit of relaxation in each local time step with the high resolution of the spatial and timescales. Therefore, we see a more accurate solution also with larger time steps than in the global linearization method.

6. Conclusions and discussion

We present a fluid model based on a delicate mixture of components. Such a model can be resolved by a MSD equation, while we can embed the complex mixture processes. The underlying problems for such a more delicate diffusion matrix are discussed. Based on a delicate nonlinear partial differential equation, which has to be reduced to a solvable linearized partial differential equation, we have to discuss two linearization approaches. The first approach deals with a global linearization, which can be controlled by the time steps. The second approach deals with a local linearization and applies an iterative scheme. Such a novel approach is more flexible and can be controlled by the time steps and additionally with the iterative steps. Therefore, we obtain the same accuracy with much more larger time steps and moderate iterative steps and reduce the computational amount. For the first test examples, we achieve more accurate results for a new local linearized scheme and optimize their computational amount with larger time steps. In future, we concentrate on numerical convergence analysis of the local linearization schemes and generalize our results to real-life applications.

Funding
The author has received no direct funding for this research.

Author details
Jürgen Geiser[1]
E-mail: juergen.geiser@ruhr-uni-bochum.de
ORCID ID: http://orcid.org/0000-0003-1093-0001
[1] The Institute of Theoretical Electrical Engineering, Ruhr University of Bochum, Universitätsstrasse 150, D-44801 Bochum, Germany.

References

Bothe, D. (2011). On the Maxwell--Stefan approach to multicomponent diffusion. *Parabolic Problems, Progress in Nonlinear Differential Equations and Their Applications, 80*, 81–93. doi:10.1007/978-3-0348-0075-4_5

Böttcher, K. (2010). Numerical solution of a multicomponent species transport problem combining diffusion and fluid flow as engineering benchmark. *International Journal of Heat and Mass Transfer, 53*, 231–240. doi:10.1016/j.ijheatmasstransfer.2009.09.038

Boudin, L., Grec, B., & Salvarani, F. (2012). A mathematical and numerical analysis of the Maxwell-Stefan diffusion equations. *Discrete and Continuous Dynamical Systems Series B, 17*, 1427–1440.

Boudin, L., Grec, B., & Salvarani, F. (2015). The Maxwell--Stefan diffusion limit for a kinetic model of mixtures. *Acta Applicandae Mathematicae, 136*, 79–90. doi:10.1007/s10440-014-9886-z

Chapman, S., & Cowling, Th. G. (1990). *The mathematical theory of non-uniform gases: An account of the kinetic theory of viscosity, thermal conduction, and diffusion in gases.* Cambridge: Cambridge University Press.

Duncan, J. B., & Toor, H. L. (1962). An experimental study of three component gas diffusion. *AIChE Journal, 8*, 38–41. doi:10.1002/aic.690080112

Fick, A. (1995). On liquid diffusion. *Journal of Membrane Science, 100*, 33–38. doi:10.1016/0376-7388(94)00230-V

Geiser, J. (2011). *Iterative splitting methods for differential equations.* Boca Raton, FL: CRC-Press.

Geiser, J. (2015). Numerical methods of the Maxwell--Stefan diffusion equations and applications in plasma and particle transport. arxiv:1501.05792. Retrieved from http://arxiv.org/abs/1501.05792

Geiser, J. (in press). *Multicomponent and multiscale systems: Theory, methods, and applications in engineering.* New York, NY: Springer International.

Kelley, C. T. (1995). *Iterative methods for linear and nonlinear equations.* Philadelphia, PA: SIAM.

Maxwell, J. C. (1867). On the dynamical theory of gases. *Philosophical Transactions of the Royal Society, 157*, 49–88. Retrieved from http://www.jstor.org/stable/108968?seq=1#page_scan_tab_contents

Peerenboom, K., van Boxtel, J., Janssen, J., & van Dijk, J. (2014). A conservative multicomponent diffusion algorithm for ambipolar plasma flows in local thermodynamic equilibrium. *Journal of Physics D: Applied Physics, 47*, 425202. doi:10.1088/0022-3727/47/42/425202

Senega, T. K., & Brinkmann, R. P. (2006). A multi-component transport model for non-equilibrium low-temperature low-pressure plasmas. *Journal of Physics D: Applied Physics, 39*, 1606–1618. doi:10.1088/0022-3727/39/8/020

Spille-Kohoff, A., Preuß, E., & Böttcher, K. (2012). Numerical solution of multi-component species transport in gases at any total number of components. *International Journal of Heat and Mass Transfer, 55*, 5373–5377. doi:10.1016/j.ijheatmasstransfer.2012.05.040

Tanaka, Y. (2004). Two-temperature chemically non-equilibrium modelling of high-power Ar-N2 inductively coupled plasmas at atmospheric pressure. *Journal of Physics D: Applied Physics, 37*, 1190–1205. doi:10.1088/0022-3727/37/8/007

5

Local convergence for deformed Chebyshev-type method in Banach space under weak conditions

Ioannis K. Argyros[1]* and Santhosh George[2]

*Corresponding author: Ioannis K. Argyros, Department of Mathematical Sciences, Cameron University, Lawton, OK 73505, USA
E-mail: iargyros@cameron.edu

Reviewing editor: Yong Hong Wu, Curtin University, Australia

Abstract: We present a local convergence analysis for deformed Chebyshev methods in order to approximate a solution of a nonlinear equation in a Banach space setting. Our methods include the Chebyshev and other high-order methods under hypotheses up to the first Fréchet derivative in contrast to earlier studies using hypotheses up to the second or third Fréchet derivative. The convergence ball and error estimates are given for these methods. Numerical examples are also provided in this study.

Subjects: Applied Mathematics; Computer Mathematics; Mathematics & Statistics; Non-Linear Systems; Science

Keywords: Chebyshev method; Banach space; convergence ball; local convergence

AMS subject classifications: 65D10; 65D99; 65G99; 47J25; 49M15

1. Introduction

Many problems in computational sciences and other disciplines can be brought in the form of

$$F(x) = 0 \tag{1.1}$$

where F is a Fréchet differentiable operator defined on a convex subset D of a Banach space X with values in a Banach space Y using mathematical modelling (Argyros, 1985, 2004, 2007; Argyros & Hilout, 2013; Gutiérrez & Hernández, 1997; Kantorovich & Akilov, 1982; Ortega & Rheinboldt, 1970).

ABOUT THE AUTHORS

Ioannis K. Argyros was born in 1956 in Athens, Greece. He received a BSc from the University of Athens, Greece; and a MSc and PhD from the University of Georgia, Athens, Georgia, USA, under the supervision of Dr Douglas N. Clark. He is currently a full professor of Mathematics at Cameron University, Lawton, OK, USA. He published more than 800 papers and 17 books/monographs in his area of research, computational mathematics.

Santhosh George was born in 1967 in Kerala, India. He received his MSc degree in Mathematics from University of Calicut and PhD degree in Mathematics from Goa University, under the supervision of M. T. Nair. He is a professor of Mathematical at National Institute of Technology, Karnataka. He has guided many MTech thesis works and five students have completed their PhD under his guidance.

PUBLIC INTEREST STATEMENT

A large number of problems in applied mathematics, mathematical physics, mathematical economics, engineering and other areas are formulated as equations using mathematical modelling. The unknowns of these equations can be functions, vectors or real or complex numbers. In the present paper, we study the convergence of a fast sequence to the solution of an equation. The new results improve the results of earlier works. Solutions of these equations are very important, since they have a physical meaning.

In this study, we are concerned with approximating a solution x^* of Equation 1.1. The solutions of these equations in general can not be found in closed form, so one has to consider some iterative methods for solving Equation 1.1. The convergence analysis of iterative methods is usually based on two types: semi-local and local convergence analyses. The semi-local convergence analysis is, based on the information around an initial point, to give conditions ensuring the convergence of the iterative procedure; while the local one is, based on the information around a solution, to find estimates of the radii of convergence balls. In particular, the practice of numerical functional analysis for finding solution x^* of Equation 1.1 is essentially connected to variants of Newton's method. This method converges quadratically to x^* if the initial guess is close enough to the solution. Iterative methods of convergence order higher than two such as Chebyshev–Halley-type methods (Amat, Busquier, & Gutiérrez, 2003; Argyros, 2007; Argyros & Hilout, 2013; Candela & Marquina, 1990a, 1990b; Chun, Stanica, & Neta, 2011; Gutiérrez & Hernández, 1997, 1998; Hernández, 2001; Hernandez & Salanova, 2000; Kantorovich, 1982; Ortega & Rheinboldt, 1970; Parida & Gupta, 2008) require the evaluation of the second Fréchet derivative, which is very expensive in general. However, there are integral equations, where the second Fréchet derivative is diagonal by blocks and inexpensive (Gutiérrez & Hernández, 1997, 1998; Hernández, 2001; Hernández & Salanova, 2000) or for quadratic equations the second Fréchet derivative is constant (Argyros, 1985; Hernández & Salanova, 2000). Moreover, in some applications involving stiff systems (Argyros, 2004; Argyros & Hilout, 2013; Chun et al., 2011), high-order methods are usefull. That is why we study the local convergence of Deformed Chebyshev Method (DCM) defined for each $n = 0, 1, 2, \ldots$ by

$$\begin{aligned} y_n &= x_n - F'(x_n)^{-1}F(x_n) \\ z_n &= x_n + \alpha F'(x_n)^{-1}F(x_n) \\ H_n &= \frac{1}{\lambda}F'(x_n)^{-1}[F'\left(x_n + \lambda(z_n - x_n)\right) - F'(x_n)] \\ x_{n+1} &= x_n + \frac{1}{2}H_n(y_n - x_n) \end{aligned} \tag{1.2}$$

where x_0 is an initial point, $\lambda \in (0, 1]$ and $\alpha \in \mathbb{R}$ are given parameters. Deformed methods have been introduced to improve on the convergence order of Newton's method or Newton-like methods (Amat et al., 2003; Argyros, 1985, 2004, 2007; Argyros & Hilout, 2012, 2013; Candela & Marquina, 1990a, 1990b; Chun et al., 2011; Gutiérrez & Hernández, 1997, 1998; Hernández, 2001; Hernández & Salanova, 2000; Kantorovich, 1982; Ortega & Rheinboldt, 1970; Parida & Gupta, 2008; Wu & Zhao, 2007). In particular, DCM was proposed in Wu and Zhao (2007) as an alternative to the famous Chebyshev method (Amat et al., 2003; Argyros & Hilout, 2012, 2013; Candela & Marquina, 1990a, 1990b; Chun et al., 2011; Gutiérrez & Hernández, 1997, 1998; Hernández, 2001; Hernández & Salanova, 2000; Kantorovich, 1982; Ortega & Rheinboldt, 1970; Parida & Gupta, 2008; Wu & Zhao, 2007) defined for each $n = 0, 1, 2, \ldots$ by

$$\begin{aligned} y_n &= x_n - F'(x_n)^{-1}F(x_n) \\ L_n &= F'(x_n)^{-1}F''(x_n)F'(x_n)^{-1}F(x_n) \\ x_{n+1} &= y_n + \frac{1}{2}L_n(y_n - x_n) \end{aligned} \tag{1.3}$$

Notice that the computation of the expensive in general second Fréchet derivative $F''(x_n)$ is required in method (1.3) but not in DCM.

The semilocal convergence analysis of DCM was given in Wu and Zhao (2007) under Lipschitz continuity conditions on up to the second Fréchet derivative in the special case when $\alpha = 1$ and $\lambda > 0$. In particular, the third order of convergence of DCM was shown in Wu and Zhao (2007) under these values of the parameters α and λ.

The usual conditions for the semi-local convergence of these methods are (C): There exist constants β, η, β_1, β_2 such that

(C_1) There exists $\Gamma_0 = F'(x_0)^{-1}$ and $\|\Gamma_0\| \leq \beta$;

C_2 $\|\Gamma_0 F(x_0)\| \leq \eta$;

C_3 $\|F''(x)\| \leq \beta_1$ for each $x \in D$;

C_4 $\|F''(x) - F''(y)\| \leq \beta_2 \|x - y\|$ for each $x, y \in D$.

The local convergence conditions are similar but x_0 is x^* in (C_1) and (C_2). There is a plethora of local and semi-local convergence results under the (C) conditions (Amat et al., 2003; Argyros, 1985, 2004, 2007; Argyros & Hilout, 2012, 2013; Candela & Marquina, 1990a, 1990b; Chun et al., 2011; Gutiérrez & Hernández, 1997, 1998; Hernández, 2001; Hernández & Salanova, 2000; Kantorovich, 1982; Ortega & Rheinboldt, 1970; Parida & Gupta, 2008; Wu & Zhao, 2007). The conditions (C_3) and (C_4) limit the applicability of these methods although only the first Fréchet derivative appears in these methods. Therefore, these usefull methods cannot be applied according to the earlier results. Therefore, the motivation for this study is to use these usefull methods (Wu & Zhao, 2007) in cases when (C_3) and (C_4) are not satisfied.

As a motivational example, let us define function f on $D = \left[-\frac{1}{2}, \frac{5}{2}\right]$ by

$$f(x) = \begin{cases} x^3 \ln x^2 + x^5 - x^4, & x \neq 0 \\ 0, & x = 0 \end{cases}$$

Choose $x^* = 1$. We have that

$$f'(x) = 3x^2 \ln x^2 + 5x^4 - 4x^3 + 2x^2$$
$$f''(x) = 6x \ln x^2 + 20x^3 - 12x^2 + 10x$$
$$f'''(x) = 6 \ln x^2 + 60x^2 - 24x + 22$$

Notice that $f''(x)$ does not satisfy (C_4) on D. Hence, the results depending on (C_4) cannot apply in this case. However, using Equations 2.7–2.10 that follow we have $f'(x^*) = 3$ and $f(x^*) = 0$, $p = 1$, $L_0 = L = 146.6629073$ and $M = 101.5578008$. Hence, the results of our Theorem 2.1 that follows can apply to solve equation $f(x) = 0$ using DCM. Hence, the applicability of DCM is expanded under our new conditions.

In the rest of this study, $U(w, q)$ and $\overline{U}(w, q)$ stand, respectively, for the open and closed ball in X with center $w \in X$ and of radius $q > 0$.

The paper is organized as follows: In Section 2, we present the local convergence of these methods. The numerical examples are given in the concluding Section 3.

2. Local convergence

In this section, we present the local convergence analysis of DCM. Let $L_0 > 0, L > 0, M > 0, \alpha \in \mathbb{R}, \lambda \in (0, 1]$ and $p \in [0, 1]$ be given parameters. It is convenient for the local convergence analysis that follows to introduce some functions and parameters.

Define functions on the interval $[\,0, \left(\frac{1}{L_0}\right)^p)$ by

$$g_1(t) = \frac{Lt^p}{(1+p)(1-L_0t^p)}$$
$$g_2(t) = g_1(t) + \frac{|1+\alpha|M}{1-L_0^p}$$
$$g_3(t) = \frac{|\alpha|^p L|\lambda|^{p-1}M^p t^p}{2(1-L_0t^p)^{1+p}}$$
$$g_4(t) = g_1(t) + \frac{|\alpha|^p L|\lambda|^{p-1}M^p t^p}{2(1-L_0t^p)^{1+p}}(1+g_1(t))$$
$$\bar{g}_4(t) = g_4(t) - 1$$

and parameters

$$r_1 = \left(\frac{1+p}{(1+p)L_0 + L}\right)^{\frac{1}{p}} < \left(\frac{1}{L_0}\right)^{\frac{1}{p}}$$

and

$$r_2 = \left(\frac{(1+p)(1-M|1+\alpha|)}{(1+p)L_0 + L}\right)^{\frac{1}{p}}$$

Suppose that

$$M|1+\alpha| < 1 \tag{2.1}$$

Then, r_2 is well defined and

$$0 < r_2 < r_1 < \left(\frac{1}{L_0}\right)^{\frac{1}{p}}$$

We also have that

$$0 \le g_1(t) < 1$$

$$0 \le g_2(t) < 1$$

and

$$g_3(t) \ge 0$$

Moreover, we have that $\bar{g}_4(0) = g_4(0) - 1 = 0 - 1 = -1 < 0$ and $\bar{g}_4(t) \to \infty$ as $t \to \left(\left(\frac{1}{L_0}\right)^{\frac{1}{p}}\right)^{-}$.

Then, it follows from the intermediate value theorem that function $\bar{g}_4$ has zeros in $\left(0, \left(\frac{1}{L_0}\right)^{\frac{1}{p}}\right)$.

Denote by r_4 the smallest such zero. Set

$$r = \min\{r_2, r_4\} \tag{2.2}$$

Then, we have that

$$0 \le g_1(t) < 1 \tag{2.3}$$

$$0 \le g_2(t) < 1 \tag{2.4}$$

$$g_3(t) \geq 0 \tag{2.5}$$

and

$$0 \leq g_4(t) < 1 \text{ for each } t \in [0, r) \tag{2.6}$$

Next, we present the local convergence result for DCM.

THEOREM 2.1 *Let $F: D \subseteq X \to Y$ be a Fréchet differentiable operator. Suppose that there exist $x^* \in D$, $L_0 > 0$, $L > 0$, $M > 0$, $\alpha \in \mathbb{R}|$, $\lambda \in (0,1]$ and $p \in (0,1]$ such that for each $x, y \in D$*

$M|1+\alpha| < 1$

$$F(x^*) = 0, \quad F'(x^*)^{-1} \in L(Y, X) \tag{2.7}$$

$$\|F'(x^*)^{-1}(F'(x) - F'(x^*))\| \leq L_0\|x - x^*\|^p \tag{2.8}$$

$$\|F'(x^*)^{-1}(F'(x) - F'(y))\| \leq L\|x - y\|^p \tag{2.9}$$

$$\|F'(x^*)^{-1}F'(x)\| \leq M \tag{2.10}$$

and

$$\bar{U}(x^*, r) \subseteq D \tag{2.11}$$

where r is given by Equation 2.2. Then, sequence $\{x_n\}$ generated by DCM for $x_0 \in U(x^*, r)$ is well defined, remains in $U(x^*, r)$ for each $n = 0, 1, 2, \ldots$ and converges to x^*. Moreover, the following estimates hold for each $n = 0, 1, 2, \ldots$.

$$\|y_n - x^*\| \leq g_1(\|x_n - x^*\|)\|x_n - x^*\| < \|x_n - x^*\| < r \tag{2.12}$$

$$\|z_n - x^*\| \leq g_2(\|x_n - x^*\|)\|x_n - x^*\| < \|x_n - x^*\| < r \tag{2.13}$$

$$\|\frac{1}{2}H_n\| \leq g_3(\|x_n - x^*\|) \tag{2.14}$$

and

$$\|x_{n+1} - x^*\| \leq g_4(\|x_n - x^*\|)\|x_n - x^*\| < \|x_n - x^*\| < r \tag{2.15}$$

where the "g" functions are defined above Theorem 2.1. Furthermore, if there exists $R \in [\, r, \frac{2}{L_0})$ such that $\bar{U}(x^*, R) \subseteq D$, then the limit point x^* is the only solution of equation $F(x) = 0$ in $U(x^*, r)$.

Proof We shall show estimates Equations 2.12–2.15 using mathematical induction. Firstly, we shall show that y_0, z_0, x_1 exist and lie inside $U(x^*, r)$. In order for us to achieve this, we must show that the inverses appearing in method DCM exist for $n = 0$. By hypothesis $x_0 \in U(x^*, r)$. Using the definition of radius r and Equation 2.8, we get that

$$\|F'(x^*)^{-1}(F'(x_0) - F'(x^*))\| \leq L_0\|x - x^*\|^p < L_0 r^p < 1 \tag{2.16}$$

It follows from Equation 2.16 and the Banach Lemma on invertible operators (Argyros, 2007; Argyros & Hilout, 2013; Kantorovich, 1982; Ortega & Rheinboldt, 1970) that $F'(x_0)^{-1} \in L(Y, X)$ and

$$\|F'(x^*)^{-1}F'(x^*)\| \leq \frac{1}{1 - L_0\|x - x^*\|^p} < \frac{1}{1 - L_0 r^p} \tag{2.17}$$

Moreover y_0, z_0 are well defined by first and second substep of DCM for $n = 0$. Using the first substep of DCM for $n = 0$, we also get that

$$y_0 - x^* = x_0 - x^* - F'(x_0)^{-1}F(x_0)$$
$$= [F'(x_0)^{-1}F'(x^*)][\int_0^1 F'(x^*)^{-1}$$
$$\times [F'(x^* + \theta(x_0 - x^*)) - F'(x_0))(x_0 - x^*)]d\theta \quad (2.18)$$

Then, by the definition of function g_1, Equations 2.3, 2.9, 2.17 and 2.18, we obtain that

$$\begin{aligned}
\|y_0 - x^*\| &\leq \|F'(x_0)^{-1}F'(x^*)\| \| \int_0^1 F'(x^*)^{-1}[F'(x^* + \theta(x_0 - x^*)) \\
&\quad - F'(x_0))d\theta\| \|x_0 - x^*\| \\
&\leq \frac{L\|x_0 - x^*\|^{1+p}}{(1+p)(1 - L_0\|x_0 - x^*\|)} \\
&\leq g_1(\|x_0 - x^*\|)\|x_0 - x^*\| \\
&< \|x_k - x^*\| < r
\end{aligned}$$

which shows Equation 2.12 for $n = 0$ and $y_0 \in U(x^*, r)$. Similarly, using the second substep of DCM for $n = 0$, we get that

$$z_0 - x^* = x_0 - x^* - F'(x_0)^{-1}F(x_0) + (1+\alpha)F'(x_0)^{-1}F(x_0) \quad (2.19)$$

Then, by Equations 2.4, 2.10, 2.17, 2.19 the definition of function g_2 and Equation 2.12 (for $n = 0$), we obtain, since $F(x_0) = \int_0^1 F'(x^* + \theta(x_0 - x^*))(x_0 - x^*)d\theta$ that,

$$\begin{aligned}
\|z_0 - x^*\| &\leq \|x_0 - x^* - F'(x_0)^{-1}F(x_0)\| \\
&\quad + |1+\alpha| \|F'(x_0)^{-1}F'(x^*)\| \\
&\quad \times \| \int_0^1 F'(x^*)^{-1}F'(x^* + \theta(x_0 - x^*))d\theta\| \|x_0 - x^*\| \\
&\leq \left[g_1(\|x_0 - x^*\|) + \frac{|1+\alpha|M}{1 - L_0\|x_0 - x^*\|} \right] \|x_0 - x^*\| \\
&= g_2(\|x_0 - x^*\|)\|x_0 - x^*\| < \|x_0 - x^*\| < r
\end{aligned}$$

which shows Equation 2.13 for $n = 0$ and $z_0 \in U(x^*, r)$. We have by the definition of λ and Equations 2.12, 2.13 (for $n = 0$) that

$$\|x_0 - x^* + \lambda(z_0 - x_0)\| \leq |1 - \lambda| \|x_0 - x^*\| + |\lambda| \|z_0 - x^*\| < (|1-\lambda| + |\lambda|)r \leq r$$

which shows that $x_0 + \lambda(z_0 - x_0) \in U(x^*, r)$ and H_0 is well defined. We need an estimate on $\|H_0\|$. Using the definition of H_0, g_3, Equations 2.17 and 2.9, we get in turn that

$$\begin{aligned}
\left\| \frac{1}{2}H_0 \right\| &\leq \frac{1}{2|\lambda|} \|F'(x_0)^{-1}F'(x^*)\| \|F'(x^*)^{-1}[F'(x^* + \lambda(z_0 - x_0) - F'(x_0)]\| \\
&\leq \frac{1}{2|\lambda|} \frac{L|\lambda|^p \|z_0 - x_0\|^p}{1 - L_0\|x_0 - x^*\|^p} \\
&\leq \frac{|\alpha|^p L |\lambda|^{p-1} (\|F'(x_0)^{-1}F'(x^*)\| \|F'(x^*)^{-1}F(x_0)\|)^p}{2(1 - L_0\|x_0 - x^*\|^p)} \\
&\leq \frac{|\alpha|^p L |\lambda|^{p-1} M^p \|x_0 - x^*\|^p}{2(1 - L_0\|x_0 - x^*\|^p)^{1+p}} \\
&= g_3(\|x_0 - x^*\|)
\end{aligned}$$

which shows Equation 2.14 for $n = 0$. Then, using the last substep of DCM for $n = 0$, we get

$$\begin{aligned}\|x_1 - x^*\| &\le \|y_0 - x^*\| + \|\frac{1}{2}H_0\|\|y_0 - x_0\| \\ &\le g_1(\|x_0 - x^*\|)\|x_0 - x^*\| + \frac{|\alpha|^p L|\lambda|^{p-1}M^p\|x_0 - x^*\|^p}{2(1 - L_0\|x_0 - x^*\|^p)^{1+p}} \\ &\quad \times (\|y_0 - x^*\| + \|x_0 - x^*\|) \\ &\le [g_1(\|x_0 - x^*\|) + \frac{|\alpha|^p L|\lambda|^{p-1}\|x_0 - x^*\|^p}{2(1 - L_0\|x_0 - x^*\|^p)^{1+p}} \\ &\quad \times (1 + g_1(\|x_0 - x^*\|)]\|x_0 - x^*\| \\ &= g_4(\|x_0 - x^*\|)\|x_0 - x^*\| < \|x_0 - x^*\| < r\end{aligned}$$

which shows Equation 2.15 for $n = 0$ and $x_1 \in U(x^*, r)$. By simply replacing x_0, y_0, z_0, x_1 by x_k, y_k, z_k, x_{k+1} in the preceding estimates we arrive at estimates Equations 2.12–2.15. Finally, using the estimate $\|x_{k+1} - x^*\| < \|x_k - x^*\| < r$, we deduce that $x_{k+1} \in U(x^*, r)$ and $\lim_{k\to\infty} x_k = x^*$.

To show the uniqueness part, let $B = \int_0^1 F'(y^* + \theta(x^* - y^*)d\theta$ for some $y^* \in \bar{U}(x^*, R)$ with $F(y^*) = 0$. Using Equation 2.8 with $p = 1$, we get that

$$\begin{aligned}|F'(x^*)^{-1}(B - F'(x^*))| &\le \int_0^1 L_0|y^* + \theta(x^* - y^*) - x^*|d\theta \\ &\le L_0\int_0^1 (1-\theta)|x^* - y^*|d\theta \le \frac{L_0}{2}R < 1\end{aligned} \tag{2.20}$$

It follows from Equation 2.30 and the Banach Lemma on invertible functions that B is invertible. Finally, from the identity $0 = F(x^*) - F(y^*) = B(x^* - y^*)$, we deduce that $x^* = y^*$. □

Remark 2.2 (a) Condition Equation 2.8 can be dropped, since this condition follows from $(\mathcal{A}_3)$. Notice, however that

$$L_0 \le L \tag{2.21}$$

holds in general and $\frac{L}{L_0}$ can be arbitrarily large (Argyros, 1985, 2004, 2007; Argyros & Hilout, 2012, 2013) (see also the examples) .

(b) In view of condition Equation 2.8 and the estimate

$$\begin{aligned}\|F'(x^*)^{-1}F'(x)\| &= \|F'(x^*)^{-1}[F'(x) - F'(x^*)] + I\| \\ &\le 1 + \|F'(x^*)^{-1}(F'(x) - F'(x^*))\| \\ &\le 1 + L_0\|x - x^*\|^p\end{aligned}$$

condition (2.10) can be dropped and M can be replaced by

$$M(t) = 1 + L_0 t^p \tag{2.22}$$

(c) The convergence ball of radius r_1 was given by us in Argyros (1985, 2004, 2007) and Argyros and Hilout (2013) for Newton's method under conditions Equations 2.8 and 2.9. Estimate $r_2 < r_1$ shows that the convergence ball of higher than two DCM methods are smaller than the convergence ball of radius r_1 of the quadratically convergent Newton's method. The convergence ball given by Ortega and Rheinboldt (1970) for Newton's method is

$$r_R = \frac{2}{3L} < r_1 \text{ (for } p = 1) \tag{2.23}$$

if $L_0 < L$ and $\frac{r_R}{r_1} \to \frac{1}{3}$ as $\frac{L_0}{L} \to 0$. Hence, we do not expect r to be larger than r_1 no matter how we choose L_0, L, M and α.

(d) The local results can be used for projection methods such as Arnoldi's method, the generalized minimum residual method, the generalized conjugate method for combined Newton/finite projection methods and in connection to the mesh independence principle in order to develop the cheapest and most efficient mesh refinement strategy (Argyros, 1985, 2004, 2007; Argyros & Hilout, 2013; Kantorovich, 1982; Ortega & Rheinboldt , 1970).

(e) The results can also be used to solve equations where the operator F' satisfies the autonomous differential equation (Argyros, 1985, 2004, 2007; Argyros & Hilout, 2013; Kantorovich, 1982; Ortega & Rheinboldt , 1970):

$$F'(x) = T(F(x))$$

where T is a known continuous operator. Since $F'(x^*) = T(F(x^*)) = T(0)$, $F''(x^*) = F'(x^*)T'(F(x^*)) = T(0)T'(0)$, we can apply the results without actually knowing the solution x^*. Let as an example $F(x) = e^x - 1$. Then, we can choose $T(x) = x + 1$ and $x^* = 0$.

(f) It is worth noticing that DCM is not changing when we use the condition of Theorem 2.1 instead of the stronger $(\mathcal{C})$ conditions used in the earlier mentioned works. We can also use the computational order of convergence (COC) (Argyros & Hilout, 2013; Hernández & Salanova, 2000) defined by

$$\xi = \ln\left(\frac{\|x_{n+1} - x^*\|}{\|x_n - x^*\|}\right) / \ln\left(\frac{\|x_n - x^*\|}{\|x_{n-1} - x^*\|}\right)$$

or the approximate COC

$$\xi_1 = \ln\left(\frac{\|x_{n+1} - x_n\|}{\|x_n - x_{n-1}\|}\right) / \ln\left(\frac{\|x_n - x_{n-1}\|}{\|x_{n-1} - x_{n-2}\|}\right)$$

since the bounds given in Theorem 2.1 may be very pessimistic. This way we also avoid the computation of the error bounds as given in the earlier studies where bounds on the Frécher-derivatives higher than one are used.

(g) The restriction $\lambda \in (0, 1]$ can be dropped, if Equation 2.11 is replaced by

$$U_1 = \bar{U}(x^*, (|\lambda| + |1 - \lambda|)r) \subseteq D \tag{2.24}$$

for $\lambda \in \mathbb{R} - \{0\}$. Indeed, we will then have

$$\begin{aligned} \|x_n + \lambda(y_n - x_n) - x^*\| &\leq |\lambda|\|x_n - x^*\| + |1 - \lambda|\|y_n - x^*\| \\ &\leq (|\lambda| + |1 - \lambda|)r \\ &\Rightarrow x_n + \lambda(y_n - x_n) \in U_1 \end{aligned}$$

3. Numerical examples

We present numerical examples where we compute the radii of the convergence balls.

Example 3.1 Let $X = Y = \mathbb{R}$. Define function F on $D = [1, 3]$ by

$$F(x) = \frac{2}{3}x^{\frac{2}{3}} - x \tag{3.1}$$

Then, $x^* = \frac{9}{4}$, $F'(x^*)^{-1} = 2$, $L_0 = 1 < L = 2$, $p = 0.5$, $\alpha = -0.6585$, $\lambda = 1$ and $M = 2(\sqrt{3} - 1)$, $r_1 = 0.7746$, $r_2 = 0.04629$, $r_4 = 0.0167$ and $r = 0.0167$.

Example 3.2 Let $X = Y = \mathbb{R}^3$, $D = \overline{U}(0, 1)$ and $B(x) = F''(x)$ for each $x \in D$. Define F on D for $v = x, y, z$ by

$$F(v) = \left(e^x - 1, \frac{e-1}{2}y^2 + y, z\right) \tag{3.2}$$

Then, the Fréchet derivative is given by:

$$F'(v) = \begin{bmatrix} e^x & 0 & 0 \\ 0 & (e-1)y+1 & 0 \\ 0 & 0 & 1 \end{bmatrix}$$

Notice that $x^* = (0,0,0)$, $F'(x^*) = F'(x^*)^{-1} = diag\{1,1,1\}$, $L_0 = e - 1 < L = M = e$, $p = 1, \alpha = -0.8161$, $\lambda = 0.5$. The values of $r_1 = 0.3245$, $r_2 = 0.1625$, $r_4 = 0.2291$ and $r = 0.1625$.

Example 3.3 Let $X = Y = C[0,1]$, the space of continuous functions defined on $[0,1]$ and be equipped with the max norm. Let $D = \overline{U}(0,1)$ and $B(x) = F''(x)$ for each $x \in D$. Define function F on D by

$$F(\varphi)(x) = \varphi(x) - 5\int_0^1 x\theta\varphi(\theta)^3 d\theta \tag{3.3}$$

we have that

$$F'(\varphi(\xi))(x) = \xi(x) - 15\int_0^1 x\theta\varphi(\theta)^2\xi(\theta)d\theta, \text{ for each } \xi \in D$$

Then, we get that $p = 1$, $x^* = 0$, $L_0 = 7.5$, $L = 15$, $\alpha = -0.9412$, $\lambda = 0.5$ and $M = M(t) = 1 + 7.5t$. The values of $r_1 = 0.0667$, $r_2 = 0.0333$, $r_4 = 0.0449$ and $r = 0.0333$.

Example 3.4 Returing back to the motivational example at the introduction of this study, we have $p = 1$, $L_0 = L = 146.6629073$, $M = 101.5578008$, $\alpha = -0.9951$, $\lambda = 0.5$. The values of $r_1 = 0.0045$, $r_2 = 0.0023$, $r_4 = 0.0034$ and $r = 0.0023$.

Example 3.5 Let us consider the nonlinear Hammerstein integral equation of the second kind (Argyros, 2007; Argyros & Hilout, 2013; Kantorovich, 1982; Ortega & Rheinboldt, 1970) given by

$$y(s) = w(s) + \int_a^b G(s,t)[y(t)^{1+p} + \mu y(t)]dt \text{ for each } s \in [a,b] \tag{3.4}$$

where w is a given continuous function such that $w(s) > 0$ for each $s \in [a,b]$, $\mu \in \mathbb{R}$, $p \in (0,1]$ and the kernel G is a nonnegative continuous function on $[a,b] \times [a,b]$. Equation 3.4 appears in many studies in mathematical physics (radiative transfer, kinetic theory of gases, neuron transport) and other applied areas (Argyros, 1985, 2007; Argyros & Hilout, 2013; Gutiérrez & Hernández, 1997, 1998; Hernández & Salanova, 2000; Hernández, 2001; Kantorovich, 1982; Ortega & Rheinboldt, 1970; Wu & Zhao, 2007). The kernel G is defined by

$$G(s,t) = \begin{cases} \frac{(b-s)(t-a)}{b-a}, & t \leq s \\ \frac{(s-a)(b-t)}{b-a}, & s \leq t \end{cases} \tag{3.5}$$

It is well known that Equation 3.4 is equivalent to the following two-point boundary value problem;

$$\begin{aligned} y'' + y^{1+p} + \mu y &= 0 \\ y(a) = v(a), y(b) &= v(b) \end{aligned} \tag{3.6}$$

Next, we shall solve Equation 3.5 by first discretizing it and using DCM for $\mu = 0$, $p = \frac{1}{2}$, $a = 0$, $b = 1$ and $y(0) = y(1) = 0$. We divide the interval $[0,1]$ into m subintervals and let $l = \frac{1}{m}$. Let us denote the points of subdivision by $u_0 = 0 < u_1 < \cdots < u_m = 1$ with the corresponding values of the function

$y_0 = y(u_0), y_1 = y(u_1), \cdots, y_m = y(u_m)$. A simple approximation for the second derivative at these points is given by

$$y_i'' \approx \frac{y_{i-1} - 2y_i + y_{i+1}}{l^2}, \; i = 1, 2, \ldots, m-1$$

Notice that $y_0 = o$ and $y_m = 0$. Therefore, we obtain the following system of nonlinear equations

$$\begin{aligned} l^2 y_1^{\frac{3}{2}} - 2y_1 + y_2 &= 0 \\ y_{i-1} + l^2 y_i^{\frac{3}{2}} - 2y_i + y_{i+1} &= 0 \\ y_{m-2} + l^2 y_{m-1}^{\frac{3}{2}} - 2y_{m-1} &= 0, \;\; i = 2, 3, \ldots, m-1 \end{aligned} \tag{3.7}$$

Hence, we have an operator $F: \mathbb{R}^{m-1} \to \mathbb{R}^{m-1}$ whose Fréchet differential can be written as:

$$F'(y) = \begin{bmatrix} \frac{3}{2} l^2 y_1^{\frac{1}{2}} - 2 & 1 & 0 & \cdots & \cdots & 0 \\ 1 & \frac{3}{2} l^2 y_2^{\frac{1}{2}} - 2 & 1 & 0 & \cdots & 0 \\ \vdots & & & & \vdots & \\ 0 & \cdots & \cdots & 1 & \frac{3}{2} l^2 y_{m-1}^{\frac{1}{2}} - 2 & 1 \end{bmatrix}$$

Notice that we may not be able to find the second Fréchet differential since it would involve terms of the form $y_i^{\frac{1}{2}}$ and they may not exist. Therefore, the famous Chebyshev method Equation 1.3 cannot be used. However, DCM method Equation 1.2 obtained from method Equation 1.3 can be used, since only the first Fréchet differential appears in this method. Moreover, the theory introduced in this paper applies. Indeed, let $x \in \mathbb{R}^{m-1}$. Define the norm to be $\|x\| = \max_{1 \le j \le m-1} \|x_j\|$. The corresponding norm on $A \in \mathbb{R}^{m-1} \times \mathbb{R}^{m-1}$ is $\|A\| = \max_{1 \le j \le m-1} \sum_{i=1}^{m-1} |a_{ji}|$. Then, for all $y, z \in \mathbb{R}^{m-1}$ for which $|y_i| > 0$, $|z_i| > 0$, $i = 1, 2, \ldots, m-1$:

$$\begin{aligned} \|F'(y) - F'(z)\| &= \left\| diag \left\{ \frac{3}{2} \left(y_i^{\frac{1}{2}} - z_i^{\frac{1}{2}} \right) \right\} \right\| \\ &= \frac{3}{2} l^2 \max_{1 \le j \le m-1} |y_i^{\frac{1}{2}} - z_i^{\frac{1}{2}}| \\ &\le \frac{3}{2} l^2 \max_{1 \le j \le m-1} |y_i - z_i|^{\frac{1}{2}} \\ &= \frac{3}{2} l^2 \|y - z\|^{\frac{1}{2}} \end{aligned}$$

That is we can choose $L_0 = L = \frac{3}{2} l^2 \|F'(x^*)^{-1}\|$. We choose $m = 10$ leading to nine equations. Since a solution would vanish at the endpoints and be positive in the interior a reasonable choice of initial approximation seems to be $130 \sin \pi x$. Hence, we obtain the following initial vector for solving system Equation 3.7

$$y^0 = \begin{bmatrix} 4.01524e+01 \\ 7.63785e+01 \\ 1.05135e+02 \\ 1.23611e+02 \\ 1.29999e+02 \\ 1.23675e+02 \\ 1.05257e+02 \\ 7.65462e+01 \\ 4.03495e+01 \end{bmatrix}$$

Then, after two iterations DCM for $\alpha = \lambda = 1$ gives us the solution x^* defined by

$$x^* = \begin{bmatrix} 3.35740e+01 \\ 6.52027e+01 \\ 9.15664e+01 \\ 1.09168e+02 \\ 1.15363e+02 \\ 1.09168e+02 \\ 9.15664e+01 \\ 6.52027e+01 \\ 3.35740e+01 \end{bmatrix}$$

Funding
The authors received no direct funding for this research.

Author details
Ioannis K. Argyros[1]
E-mail: iargyros@cameron.edu
Santhosh George[2]
E-mail: sgeorge@nitk.ac.in
ORCID ID: http://orcid.org/0000-0002-3530-5539
[1] Department of Mathematical Sciences, Cameron University, Lawton, OK 73505, USA.
[2] Department of Mathematical and Computational Sciences, National Institute of Technology Karnataka, Mangaluru, Karnataka 757 025, India.

References

Amat, S., Busquier, S., & Gutiérrez, J. M. (2003). Geometric constructions of iterative functions to solve nonlinear equations. *Journal of Computational and Applied Mathematics, 157*, 197–205.

Argyros, I. K. (1985). Quadratic equations and applications to Chandrasekhar's and related equations. *Bulletin of the Australian Mathematical Society, 32*, 275–292.

Argyros, I. K. (2004). A unifying local-semilocal convergence analysis and applications for two-point Newton-like methods in Banach spaces. *Journal of Mathematical Analysis and Applications, 20*, 373–397.

Argyros, I. K. (2007). Studies in computational mathematics. In C. K. Chui & L. Wuytack (Eds.), *Computational theory of iterative methods* (Chapter. 15). New York, NY: Elsevier.

Argyros, I. K., & Hilout, S. (2012). Weaker conditions for the convergence of Newton's method. *Journal of Complexity, 28*, 364–387.

Argyros, I. K., & Hilout, S. (2013). *Numerical methods in nonlinear analysis*. London: World Scientific Publishing.

Candela, V., & Marquina, A. (1990a). Recurrence relations for rational cubic methods I: The Halley method. *Computing, 44*, 169–184.

Candela, V., & Marquina, A. (1990b). Recurrence relations for rational cubic methods II: The Chebyshev method. *Computing, 45*, 355–367.

Chun, C., Stanica, P., & Neta, B. (2011). Third order family of methods in Banach spaces. *Computers and Mathematics with Applications, 61*, 1665–1675.

Gutiérrez, J. M., & Hernández, M. A. (1997). Third-order iterative methods for operators with bounded second derivative. *Journal of Computational and Applied Mathematics, 82*, 171–183.

Gutiérrez, J. M., & Hernández, M. A. (1998). Recurrence relations for the super-Halley method. *Computers & Mathematics with Applications, 36*, 1–8.

Hernández, M. A. (2001). Chebyshev's approximation algorithms and applications. *Computers & Mathematics with Applications, 41*, 433–455.

Hernández, M. A., & Salanova, M. A. (2000). Modification of the Kantorovich assumptions for semilocal convergence of the Chebyshev method. *Journal of Computational and Applied Mathematics, 126*, 131–143.

Kantorovich, L. V., & Akilov, G. P. (1982). *Functional analysis*. Oxford: Pergamon Press.

Ortega, J. M., & Rheinboldt, W. C. (1970). *Iterative solution of nonlinear equations in several variables*. New York, NY: Academic Press.

Parida, P. K., & Gupta, D. K. (2008). Recurrence relations for semi-local convergence of a Newton-like method in Banach spaces. *Journal of Mathematical Analysis and Applications, 345*, 350–361.

Wu, Q., & Zhao, Y. (2007). Newton–Kantorovich type convergence theorem for a family of new deformed Chebyshev method. *Applied Mathematics and Computation, 192*, 405–412.

6

An asymptotic analysis for an integrable variant of the Lotka–Volterra prey–predator model via a determinant expansion technique

Masato Shinjo[1], Masashi Iwasaki[2*], Akiko Fukuda[3], Emiko Ishiwata[4], Yusaku Yamamoto[5] and Yoshimasa Nakamura[1]

*Corresponding author: Masashi Iwasaki, Faculty of Life and Environmental Sciences, Kyoto Prefectural University, Kyoto 606-8522, Japan
E-mail: imasa@kpu.ac.jp

Reviewing editor: Kok Lay Teo, Curtin University, Australia

Abstract: The Hankel determinant appears in representations of solutions to several integrable systems. An asymptotic expansion of the Hankel determinant thus plays a key role in the investigation of asymptotic analysis of such integrable systems. This paper presents an asymptotic expansion formula of a certain Casorati determinant as an extension of the Hankel case. This Casorati determinant is then shown to be associated with the solution to the discrete hungry Lotka–Volterra (dhLV) system, which is an integrable variant of the famous prey–predator model in mathematical biology. Finally, the asymptotic behavior of the dhLV system is clarified using the expansion formula for the Casorati determinant.

Subjects: Applied Mathematics; Dynamical Systems; Mathematical Biology; Mathematics & Statistics; Science

Keywords: Casorati determinant; discrete hungry Lotka–Volterra system; asymptotic expansion

AMS subject classification: 39A12; 34E05; 15A15

1. Introduction

Integrable systems are often classified as nonlinear dynamical systems whose solutions can be explicitly expressed. Such an integrable system is the Toda equation which describes the current-voltage function in an electric circuit. A time discretization, called the discrete Toda equation (Hirota, 1981), is simply equal to the recursion formula of the qd algorithm for computing eigenvalues of a symmetric tridiagonal matrix (Henrici, 1988; Rutishauser, 1990) and singular values of a bidiagonal matrix (Parlett, 1995).

Another commonly investigated integrable system is the integrable Lotka–Volterra (LV) system, which is a prey–predator model in mathematical biology (Yamazaki, 1987). The discrete LV (dLV) system was shown in Iwasaki and Nakamura (2002) to be applicable to computing for bidiagonal

ABOUT THE AUTHOR

Masato Shinjo is a doctoral student in the Department of Applied Mathematics and Physics, Graduate School of Informatics, Kyoto University. He studies asymptotic analysis of nonlinear dynamical systems known as integrable systems.

PUBLIC INTEREST STATEMENT

In this paper, we present a powerful new technique for an asymptotic expansion of the Casorati determinant. The Casorati determinant is associated with several difference equations appearing in mathematical physics, and plays a role similar to the Wronskian in the theory of differential equations. Our technique will be useful for asymptotically analyzing not only the discrete hungry Lotka-Volterra system, but also other dynamical systems associated with the Casorati determinant.

singular values. The hungry Lotka–Volterra (hLV) system is a variant that captures a more complicated prey-predator relationship in comparison with the original LV system (Bogoyavlensky, 1988; Itoh, 1987). Time discretization of this system leads to the discrete hungry Lotka–Volterra (dhLV) system. It was shown in Fukuda, Ishiwata, Iwasaki, and Nakamura (2009), Fukuda, Ishiwata, Yamamoto, Iwasaki, and Nakamura (2013), Yamamoto, Fukuda, Iwasaki, Ishiwata, and Nakamura (2010) that the dhLV system can generate LR matrix transformations for computing eigenvalues of a banded totally nonnegative (TN) matrix whose minors are all nonnegative.

The determinant solutions to both the discrete Toda equation and the dLV system can be expressed using the Hankel determinant,

$$H_0^{(n)} := 1, \quad H_j^{(n)} := \begin{vmatrix} a^{(n)} & a^{(n+1)} & \cdots & a^{(n+j-1)} \\ a^{(n+1)} & a^{(n+2)} & \cdots & a^{(n+j)} \\ \vdots & \vdots & \ddots & \vdots \\ a^{(n+j-1)} & a^{(n+j)} & \cdots & a^{(n+2j-2)} \end{vmatrix}, \quad j = 1, 2, \ldots \tag{1.1}$$

where j and n correspond to the discrete spatial and discrete time variables, respectively (Tsujimoto, 2001). Here, the formal power series $f(z) = \sum_{n=0}^{\infty} a^{(n)} z^n$ associated with $H_j^{(n)}$ is assumed to be analytic at $z = 0$ and meromorphic in the disk $D = \{z \mid |z| < \zeta\}$. The finite or infinite types of poles $u_1^{-1}, u_2^{-1}, \ldots$ of $f(z)$ are ordered such that $0 < |u_1^{-1}| < |u_2^{-1}| < \cdots < \zeta$. Then, there exists a nonzero constant c_j independent of n such that, for some ϱ satisfying $|u_j| > \varrho > |u_{j+1}|$,

$$H_j^{(n)} = c_j (u_1 u_2 \cdots u_j)^n \left(1 + O\left(\left(\frac{\varrho}{|u_j|} \right)^n \right) \right) \tag{1.2}$$

as $n \to \infty$ (Henrici, 1988). The asymptotic expansion (1.2) as $n \to \infty$ enables the asymptotic analysis of the discrete Toda equation and the dLV system as in Henrici (1988), Rutishauser (1990) and in Iwasaki and Nakamura (2002), respectively.

A generalization of the Hankel determinant $H_j^{(n)}$ is given in the below determinant of a nonsymmetric square matrix of order j,

$$C_{i,0}^{(n)} := 1, \quad C_{i,j}^{(n)} := \begin{vmatrix} a_i^{(n)} & a_{i+1}^{(n)} & \cdots & a_{i+j-1}^{(n)} \\ a_i^{(n+1)} & a_{i+1}^{(n+1)} & \cdots & a_{i+j-1}^{(n+1)} \\ \vdots & \vdots & \ddots & \vdots \\ a_i^{(n+j-1)} & a_{i+1}^{(n+j-1)} & \cdots & a_{i+j-1}^{(n+j-1)} \end{vmatrix}, \quad i = 0, 1, \ldots, \quad j = 1, 2, \ldots \tag{1.3}$$

which is called the Casorati determinant or Casoratian. The Casorati determinant is useful in the theory of difference equations, particularly in mathematical physics, and plays a role similar to the Wronskian in the theory of differential equations (Vein & Dale, 1999). No one wonder here that the formal power series $f_i(z) = \sum_{n=0}^{\infty} a_i^{(n)} z^n$ is associated with the Casorati determinant $C_{i,j}^{(n)}$ for each i. The formal power series $f_i(z)$ differs from $f(z)$ in that not only the superscript, but also the subscript, appears in the coefficients.

To the best of our knowledge, from the viewpoint of the formal power series $f_i(z)$, the asymptotic analysis for the Casorati determinant $C_{i,j}^{(n)}$ has not yet been discussed in the literature. The first purpose of this paper is to present an asymptotic expansion of the Casorati determinant $C_{i,j}^{(n)}$ as $n \to \infty$. The asymptotic behavior of the dhLV system was discussed in Fukuda et al. (2009, 2013) in the case where the discretization parameter $\delta^{(n)}$ is restricted to be positive. However, it was suggested in Yamamoto et al. (2010) that the choice $\delta^{(n)} < 0$ in the dhLV system yields a convergence acceleration of the LR transformations. The discrete time evolution in the dhLV system with $\delta^{(n)} < 0$ corresponds to a reverse of the continuous-time evolution in the hLV system. It is interesting to note that such artificial dynamics are useful for computing eigenvalues of a TN matrix. The second purpose of this paper is to provide an asymptotic analysis for the dhLV system without being limited by the sign of $\delta^{(n)}$.

The remainder of this paper is organized as follows. In Section 2, we first observe that the entries in $C_{i,j}^{(n)}$ can be expressed using poles of $f_i(z)$. We then give an asymptotic expansion of the Casorati determinant in terms of the poles of $f_i(z)$ as $n \to \infty$ by expanding the theorem analyticity for the Hankel determinant given in Henrici (1988). In Section 3, we find the determinant solution to the dhLV system through relating the dhLV system to a three-term recursion formula. With the help of the resulting theorem for the Casorati determinant $C_{i,j}^{(n)}$, we explain in Section 4 that the determinant solution to the dhLV system can be rewritten using the Casorati determinant $C_{i,j}^{(n)}$, and we clarify the asymptotic behavior of the solution to the dhLV system. Finally, we give concluding remarks in Section 5.

2. An asymptotic expansion of the Casorati determinant

In this section, we first give an expression of the entries of the Casorati determinant $C_{i,j}^{(n)}$ in terms of poles of the formal power series $f_i(z)$ associated with $C_{i,j}^{(n)}$. Referring to the theorem on analyticity for the Hankel determinant given in Henrici (1988), we present an asymptotic expansion of the Casorati determinant $C_{i,j}^{(n)}$ as $n \to \infty$ using the poles of $f_i(z)$. We also describe the case where some restriction is imposed on the poles of $f_i(z)$.

Let $f_i(z) = \sum_{n=0}^{\infty} a_i^{(n)} z^n$, which is the formal power series associated with $C_{i,j}^{(n)}$ for $i = 0, 1, \ldots,$ be analytic at $z = 0$ and meromorphic in the disk $D = \{z \mid |z| < \zeta\}$. Moreover, let $r_{i,1}^{-1}, r_{i,2}^{-1}, \ldots,$ denote the poles of $f_i(z)$ such that $|r_{i,1}^{-1}| < |r_{i,2}^{-1}| < \cdots < \zeta$. By extracting the principal parts in $f_i(z)$, we derive

$$f_i(z) = \frac{\alpha_{i,1}}{r_{i,1}^{-1} - z} + \frac{\alpha_{i,2}}{r_{i,2}^{-1} - z} + \cdots + \frac{\alpha_{i,p}}{r_{i,p}^{-1} - z} + \sum_{n=0}^{\infty} b_i^{(n)} z^n \tag{2.1}$$

where p is an arbitrary positive integer, $\alpha_{i,1}, \alpha_{i,2}, \ldots, \alpha_{i,p}$ are some nonzero constants, and $b_i^{(n)}$, which contains the terms with respect to $r_{i,p+1}^{-1}, r_{i,p+2}^{-1}, \ldots$, satisfies

$$|b_i^{(n)}| \leq \mu_i \rho_i^n \tag{2.2}$$

for some nonzero positive constants μ_i and ρ_i with $|r_{i,p+1}| < \rho_i < |r_{i,p}|$. The proof of (2.2) is given in Henrici (1988) utilizing the Cauchy coefficient estimate. We now give a lemma for an expression of $a_i^{(n)}$ appearing in $f_i(z) = \sum_{n=0}^{\infty} a_i^{(n)} z^n$.

LEMMA 2.1 *Let us assume that the poles $r_{i,1}^{-1}, r_{i,2}^{-1}, \ldots, r_{i,p}^{-1}$ of $f_i(z)$ are not multiple. Then, $a_i^{(n)}$ can be expressed using $r_{i,1}, r_{i,2}, \ldots, r_{i,p}$ as*

$$a_i^{(n)} = \sum_{\ell=1}^{p} c_{i,\ell} r_{i,\ell}^n + b_i^{(n)} \tag{2.3}$$

where $c_{i,1}, c_{i,2}, \ldots, c_{i,p}$ are some nonzero constants.

Proof The crucial element is the replacement $\alpha_{i,1} = c_{i,1} r_{i,1}^{-1}, \alpha_{i,2} = c_{i,2} r_{i,2}^{-1}, \ldots, \alpha_{i,p} = c_{i,p} r_{i,p}^{-1}$ in (2.1), namely,

$$f_i(z) = \frac{c_{i,1} r_{i,1}^{-1}}{r_{i,1}^{-1} - z} + \frac{c_{i,2} r_{i,2}^{-1}}{r_{i,2}^{-1} - z} + \cdots + \frac{c_{i,p} r_{i,p}^{-1}}{r_{i,p}^{-1} - z} + \sum_{n=0}^{\infty} b_i^{(n)} z^n \tag{2.4}$$

Since each $c_{i,\ell} r_{i,\ell}^{-1} / (r_{i,\ell}^{-1} - z)$ in (2.4) can be regarded as the summation of a geometric series, we obtain

$$\begin{aligned} f_i(z) &= \sum_{n=0}^{\infty} c_{i,1} r_{i,1}^n z^n + \sum_{n=0}^{\infty} c_{i,2} r_{i,2}^n z^n + \cdots + \sum_{n=0}^{\infty} c_{i,p} r_{i,p}^n z^n + \sum_{n=0}^{\infty} b_i^{(n)} z^n \\ &= \sum_{n=0}^{\infty} \left[\left(\sum_{\ell=1}^{p} c_{i,\ell} r_{i,\ell}^n \right) + b_i^{(n)} \right] z^n \end{aligned}$$

which implies (2.3).

Similarly to the asymptotic expansion as $n \to \infty$ of the Hankel determinant $H_j^{(n)}$ given in Henrici (1988), we have the following theorem for the Casorati determinant $C_{i,j}^{(n)}$.

THEOREM 2.2 *Let us assume that the poles $r_{i,1}^{-1}, r_{i,2}^{-1}, \dots, r_{i,p}^{-1}$ of $f_i(z)$ are not multiple. Then there exists some constant $c_{i,\sigma(\kappa_1,\kappa_2,\dots,\kappa_j)}$ independently of n such that, as $n \to \infty$,*

$$C_{i,j}^{(n)} = \sum_{\sigma} \left[c_{i,\sigma(\kappa_1,\kappa_2,\dots,\kappa_j)} \left(r_{i,\kappa_1} r_{i+1,\kappa_2} \cdots r_{i+j-1,\kappa_j} \right)^n \left(1 + \sum_{\ell=1}^{j} O\left(\left(\frac{\rho_{i+\ell-1}}{|r_{i+\ell-1,\kappa_\ell}|} \right)^n \right) \right) \right] \tag{2.5}$$

where σ denotes the mapping from $\{\kappa_1, \kappa_2, \dots, \kappa_j\}$ to $\{1, 2, \dots, p\}$ and $\rho_{i+\ell-1}$ is some constant such that $|r_{i+\ell-1,p+1}| < \rho_{i+\ell-1} < |r_{i+\ell-1,p}|$.

Proof By applying Lemma 2.1 and the addition formula of determinants to the Casorati determinant $C_{i,j}^{(n)}$, we derive

$$C_{i,j}^{(n)} = \sum_{\sigma} D_{i,\sigma(\kappa_1,\kappa_2,\dots,\kappa_j)}^{(n)} + \sum_{\sigma} \hat{D}_{i,\sigma(\kappa_1,\kappa_2,\dots,\kappa_j)}^{(n)} \tag{2.6}$$

where in the first summation

$$D_{i,\sigma(\kappa_1,\kappa_2,\dots,\kappa_j)}^{(n)} := \begin{vmatrix} c_{i,\kappa_1} r_{i,\kappa_1}^{n} & c_{i+1,\kappa_2} r_{i+1,\kappa_2}^{n} & \cdots & c_{i+j-1,\kappa_j} r_{i+j-1,\kappa_j}^{n} \\ c_{i,\kappa_1} r_{i,\kappa_1}^{n+1} & c_{i+1,\kappa_2} r_{i+1,\kappa_2}^{n+1} & \cdots & c_{i+j-1,\kappa_j} r_{i+j-1,\kappa_j}^{n+1} \\ \vdots & \vdots & \ddots & \vdots \\ c_{i,\kappa_1} r_{i,\kappa_1}^{n+j-1} & c_{i+1,\kappa_2} r_{i+1,\kappa_2}^{n+j-1} & \cdots & c_{i+j-1,\kappa_j} r_{i+j-1,\kappa_j}^{n+j-1} \end{vmatrix}$$

and $\hat{D}_{i,\sigma(\kappa_1,\kappa_2,\dots,\kappa_j)}^{(n)}$ in the second summation denotes a determinant of the same form as $D_{i,\sigma(\kappa_1,\kappa_2,\dots,\kappa_j)}^{(n)}$ except that the ℓth column is replaced with $\mathbf{b}_\ell := (b_{i+\ell-1}^{(n)}, b_{i+\ell-1}^{(n+1)}, \dots, b_{i+\ell-1}^{(n+j-1)})^\top$ for at least one of ℓ. Evaluating the first summation in (2.6), we obtain

$$\sum_{\sigma} D_{i,\sigma(\kappa_1,\kappa_2,\dots,\kappa_j)}^{(n)} = \sum_{\sigma} c_{i,\sigma(\kappa_1,\kappa_2,\dots,\kappa_j)} \left(r_{i,\kappa_1} r_{i+1,\kappa_2} \cdots r_{i+j-1,\kappa_j} \right)^n \tag{2.7}$$

where

$$c_{i,\sigma(\kappa_1,\kappa_2,\dots,\kappa_j)} := \begin{vmatrix} c_{i,\kappa_1} & c_{i+1,\kappa_2} & \cdots & c_{i+j-1,\kappa_j} \\ c_{i,\kappa_1} r_{i,\kappa_1} & c_{i+1,\kappa_2} r_{i+1,\kappa_2} & \cdots & c_{i+j-1,\kappa_j} r_{i+j-1,\kappa_j} \\ \vdots & \vdots & \ddots & \vdots \\ c_{i,\kappa_1} r_{i,\kappa_1}^{j-1} & c_{i+1,\kappa_2} r_{i+1,\kappa_2}^{j-1} & \cdots & c_{i+j-1,\kappa_j} r_{i+j-1,\kappa_j}^{j-1} \end{vmatrix} \tag{2.8}$$

To estimate the second summation in (2.6), for example, we consider the case where the 1st column is replaced with $\mathbf{b}_1$. It immediately follows from (2.2) that

$$\begin{vmatrix} b_i^{(n)} & c_{i+1,\kappa_2} r_{i+1,\kappa_2}^{n} & \cdots & c_{i+j-1,\kappa_j} r_{i+j-1,\kappa_j}^{n} \\ b_i^{(n+1)} & c_{i+1,\kappa_2} r_{i+1,\kappa_2}^{n+1} & \cdots & c_{i+j-1,\kappa_j} r_{i+j-1,\kappa_j}^{n+1} \\ \vdots & \vdots & \ddots & \vdots \\ b_i^{(n+j-1)} & c_{i+1,\kappa_2} r_{i+1,\kappa_2}^{n+j-1} & \cdots & c_{i+j-1,\kappa_j} r_{i+j-1,\kappa_j}^{n+j-1} \end{vmatrix} = O\left(\left(\rho_i r_{i+1,\kappa_2} \cdots r_{i+j-1,\kappa_j} \right)^n \right)$$

It is also easy to check $O((r_{i,\kappa_1} r_{i+1,\kappa_2} \cdots r_{i+\ell-2,\kappa_{\ell-1}} \rho_{i+\ell-1} r_{i+\ell,\kappa_{\ell+1}} \cdots r_{i+j-1,\kappa_j})^n)$ if the ℓth column is replaced with $\mathbf{b}_\ell$. Similarly, by examining the case where some columns are replaced with some of $\mathbf{b}_1, \mathbf{b}_2, \dots, \mathbf{b}_j$, we can see that

$$\sum_{\sigma} \hat{D}_{i,\sigma(\kappa_1,\kappa_2,\dots,\kappa_j)} = \sum_{\sigma} c_{i,\sigma(\kappa_1,\kappa_2,\dots,\kappa_j)} (r_{i,\kappa_1} r_{i+1,\kappa_2} \cdots r_{i+j-1,\kappa_j})^n \sum_{\ell=1}^{j} O\left(\left(\frac{\rho_{i+\ell-1}}{|r_{i+\ell-1,\kappa_\ell}|} \right)^n \right) \tag{2.9}$$

Thus, from (2.7)–(2.9), we obtain (2.5)

Now, let us consider the restriction $r_{i,1} = r_1, r_{i,2} = r_2, \dots, r_{i,j} = r_j$ in $f_i(z)$. Then, by replacing $r_{i,\ell}$ with r_ℓ in (2.3), we easily obtain

$$a_i^{(n)} = \sum_{\ell=1}^{p} c_{i,\ell} r_\ell^n + b_i^{(n)} \tag{2.10}$$

As a specialization of Theorem 2.2, we derive the following theorem for an asymptotic expansion of the Casorati determinant $C_{i,j}^{(n)}$ with restricted $a_i^{(n)}$ as $n \to \infty$.

THEOREM 2.3 *Let us assume that the poles $r_1^{-1}, r_2^{-1}, \dots, r_j^{-1}$ of $f_i(z)$ are not multiple. Then there exists some constant $c_{i,j} \neq 0$ independently of n such that, for $|r_{j+1}| < \rho_i < |r_j|$, as $n \to \infty$,*

$$C_{i,j}^{(n)} = c_{i,j}\left(r_1 r_2 \dots r_j\right)^n \left(1 + \sum_{\ell=1}^{j} O\left(\left(\frac{\rho_{i+\ell-1}}{|r_j|}\right)^n\right)\right) \tag{2.11}$$

Proof Replacing $r_{i,1} = r_1, r_{i,2} = r_2, \dots, r_{i,p} = r_p$ in (2.8) gives

$$c_{i,\sigma(\kappa_1,\kappa_2,\dots,\kappa_j)} = c_{i,\kappa_1} c_{i+1,\kappa_2} \cdots c_{i+j-1,\kappa_j} \begin{vmatrix} 1 & 1 & \cdots & 1 \\ r_{\kappa_1} & r_{\kappa_2} & \cdots & r_{\kappa_j} \\ \vdots & \vdots & \ddots & \vdots \\ r_{\kappa_1}^{j-1} & r_{\kappa_2}^{j-1} & \cdots & r_{\kappa_j}^{j-1} \end{vmatrix} \tag{2.12}$$

Thus, by taking into account that $c_{\sigma(\kappa_1,\kappa_2,\dots,\kappa_j)} \neq 0$ only in the case where $\kappa_1, \kappa_2, \dots, \kappa_j$ are distinct to each other, we can simplify (2.7) as

$$\sum_{\pi} D_{i,\pi(\kappa_1,\kappa_2,\dots,\kappa_j)}^{(n)} = (r_1 r_2 \dots r_j)^n \sum_{\pi} c_{i,\kappa_1} c_{i+1,\kappa_2} \cdots c_{i+j-1,\kappa_j} \begin{vmatrix} 1 & 1 & \cdots & 1 \\ r_{\kappa_1} & r_{\kappa_2} & \cdots & r_{\kappa_j} \\ \vdots & \vdots & \ddots & \vdots \\ r_{\kappa_1}^{j-1} & r_{\kappa_2}^{j-1} & \cdots & r_{\kappa_j}^{j-1} \end{vmatrix} \tag{2.13}$$

where π denotes the bijection from $\{\kappa_1, \kappa_2, \dots, \kappa_j\}$ to $\{1, 2, \dots, j\}$. It is noted here that the bijection π is equal to the mapping σ with $p = j$. Moreover, there exists a constant ρ_i, which is not equal to one in Theorem 2.2, such that $|r_{j+1}| < \rho_i < |r_j|$. This is because ρ_i and ρ_{i+1} do not always satisfy $\rho_i = \rho_{i+1}$ even if $r_{i,1} = r_1, r_{i,2} = r_2, \dots, r_{i,j} = r_j$ in Theorem 2.2. Thus, (2.9) becomes

$$\sum_{\ell=1}^{j} O\left(\left(r_1 r_2 \dots r_{j-1} \rho_{i+\ell-1}\right)^n\right) \tag{2.14}$$

Therefore, from (2.13) and (2.14), we obtain (2.11).

Theorem 2.3 covers an asymptotic expansion of the Hankel determinant $H_j^{(n)}$. Theorems 2.2 and 2.3 should be useful for the asymptotic analysis of dynamical systems with solutions expressed in terms of the Casorati determinant $C_{i,j}^{(n)}$.

3. The dhLV system and its determinant solution

In this section, similarly to work in Tsujimoto and Kondo (2000), Spiridonov and Zhedanov (1997), we derive the dhLV system from a three-term recursion formula, and then clarify the determinant expression of an auxiliary variable in the solution to the dhLV system through investigating the three-term recursion formula.

Let us consider a three-term recursion formula with respect to the polynomials $T_0^{(n)}(x), T_1^{(n)}(x), \cdots$ at the discrete time n,

$$\begin{cases} T_{k+1}^{(n)}(x) = xT_k^{(n)}(x) - v_k^{(n)}T_{k-M}^{(n)}(x), & k = M, M+1, \dots, \\ T_0^{(n)}(x) := 1, \quad T_1^{(n)}(x) := x, \quad \dots, & T_M^{(n)}(x) := x^M \end{cases} \tag{3.1}$$

where M is a positive integer and $v_M^{(n)}, v_{M+1}^{(n)}, \dots$ do not depend on x. Accordingly, $T_0^{(n)}(x), T_1^{(n)}(x), \dots,$ are all monic. Moreover, let us prepare a time evolution from n to $n+1$,

$$T_k^{(n+1)}(x) = \frac{1}{x^{M+1} - \left(\delta^{(n)}\right)^{M+1}} \left(T_{k+M+1}^{(n)}(x) - V_k^{(n)} T_k^{(n)}(x) \right) \tag{3.2}$$

where $V_k^{(n)} := T_{k+M+1}^{(n)}(\delta^{(n)})/T_k^{(n)}(\delta^{(n)})$. Then, by replacing n with $n+1$ in (3.1) and using (3.2), we obtain

$$T_{k+M+2}^{(n)}(x) - V_{k+1}^{(n)} T_{k+1}^{(n)}(x) = x\left(T_{k+M+1}^{(n)}(x) - V_k^{(n)} T_k^{(n)}(x) \right) - v_k^{(n+1)} \left(T_{k+1}^{(n)}(x) - V_{k-M}^{(n)} T_{k-M}^{(n)}(x) \right) \tag{3.3}$$

By using (3.1) again for deleting except for terms with respect to $T_{k+1}^{(n)}(x)$ and $T_{k-M}^{(n)}(x)$ in (3.3), we derive

$$\left(V_k^{(n)} + v_k^{(n+1)} - v_{k+M+1}^{(n)} - V_{k+1}^{(n)} \right) T_{k+1}^{(n)}(x) = \left(V_k^{(n)} v_k^{(n)} - v_k^{(n+1)} V_{k-M}^{(n)} \right) T_{k-M}^{(n)}(x)$$

Thus, it is observed that

$$V_k^{(n)} + v_k^{(n+1)} = v_{k+M+1}^{(n)} + V_{k+1}^{(n)} \tag{3.4}$$

$$v_k^{(n)} V_k^{(n)} = v_k^{(n+1)} V_{k-M}^{(n)} \tag{3.5}$$

Let us introduce a new variable $u_k^{(n)}$ such that

$$v_k^{(n)} = u_{k-M}^{(n)} \prod_{j=1}^{M} \left(\delta^{(n)} + u_{k-j-M}^{(n)} \right) \tag{3.6}$$

$$V_k^{(n)} = -\prod_{j=0}^{M} \left(\delta^{(n)} + u_{k-j}^{(n)} \right) \tag{3.7}$$

Then, it follows from (3.5)-(3.7) that

$$v_k^{(n+1)} = u_{k-M}^{(n)} \prod_{j=1}^{M} \left(\delta^{(n)} + u_{k-j+1}^{(n)} \right) \tag{3.8}$$

Moreover, from (3.6) and (3.8), we see that

$$v_k^{(n+1)} - v_{k+M+1}^{(n)} = \prod_{j=0}^{M} \left(\delta^{(n)} + u_{k-j}^{(n)} \right) - \prod_{j=0}^{M} \left(\delta^{(n)} + u_{k-j+1}^{(n)} \right) \tag{3.9}$$

It is obvious from (3.7) that the right-hand side of (3.9) is equal to $V_{k+1}^{(n)} - V_k^{(n)}$. This implies that $v_k^{(n+1)}$ in (3.8) also satisfies (3.4). Consequently, by combining (3.6) and (3.8), noting that $\prod_{j=1}^{M}(\delta^{(n)} + u_{k+1-j}^{(n)}) = \prod_{j=1}^{M}(\delta^{(n)} + u_{k-M+j}^{(n)})$ and replacing $k - M$ with k, we have the discrete system

$$u_k^{(n+1)} \prod_{j=1}^{M} \left(\delta^{(n+1)} + u_{k-j}^{(n+1)} \right) = u_k^{(n)} \prod_{j=1}^{M} \left(\delta^{(n)} + u_{k+j}^{(n)} \right) \tag{3.10}$$

Equation (3.10) can be regarded as a discretization of the hLV system which differs from the simple LV system in that more than one food exists for each species. Thus, (3.10) is the dhLV system and M

corresponds to the number of the species of foods for each species. Clearly, from the definition, (3.10) with $M = 1$ is simply equal to the dLV system. The dhLV system (3.10) is essentially equal to the dhLV system in Fukuda et al. (2009),

$$u_k^{(n+1)} \prod_{j=1}^{M} \left(1 + \delta^{(n+1)} u_{k-j}^{(n+1)}\right) = u_k^{(n)} \prod_{j=1}^{M} \left(1 + \delta^{(n)} u_{k+j}^{(n)}\right) \tag{3.11}$$

This is because (3.11) is derived by replacing $u_k^{(n)}$ with $[1/(\delta^{(n)})^M]u_k^{(n)}$ and $1/(\delta^{(n)})^{M+1}$ with $\delta^{(n)}$ for $n = 0, 1, \ldots$ in (3.10).

Let $\tilde{T}_0^{(n)}, \tilde{T}_1^{(n)}, \ldots$ be polynomials satisfying a three-term recursion formula,

$$\begin{cases} \tilde{T}_{k+1}^{(n)}(x) = x^M \tilde{T}_{k-M+1}^{(n)}(x) - w_k^{(n)} \tilde{T}_{k-M}^{(n)}(x), \quad k = M, M+1, \ldots, \\ \tilde{T}_0^{(n)}(x) := 1, \quad \tilde{T}_1^{(n)}(x) := x, \quad \ldots, \quad \tilde{T}_M^{(n)}(x) := x^M \end{cases} \tag{3.12}$$

where $w_M^{(n)}, w_{M+1}^{(n)}, \ldots$, do not depend on x. It is obvious from (3.12) that $\tilde{T}_M^{(n)}(x), \tilde{T}_{M+1}^{(n)}(x), \ldots$ are also all monic. Moreover, let us introduce a linear functional (form) $\mathcal{L}^{(n)}$,

$$\begin{aligned} \mathcal{L}^{(n)}[T_k^{(n)}(x^M)\tilde{T}_\ell^{(n)}(x)] :&= \int_{\mathbb{R}} T_k^{(n)}(x^M)\tilde{T}_\ell^{(n)}(x)\omega^{(n)}(x)dx \\ &= \begin{cases} h_k^{(n)} & (k = \ell) \\ 0 & (k \neq \ell) \end{cases} \end{aligned} \tag{3.13}$$

where $\omega^{(n)}(x)$ is a weight function. The linear functional $\mathcal{L}^{(n)}$ with $M = 1$ is equivalent to that in Chihara (1978). Further, $\mathcal{L}^{(n)}$ with arbitrary M is a specialization of a linear function appearing in Maeda, Miki, and Tsujimoto (2013). Since it follows from (3.1), (3.12), and (3.13) that $\mathcal{L}[T_k^{(n)}(x^M) x^M \tilde{T}_{k-M}^{(n)}(x)] = h_k^{(n)}$ and $\mathcal{L}[x^M T_k^{(n)}(x^M)\tilde{T}_{k-M}^{(n)}(x)] = v_k^{(n)} h_{k-M}^{(n)}$, we easily derive

$$v_k^{(n)} = \frac{h_k^{(n)}}{h_{k-M}^{(n)}} \tag{3.14}$$

Let $\mu_k^{(n)} := \mathcal{L}^{(n)}[x^k]$ for $k = 0, 1, \ldots$. From (3.12), it turns out that $\tilde{T}_k^{(n)}(x)$ is expressed as $\tilde{T}_k^{(n)}(x) = s_{k,0}^{(n)} + \cdots + s_{k,k-1}^{(n)} x^{k-1} + x^k$ where $s_{k,0}^{(n)}, \ldots, s_{k,k-1}^{(n)}$ are some constants at each k and each n. Since it is clear from (3.1) that $T_\ell^{(n)}(x)$ can be given as the summation of x^ℓ and the linear combination of $T_0^{(n)}(x), T_1^{(n)}(x), \ldots, T_{\ell-1}^{(n)}(x)$, we see from (3.13) that $\mathcal{L}^{(n)}[T_\ell^{(n)}(x^M)\tilde{T}_k^{(n)}(x)] = \mathcal{L}^{(n)}[x^{\ell M}\tilde{T}_k^{(n)}(x)]$. Thus, it follows that

$$\begin{aligned} &\mathcal{L}^{(n)}[T_0^{(n)}(x^M)\tilde{T}_k^{(n)}(x)] = s_{k,0}^{(n)}\mu_0^{(n)} + \cdots + s_{k,k-1}^{(n)}\mu_{k-1}^{(n)} + \mu_k^{(n)} \\ &\quad\vdots \\ &\mathcal{L}^{(n)}[T_{k-1}^{(n)}(x^M)\tilde{T}_k^{(n)}(x)] = s_{k,0}^{(n)}\mu_{(k-1)M}^{(n)} + \cdots + s_{k,k-1}^{(n)}\mu_{(k-1)(M+1)}^{(n)} + \mu_{(k-1)M+k}^{(n)} \\ &\mathcal{L}^{(n)}[T_k^{(n)}(x^M)\tilde{T}_k^{(n)}(x)] = s_{k,0}^{(n)}\mu_{kM}^{(n)} + \cdots + s_{k,k-1}^{(n)}\mu_{k(M+1)-1}^{(n)} + \mu_{k(M+1)}^{(n)} \end{aligned}$$

By combining the above with (3.13), we derive a system of linear equations

$$\begin{pmatrix} \mu_0^{(n)} & \cdots & \mu_{k-1}^{(n)} & \mu_k^{(n)} \\ \vdots & \ddots & \vdots & \vdots \\ \mu_{(k-1)M}^{(n)} & \cdots & \mu_{(k-1)(M+1)}^{(n)} & \mu_{(k-1)M+k}^{(n)} \\ \mu_{kM}^{(n)} & \cdots & \mu_{k(M+1)-1}^{(n)} & \mu_{k(M+1)}^{(n)} \end{pmatrix} \begin{pmatrix} s_{k,0}^{(n)} \\ \vdots \\ s_{k,k-1}^{(n)} \\ 1 \end{pmatrix} = \begin{pmatrix} 0 \\ \vdots \\ 0 \\ h_k^{(n)} \end{pmatrix} \tag{3.15}$$

Since $s_{k,0}^{(n)}, \dots, s_{k,k-1}^{(n)}$ are uniquely determined, the coefficient matrix in (3.15) is nonsingular. This suggests that (3.15) can be transformed into

$$\begin{pmatrix} s_{k,0}^{(n)} \\ \vdots \\ s_{k,k-1}^{(n)} \\ 1 \end{pmatrix} = \frac{1}{\tau_{k+1}^{(n)}} \begin{pmatrix} \hat{\mu}_0^{(n)} & \cdots & \hat{\mu}_{k-1}^{(n)} & \hat{\mu}_k^{(n)} \\ \vdots & \ddots & \vdots & \vdots \\ \hat{\mu}_{(k-1)M}^{(n)} & \cdots & \hat{\mu}_{(k-1)(M+1)}^{(n)} & \hat{\mu}_{(k-1)M+k}^{(n)} \\ \hat{\mu}_{kM}^{(n)} & \cdots & \hat{\mu}_{k(M+1)-1}^{(n)} & \hat{\mu}_{k(M+1)}^{(n)} \end{pmatrix} \begin{pmatrix} 0 \\ \vdots \\ 0 \\ h_k^{(n)} \end{pmatrix} \tag{3.16}$$

where the hat denotes cofactors of the coefficient matrix in (3.15) and

$$\tau_{k+1}^{(n)} := \begin{vmatrix} \mu_0^{(n)} & \cdots & \mu_{k-1}^{(n)} & \mu_k^{(n)} \\ \vdots & \ddots & \vdots & \vdots \\ \mu_{(k-1)M}^{(n)} & \cdots & \mu_{(k-1)(M+1)}^{(n)} & \mu_{(k-1)M+k}^{(n)} \\ \mu_{kM}^{(n)} & \cdots & \mu_{k(M+1)-1}^{(n)} & \mu_{k(M+1)}^{(n)} \end{vmatrix} \tag{3.17}$$

It is of significance to note that $\hat{\mu}_{k(M+1)}^{(n)} = \tau_k^{(n)}$. Thus, by examining the last row for both sides of (3.16), we find

$$h_k^{(n)} = \frac{\tau_{k+1}^{(n)}}{\tau_k^{(n)}} \tag{3.18}$$

Equations (3.14) and (3.18) therefore lead to

$$v_k^{(n)} = \frac{\tau_{k+1}^{(n)} \tau_{k-M}^{(n)}}{\tau_k^{(n)} \tau_{k-M+1}^{(n)}} \tag{3.19}$$

Since we can easily obtain the solution to the dhLV system (3.10), by combining (3.6) with (3.19), the determinant expression of $v_k^{(n)}$ is important for the asymptotic analysis of the dhLV system (3.10) in the next section.

Let us define the time evolution of the linear functional from $\mathcal{L}^{(n)}$ to $\mathcal{L}^{(n+1)}$ by

$$\mathcal{L}^{(n+1)}[P(x)] = \mathcal{L}^{(n)} \left[\left(x^{M(M+1)} - (\delta^{(n)})^{M+1} \right) P(x) \right] \tag{3.20}$$

where $P(x)$ is an arbitrary polynomial. Then, it is easy to check that $T_k^{(n+1)}(x^M)$ and $\tilde{T}_\ell^{(n)}(x)$ are orthogonal to each other with respect to $\mathcal{L}^{(n+1)}$. Equation (3.20) yields a time evolution with respect to μ's,

$$\mu_k^{(n+1)} = \mu_{k+M(M+1)}^{(n)} - (\delta^{(n)})^{M+1} \mu_k^{(n)} \tag{3.21}$$

Noting (3.1) and (3.12), we find that $\mathcal{L}^{(n)}[T_k^{(n)}(x^M)\tilde{T}_\ell^{(n)}(x)]$ with $k = \ell$ can be expressed as the linear combination of $\mu_0^{(n)}, \mu_{(M+1)}^{(n)}, \dots, \mu_{k(M+1)}^{(n)}$. Thus, by combining it with (3.13), we derive

$$\mu_{j(M+1)}^{(n)} \neq 0, \quad j = 0, 1, \dots, k \tag{3.22}$$

Similarly, in the case where $\mathcal{L}^{(n)}[T_k^{(n)}(x^M)\tilde{T}_\ell^{(n)}(x)]$ with $k \neq \ell$, we have

$$\mu_{i+j(M+1)}^{(n)} = 0, \quad i = 1, 2, \dots, M, \quad j = 0, 1, \dots, k \tag{3.23}$$

Taking into account that the sequence $\{\mu_{j(M+1)}^{(n)}\}_{n=0,1,\dots}$ with (3.21) is a specialization of the sequence $\{a_j^{(n)}\}_{n=0,1,\dots}$ appearing in the previous section, we may replace $\mu_{j(M+1)}^{(n)}$ with $a_j^{(n)}$ in the following discussion. Thus, we can rewrite $\tau_k^{(n)}$ as

$$\tau_0^{(n)} := 1, \quad \tau_{j(M+1)}^{(n)} := \begin{vmatrix} \tau_{0,M}^{(n)} & \tau_{1,M}^{(n)} & \cdots & \tau_{j-1,M}^{(n)} \\ \tau_{M,M}^{(n)} & \tau_{M+1,M}^{(n)} & \cdots & \tau_{M+j-1,M}^{(n)} \\ \vdots & \vdots & \ddots & \vdots \\ \tau_{(j-1)M,M}^{(n)} & \tau_{(j-1)M+1,M}^{(n)} & \cdots & \tau_{(j-1)(M+1)-1,M}^{(n)} \end{vmatrix}$$

$$\tau_{i+j(M+1)}^{(n)} := \begin{vmatrix} \tau_{0,M}^{(n)} & \tau_{1,M}^{(n)} & \cdots & \tau_{j-1,M}^{(n)} & \tau_{j,i-1}^{(n)} \\ \tau_{M,M}^{(n)} & \tau_{M+1,M}^{(n)} & \cdots & \tau_{M+j-1,M}^{(n)} & \tau_{M+j,i-1}^{(n)} \\ \vdots & \vdots & \ddots & \vdots & \vdots \\ \tau_{(j-1)M,M}^{(n)} & \tau_{(j-1)M+1,M}^{(n)} & \cdots & \tau_{(j-1)(M+1)-1,M}^{(n)} & \tau_{(j-1)(M+1),i-1}^{(n)} \\ \tau_{jM,i-1}^{(n)} & \tau_{jM+1,i-1}^{(n)} & \cdots & \tau_{j(M+1)-1,i-1}^{(n)} & \tau_{j(M+1),i-1}^{(n)} \end{vmatrix}, \quad i = 1, 2, \ldots, M$$

where $\tau_{s,t}^{(n)} := \mathrm{diag}(a_s^{(n)}, a_{s+1}^{(n)}, \ldots, a_{s+t}^{(n)})$ is an $(t+1)$-by-$(t+1)$ diagonal matrix with the relationship concerning the evolution from n to $n+1$,

$$a_k^{(n+1)} = a_{k+M}^{(n)} - (\delta^{(n)})^{M+1} a_k^{(n)} \tag{3.24}$$

4. Asymptotic analysis of the dhLV system

This section begins by explaining that the auxiliary variable in the dhLV system can be rewritten in terms of the Casorati determinant. By using Theorem 2.2, we clarify the asymptotic behavior of the dhLV variables as $n \to \infty$.

The 1st, 2nd, ⋯, $(j-1)$th row and column blocks in $\tau_{i+j(M+1)}^{(n)}$ are M-by-M matrices, but the jth row and column blocks are $(i-1)$-by-$(i-1)$ matrices. The following lemma gives the representation of $v_k^{(n)}$ in terms of the $C_{i,j}^{(n)}$ appearing in Section 1.

LEMMA 4.1 *The auxiliary variable $v_k^{(n)}$ is expressed as*

$$v_{i+j(M+1)}^{(n)} = \frac{C_{i,j+1}^{(n)} C_{i+1,j-1}^{(n)}}{C_{i,j}^{(n)} C_{i+1,j}^{(n)}}, \qquad i = 0, 1, \ldots, M-1, \qquad j = 1, 2, \ldots, m-1 \tag{4.1}$$

$$v_{M+j(M+1)}^{(n)} = \frac{C_{M,j+1}^{(n)} C_{0,j}^{(n)}}{C_{M,j}^{(n)} C_{0,j+1}^{(n)}}, \qquad j = 0, 1, \ldots, m-1 \tag{4.2}$$

Proof Let us introduce a new determinant of a square matrix of order j,

$$G_{i,0}^{(n)} := 1, \quad G_{i,j}^{(n)} := \begin{vmatrix} a_i^{(n)} & a_{i+1}^{(n)} & \cdots & a_{i+j-1}^{(n)} \\ a_{i+M}^{(n)} & a_{i+M+1}^{(n)} & \cdots & a_{i+M+j-1}^{(n)} \\ \vdots & \vdots & \ddots & \vdots \\ a_{i+M(j-1)}^{(n)} & a_{i+M(j-1)+1}^{(n)} & \cdots & a_{i+(M+1)(j-1)}^{(n)} \end{vmatrix}, \quad j = 1, 2, \ldots \tag{4.3}$$

We begin by showing that $\tau_{j(M+1)}^{(n)}$ can be transformed into a block diagonal determinant with respect to $G_{0,j}^{(n)}, G_{1,j}^{(n)}, \ldots, G_{M,j}^{(n)}$. By interchanging the 2nd, 3rd, ⋯jth rows and columns with the $[1+(M+1)]$th, $[1+2(M+1)]$th, ⋯, $[1+(j-1)(M+1)]$th rows and columns in $\tau_{j(M+1)}^{(n)}$, we observe that the same form of $G_{0,j}^{(n)}$ appears in the 1st diagonal block of $\tau_{j(M+1)}^{(n)}$. The entries in the 1st, 2nd, ⋯, jth rows and columns in $\tau_{j(M+1)}^{(n)}$ are simultaneously all 0, except for those in the diagonal block section. Permutations similar to the above provide the forms of $G_{1,j}^{(n)}, G_{2,j}^{(n)}, \ldots, G_{M,j}^{(n)}$ as the 2nd, 3rd, ⋯, $(M+1)$th blocks in $\tau_{j(M+1)}^{(n)}$. Thus, $\tau_{j(M+1)}^{(n)}$ can be expressed in terms of $G_{0,j}^{(n)}, G_{1,j}^{(n)}, \ldots, G_{M,j}^{(n)}$ as

$$\tau_{j(M+1)}^{(n)} = \prod_{\ell=0}^{M} G_{\ell,j}^{(n)} \tag{4.4}$$

Similarly, $\tau^{(n)}_{i+j(M+1)}$ can be transformed into the determinant of a block diagonal matrix whose $M+1$ blocks are $G^{(n)}_{0,j+1}, G^{(n)}_{1,j+1}, \ldots, G^{(n)}_{i-1,j+1}$ and $G^{(n)}_{i,j}, G^{(n)}_{i+1,j}, \ldots, G^{(n)}_{M,j}$. Thus, it follows that

$$\tau^{(n)}_{i+j(M+1)} = \left(\prod_{\ell=0}^{i-1} G^{(n)}_{\ell,j+1}\right)\left(\prod_{\ell=i}^{M} G^{(n)}_{\ell,j}\right) \tag{4.5}$$

The cases where $k = i + j(M+1)$ and $k = M + j(M+1)$ in (3.19) become

$$\begin{aligned} v^{(n)}_{i+j(M+1)} &= \frac{\tau^{(n)}_{i+j(M+1)+1}\tau^{(n)}_{i+(j-1)(M+1)+1}}{\tau^{(n)}_{i+j(M+1)}\tau^{(n)}_{i+(j-1)(M+1)+2}} \\ v^{(n)}_{M+j(M+1)} &= \frac{\tau^{(n)}_{(j+1)(M+1)}\tau^{(n)}_{j(M+1)}}{\tau^{(n)}_{M+j(M+1)}\tau^{(n)}_{j(M+1)+1}} \end{aligned} \tag{4.6}$$

By combining them with (4.4) and (4.5), we obtain

$$\begin{aligned} v^{(n)}_{i+j(M+1)} &= \frac{G^{(n)}_{i,j+1}G^{(n)}_{i+1,j-1}}{G^{(n)}_{i,j}G^{(n)}_{i+1,j}}, \qquad i = 0, 1, \ldots, M-1 \\ v^{(n)}_{M+j(M+1)} &= \frac{G^{(n)}_{M,j+1}G^{(n)}_{0,j}}{G^{(n)}_{M,j}G^{(n)}_{0,j+1}} \end{aligned} \tag{4.7}$$

The entries in the jth row of $G^{(n)}_{i,j}$ are given by the linear combination $a^{(n)}_{i+M(j-1)+\ell} = a^{(n+1)}_{i+M(j-2)+\ell} + (\delta^{(n)})^{M+1}a^{(n)}_{i+M(j-2)+\ell}$ for $\ell = 0, 1, \ldots, j-1$. By multiplying the $(j-1)$th row by $-(\delta^{(n)})^{M+1}$ and then adding it to the jth, we get row $(a^{(n+1)}_{i+M(j-2)}, a^{(n+1)}_{i+M(j-2)+1}, \ldots, a^{(n+1)}_{i+(M+1)(j-2)+1})$ as the new jth row. Similarly, for the $(j-1)$th, $(j-2)$th, …, 2nd rows, it follows that

$$G^{(n)}_{i,j} = \begin{vmatrix} a^{(n)}_{i} & a^{(n)}_{i+1} & \cdots & a^{(n)}_{i+j-1} \\ a^{(n+1)}_{i} & a^{(n+1)}_{i+1} & \cdots & a^{(n+1)}_{i+j-1} \\ \vdots & \vdots & \ddots & \vdots \\ a^{(n+1)}_{i+M(j-2)} & a^{(n+1)}_{i+M(j-2)+1} & \cdots & a^{(n+1)}_{i+(M+1)(j-2)+1} \end{vmatrix}$$

It is worth noting here that the subscript M can be regarded as be transformed into the superscript 1. Thus, $G^{(n)}_{i,j}$ in (4.3) is equal to the Casorati determinant $C^{(n)}_{i,j}$ in (1.3). Then, by accounting for it in (4.6) and (4.7), we have (4.1) and (4.2).

Lemma 4.1 with Theorem 2.2 leads to the following theorem for asymptotic behavior of $v^{(n)}_{M+j(M+1)}$ as $n \to \infty$.

THEOREM 4.2 *The auxiliary variable $v^{(n)}_{M+j(M+1)}$ converges to some constant $\hat{c}_j$ as $n \to \infty$.*

Proof Let σ^* be the mapping from $\{\kappa_1, \kappa_2, \ldots, \kappa_j\}$ to $\{\kappa_1^*, \kappa_2^*, \ldots, \kappa_j^*\}$ where $\kappa_1^*, \kappa_2^*, \ldots, \kappa_j^*$ are positive integers such that $r_{i,\kappa_1^*}r_{i+1,\kappa_2^*}\cdots r_{i+j-1,\kappa_j^*} = \max_{\sigma}(r_{i,\kappa_1}r_{i+1,\kappa_2}\cdots r_{i+j-1,\kappa_j})$. Then, it follows from Theorem 2.2 that

$$\begin{aligned} &\lim_{n\to\infty} \frac{C^{(n)}_{i,j}}{\left(r_{i,\kappa_1^*}r_{i+1,\kappa_2^*}\cdots r_{i+j-1,\kappa_j^*}\right)^n} \\ &= \lim_{n\to\infty}\left\{ c_{i,\sigma^*(\kappa_1,\kappa_2,\ldots,\kappa_j)}\left(1 + \sum_{\ell=1}^{j} O\left(\left(\frac{\rho_{i+\ell-1}}{|r_{i+\ell-1,\kappa_\ell^*}|}\right)^n\right)\right)\right. \\ &\quad \left. + \sum_{\sigma\backslash\sigma^*}\left[c_{i,\sigma(\kappa_1,\kappa_2,\ldots,\kappa_j)}\left(\frac{r_{i,\kappa_1}r_{i+1,\kappa_2}\cdots r_{i+j-1,\kappa_j}}{r_{i,\kappa_1^*}r_{i+1,\kappa_2^*}\cdots r_{i+j-1,\kappa_j^*}}\right)^n\left(1 + \sum_{\ell=1}^{j} O\left(\left(\frac{\rho_{i+\ell-1}}{|r_{i+\ell-1,\kappa_\ell}|}\right)^n\right)\right)\right]\right\} \\ &= c_{i,\sigma^*(\kappa_1,\kappa_2,\ldots,\kappa_j)} \end{aligned} \tag{4.8}$$

It is of significance to note the relationship between $f_i(z)$ and $f_{i+M}(z)$ is derived from (3.24),

$$f_{i+M}(z) = \frac{\left[1 + (\delta^{(n)})^{M+1} z\right] f_i(z) - a_i^{(0)}}{z} \tag{4.9}$$

Equation (4.9) implies that the poles of $f_i(z)$ and $f_{i+M}(z)$ are equal to each other, namely, $r_{i,1} = r_{i+M,1}$, $r_{i,2} = r_{i+M,2}, \dots$. Thus, by combining them with Theorem 2.2, we derive

$$\lim_{n\to\infty} \frac{C_{i+M,j}^{(n)}}{\left(r_{i,\kappa_1^*} r_{i+1,\kappa_2^*} \cdots r_{i+j-1,\kappa_j^*}\right)^n} = \lim_{n\to\infty} \frac{C_{i+M,j}^{(n)}}{\left(r_{i+M,\kappa_1^*} r_{i+M+1,\kappa_2^*} \cdots r_{i+M+j-1,\kappa_j^*}\right)^n}$$
$$= c_{i+M,\sigma^*(\kappa_1,\kappa_2,\dots,\kappa_j)} \tag{4.10}$$

Since (4.8) and (4.10) imply that $C_{0,j}^{(n)}/C_{M,j}^{(n)} \to c_{0,\sigma^*(\kappa_1,\kappa_2,\dots,\kappa_j)}/c_{M,\sigma^*(\kappa_1,\kappa_2,\dots,\kappa_j)}$ as $n \to \infty$, we can conclude that $v_{M+j(M+1)}^{(n)} \to \hat{c}_j = (c_{0,\sigma^*(\kappa_1,\kappa_2,\dots,\kappa_j)} c_{M,\sigma^*(\kappa_1,\kappa_2,\dots,\kappa_{j+1})})/(c_{M,\sigma^*(\kappa_1,\kappa_2,\dots,\kappa_j)} c_{0,\sigma^*(\kappa_1,\kappa_2,\dots,\kappa_{j+1})})$ as $n \to \infty$.

By considering the positivity of $v_{j(M+1)}^{(n)}, v_{1+j(M+1)}^{(n)}, \dots, v_{M-1+j(M+1)}^{(n)}$, we derive the following theorem for the asymptotic behavior of $v_{j(M+1)}^{(n)}, v_{1+j(M+1)}^{(n)}, \dots, v_{M-1+j(M+1)}^{(n)}$ as $n \to \infty$.

THEOREM 4.3 *Let us assume that* $v_{j(M+1)}^{(n)} > 0, v_{1+j(M+1)}^{(n)} > 0, \dots, v_{M-1+j(M+1)}^{(n)} > 0$ *for* $n = 0, 1, \dots$. *Then* $v_{j(M+1)}^{(n)}, v_{1+j(M+1)}^{(n)}, \dots, v_{M-1+j(M+1)}^{(n)}$ *converge to 0 as* $n \to \infty$.

Proof From the Jacobi determinant identity (Hirota, 2003), it follows that

$$C_{i,j+1}^{(n)} C_{i+1,j-1}^{(n+1)} = C_{i,j}^{(n)} C_{i+1,j}^{(n+1)} - C_{i,j}^{(n+1)} C_{i+1,j}^{(n)} \tag{4.11}$$

Equation (4.11) allows us to simplify $\sum_{i=0}^{M-1} v_{i+j(M+1)}^{(n)}$ as

$$\sum_{i=0}^{M-1} v_{i+j(M+1)}^{(n)} = \sum_{i=0}^{M-1} \left(\frac{C_{i+1,j}^{(n+1)}}{C_{i+1,j}^{(n)}} - \frac{C_{i,j}^{(n+1)}}{C_{i,j}^{(n)}} \right)$$
$$= \frac{C_{M,j}^{(n+1)}}{C_{M,j}^{(n)}} - \frac{C_{0,j}^{(n+1)}}{C_{0,j}^{(n)}} \tag{4.12}$$

From (4.8), we derive

$$\lim_{n\to\infty} \frac{C_{i,j}^{(n+1)}}{C_{i,j}^{(n)}} = r_{i,\kappa_1^*} r_{i+1,\kappa_2^*} \cdots r_{i+j-1,\kappa_j^*} \tag{4.13}$$

Thus, by combining (4.13) and $r_{M,\kappa_1^*} = r_{0,\kappa_1^*}, r_{M+1,\kappa_2^*} = r_{1,\kappa_2^*}, \dots, r_{M+j-1,\kappa_j^*} = r_{j-1,\kappa_j^*}$ with (4.12), we have

$$\lim_{n\to\infty} \sum_{i=0}^{M-1} v_{i+j(M+1)}^{(n)} = 0 \tag{4.14}$$

Therefore, by taking into account that $v_{j(M+1)}^{(n)} > 0, v_{1+j(M+1)}^{(n)} > 0, \dots, v_{M-1+j(M+1)}^{(n)} > 0$ in (4.14), we find that $v_{j(M+1)}^{(n)} \to 0, v_{1+j(M+1)}^{(n)} \to 0, \dots, v_{M-1+j(M+1)}^{(n)} \to 0$ as $n \to \infty$

By recalling the relationship of the dhLV variable $u_k^{(n)}$ to the auxiliary variable $v_k^{(n)}$ in (3.6), we have the following theorem concerning an asymptotic convergence of $u_k^{(n)}$ as $n \to \infty$.

THEOREM 4.4 *As* $n \to \infty$, *the dhLV variable* $u_{j(M+1)}^{(n)}$ *converges to some nonzero constant* $\bar{c}_j$, *and* $u_{1+j(M+1)-M}^{(n)}, u_{2+j(M+1)-M}^{(n)}, \dots, u_{M+j(M+1)}^{(n)}$ *go to 0, provided that* $\delta^{(n)}$ *satisfy* $u_{k-M}^{(n)} \prod_{j=1}^{M} (\delta^{(n)} + u_{k-M-j}^{(n)}) > 0$ *for* $n = 0, 1, \dots$ *and the limit of* $\delta^{(n)}$ *as* $n \to \infty$ *exists.*

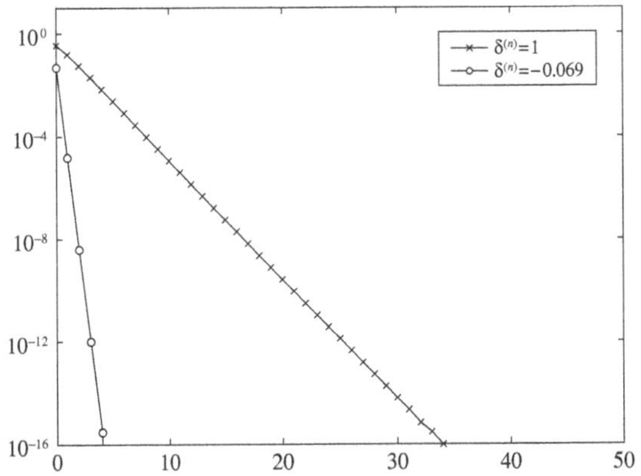

Figure 1. A graph of the discrete time n (x-axis) and the value of $|u_7^{(n)}|$ (y-axis) in the dhLV system (3.10) with $M = 3$ and $m = 3$. Cross : $\delta^{(n)} = 1$, Circle : $\delta^{(n)} = -0.069$.

Proof The proof is given by induction for j. Without loss of generality, let us assume that $\lim_{n\to\infty} \delta^{(n)} = \delta$ where δ denotes some constant. From (3.6), it holds that

$$u_k^{(n)} = \frac{v_{k+M}^{(n)}}{\prod_{\ell=1}^{M}\left(\delta^{(n)} + u_{k-\ell}^{(n)}\right)} \tag{4.15}$$

By taking the limit as $n \to \infty$ of both sides of (4.15) with $k = 0$ and using $v_M^{(n)} \to \hat{c}_0$ as $n \to \infty$, we obtain

$$\lim_{n\to\infty} u_0^{(n)} = \bar{c}_0 \tag{4.16}$$

where $\bar{c}_0 = \hat{c}_0/\delta^M$. By considering Theorem 4.2 with (4.16) in the case where $k = 1, 2, \ldots, M$ in (4.15), we successively check that $u_1^{(n)} \to 0, u_2^{(n)} \to 0, \ldots, u_M^{(n)} \to 0$ as $n \to \infty$.

Let us assume that $u_{j(M+1)}^{(n)} \to \bar{c}_j$ and $u_{1+j(M+1)}^{(n)} \to 0, u_{2+j(M+1)}^{(n)} \to 0, \ldots, u_{M+j(M+1)}^{(n)} \to 0$ as $n \to \infty$. Equation (4.15) with $k = (j+1)(M+1)$ becomes

$$u_{(j+1)(M+1)}^{(n)} = \frac{v_{M+(j+1)(M+1)}^{(n)}}{\prod_{\ell=1}^{M}\left(\delta^{(n)} + u_{(j+1)(M+1)-\ell}^{(n)}\right)} \tag{4.17}$$

It is clear that the denominator on the right-hand side of (4.17) converges to δ^M as $n \to \infty$ under this assumption. By combining it with $v_{M+(j+1)(M+1)}^{(n)} \to \hat{c}_{j+1}$ as $n \to \infty$, we observe that $u_{(j+1)(M+1)}^{(n)} \to \bar{c}_{j+1} = \hat{c}_{j+1}/\delta^M$ as $n \to \infty$. Moreover, it follows that

$$\lim_{n\to\infty} u_{i+(j+1)(M+1)+1}^{(n)} = \lim_{n\to\infty} \frac{v_{i+(j+2)(M+1)}^{(n)}}{\prod_{\ell=1}^{M}\left(\delta^{(n)} + u_{i+(j+1)(M+1)+1-\ell}^{(n)}\right)} = 0, \quad i = 0, 1, \ldots, M-1 \tag{4.18}$$

since $\prod_{\ell=1}^{M}(\delta^{(n)} + u_{i+(j+1)(M+1)+1-\ell}^{(n)}) \to \delta^{M-1}(\delta + \bar{c}_{j+1})$ and $v_{i+(j+2)(M+1)}^{(n)} \to 0$ as $n \to \infty$

The convergence theorem concerning the dhLV system (3.10) in Fukuda et al. (2009) is restricted to the case where the dhLV variable $u_k^{(n)}$ is positive and the discretization parameter $\delta^{(n)}$ is fixed positive for every n. Theorem 4.4 claims that the $[j(M+1)]$th species survives and the $[1+j(M+1)]$th, $[2+j(M+1)]$th, ... , $[M+j(M+1)]$th species vanish as $n \to \infty$ even in the case where $\delta^{(n)}$ is a changeable negative for each n. Although the case of negative $u_k^{(n)}$ is not longer recognized as a valid biological model, we note that the convergence is not different from the positive case if the values of $\delta^{(n)}$ are suitable for $n = 0, 1, \ldots$.

To observe the asymptotic convergence numerically, we consider two cases where $\delta^{(n)} = 1$ and $\delta^{(n)} = -0.069$ in the dhLV system (3.10). The initial values are set as $u_k^{(0)} = (\delta^{(n)})^M / \prod_{\ell=1}^{M}(\delta^{(n)} + u_{k-\ell}^{(0)})$ for $k = 0, 1, \ldots, 8$ in the dhLV system (3.10) with $M = 3$ and $m = 3$. Figure 1 shows the behavior of $u_7^{(n)}$ for $n = 0, 1, \ldots, 50$ in the case where $\delta^{(n)} = 1$ and $\delta^{(n)} = -0.069$. This figure demonstrates that $u_7^{(n)}$ tends to 0 as n grows larger even if $\delta^{(n)} < 0$. We also see that the case where $\delta^{(n)} = -0.069$ has a superior convergence speed in comparison with the case where $\delta^{(n)} = 1$. Similarly, the asymptotic behavior of $u_0^{(n)}, u_1^{(n)}, \ldots, u_8^{(n)}$ can be seen to follow Theorem 4.4.

5. Concluding remarks

In this paper, we associated a formal power series $f_i(z) = \sum_{n=0}^{\infty} a_i^{(n)} z^n$ with the Casorati determinant $C_{i,j}^{(n)}$, and gave asymptotic expansions of the Casorati determinants as $n \to \infty$ in Theorems 2.2 and 2.3. By making use of Theorem 2.2, we then clarified the asymptotic behavior of the dhLV variables as $\to \infty$ in Theorem 4.4.

Theorems 2.2 and 2.3 may contribute to asymptotic analysis for other discrete integrable systems. One possible application is the discrete hungry Toda (dhToda) equation derived from the numbered box and ball system through inverse ultra-discretization (Tokihiro, Nagai, & Satsuma, 1999). The dhToda equation has a relationship of variables to the dhLV system whose solution is given in the Casorati determinant (Fukuda, Yamamoto, Iwasaki, Ishiwata, & Nakamura, 2011). The Casorati determinant directly appears in, for example, the solution to the discrete Darboux–Pöschl–Teller equation which is a discretization of a dynamical system concerning a special class of potentials for the 1-dimensional Schrödinger equation (Gaillard & Matveev, 2009).

It was proved in Fukuda et al. (2013) that the dhLV system (3.10) with a fixed positive $\delta^{(n)}$ is associated with the LR transformation for a TN matrix. The paper (Yamamoto et al., 2010) also suggested that the dhLV system (3.10) with changeable negative $\delta^{(n)}$ generates the shifted LR transformation for a TN matrix. Eigenvalues of an m-by-m TN matrix correspond to the constants $\hat{c}_1 = \delta^M \bar{c}_1$, $\hat{c}_2 = \delta^M \bar{c}_2, \ldots, \hat{c}_m = \delta^M \bar{c}_m$ in Theorem 4.4. Theorems 4.2–4.4 will be useful for investigating the convergence of the sequence of the shifted LR transformations based on the dhLV system (3.10) in the changeable negative case of $\delta^{(n)}$.

Acknowledgements
The authors would like to thank Prof. S. Tsujimoto for helpful discussions on the determinant expression. The authors also thank the reviewers for their careful reading and insightful suggestions.

Funding
This work is supported by the Grant-in-Aid for Scientific Research (C) [grant number 26400208] from the Japan Society for the Promotion of Science.

Author details
Masato Shinjo[1]
E-mail: mshinjo@amp.i.kyoto-u.ac.jp
Masashi Iwasaki[2]
E-mail: imasa@kpu.ac.jp
Akiko Fukuda[3]
E-mail: afukuda@shibaura-it.ac.jp
Emiko Ishiwata[4]
E-mail: ishiwata@rs.tus.ac.jp
Yusaku Yamamoto[5]
E-mail: yusaku.yamamoto@uec.ac.jp
ORCID ID: http://orcid.org/0000-0001-5682-3434
Yoshimasa Nakamura[1]
E-mail: ynaka@i.kyoto-u.ac.jp
[1] Graduate School of Informatics, Kyoto University, Kyoto, 606-8501, Japan.
[2] Faculty of Life and Environmental Sciences, Kyoto Prefectural University, Kyoto, 606-8522, Japan.
[3] College of Systems Engineering and Science, Shibaura Institute of Technology, Saitama, 337-8570, Japan.
[4] Department of Mathematical Information Science, Tokyo University of Science, Shinjuku, Tokyo, 162-8601, Japan.
[5] Department of Communication Engineering and Informatics, The University of Electro-Communications, Chofu, Tokyo, 182-8585, Japan.

References

Bogoyavlensky, O. I. (1988). Integrable discretizations of the KdV equation. *Physics Letters A, 134*, 34–38.
Chihara, T. S. (1978). *An introduction to orthogonal polynomials*. New York, NY: Golden and Breach Science Publisher.
Fukuda, A., Ishiwata, E., Iwasaki, M., & Nakamura, Y. (2009). The discrete hungry Lotka–Volterra system and a new algorithm for computing matrix eigenvalues. *Inverse Problems, 25*, 015007.
Fukuda, A., Ishiwata, E., Yamamoto, Y., Iwasaki, M., & Nakamura, Y. (2013). Integrable discrete hungry

system and their related matrix eigenvalues. *Annali di Matematica Pura ed Applicata, 192*, 423–445.

Fukuda, A., Yamamoto, Y., Iwasaki, M., Ishiwata, E., & Nakamura, Y. (2011). A Bäcklund transformation between two integrable discrete hungry systems. *Physics Letters A, 375*, 303–308.

Gaillard, P., & Matveev, V. B. (2009). Wronskian and Casorati determinant representations for Darboux-Pöschl-Teller potentials and their difference extensions. *Journal of Physics A: Mathematical and Theoretical, 42*, 404009.

Henrici, P. (1988). *Applied and computational complex analysis* (Vol. 1). New York, NY: Wiley.

Hirota, R. (1981). Discrete analogue of a generalized Toda equation. *Journal of the Physical Society of Japan, 50*, 3785–3791.

Hirota, R. (2003). Determinant and Pfaffians. *Sūrikaisekikenkyūsho Kōkyūroku, 1302*, 220–242.

Itoh, Y. (1987). Integrals of a Lotka–Volterra system of odd number of variables. *Progress of Theoretical Physics, 78*, 507–510.

Iwasaki, M., & Nakamura, Y. (2002). On the convergence of a solution of the discrete Lotka–Volterra system. *Inverse Problems, 18*, 1569–1578.

Maeda, K., Miki, H., & Tsujimoto, S. (2013). From orthogonal polynomials to integrable systems [in Japanese]. *Transactions of the Japan Society for Industrial and Applied Mathematics, 23*, 341–380.

Parlett, B. N. (1995). The new qd algorithm. *Acta Numerica, 4*, 459–491.

Rutishauser, H. (1990). *Lectures on numerical mathematics.* Boston: Birkhäuser.

Spiridonov, V., & Zhedanov, A. (1997). Discrete-time Volterra chain and classical orthogonal polynomials. *Journal of Physics A: Mathematical and General, 30*, 8727–8737.

Tokihiro, T., Nagai, A., & Satsuma, J. (1999). Proof of solitonical nature of box and ball systems by means of inverse ultra-discretization. *Inverse Problems, 15*, 1639–1662.

Tsujimoto, S., & Kondo, K. (2000). Molecule solutions to discrete equations and orthogonal polynomials [in Japanese]. *Sūrikaisekikenkyūsho Kōkyūroku, 1170*, 1–8.

Tsujimoto, S., Nakamura, Y., & Iwasaki, M. (2001). The discrete Lotka–Volterra system computes singular values. *Inverse Problems, 17*, 53–58.

Vein, R., & Dale, P. (1999). *Determinants and their applications in mathematical physics (Applied mathematical sciences,* Vol. 134). New York, NY: Springer.

Yamamoto, Y., Fukuda, A., Iwasaki, M., Ishiwata, E., & Nakamura, Y. (2010). On a variable transformation between two integrable systems: The discrete hungry Toda equation and the discrete hungry Lotka–Volterra system. *AIP Conference Proceedings, 1281*, 2045–2048.

Yamazaki, S. (1987). On the system of non-linear differential equations $\dot{y}_k = y_k(y_{k+1} - y_{k-1})$. *Journal of Physics A: Mathematical and General, 20*, 6237–6241.

7

Review of statistical actuarial risk modelling

Hiroshi Shiraishi[1]*

*Corresponding author: Hiroshi Shiraishi, Department of Mathematics, Keio University, Yokohama, Kanagawa, Japan
E-mail: shiraishi@math.keio.ac.jp

Reviewing editor: Zudi Lu, University of Southampton, UK

Abstract: In this paper, we review some results for insurance risk theory. We first introduce a variety of the insurance risk models proposed thus far. Then, we show that the expected discounted penalty function (the so-called Gerber–Shiu function) can describe some risk indicators. Next, the dividend problem is discussed; more precisely, the (approximated) optimal dividend barrier is derived and other extended dividend strategies introduced. In addition, some modified models depending on reinsurance or tax are introduced. Finally, we discuss the statistical estimation of the ruin probability and the Gerber–Shiu function.

Subjects: Actuarial & Accounting Mathematics; Applied Mathematics; Mathematical Finance; Mathematics & Statistics; Science; Statistical Theory & Methods; Statistics; Statistics & Probability

Keywords: ruin theory; Gerber–Shiu function; dividend; statistical estimation

1. Introduction

In the actuarial literature, the classical approach is to represent an insurance risk as a stochastic process with a certain number of control variables. The term "insurance risk" here means the risk for non-life insurance such as motor vehicle insurance, home and contents insurance, and travel insurance rather than for life or health insurance, for example. Once an insurance policy is insured against the insurance risk, the insured party pays an appropriate premium to the insurer at the start of the period of insurance cover. Then, the insured party will make a claim under the insurance policy each time he or she has an insurance accident. From an insurer's point of view, the surplus is generally defined by

$$\text{initial surplus} \quad + \quad \text{(aggregated) premium income} \quad - \quad \text{(aggregated) claim payment.}$$

ABOUT THE AUTHOR

Hiroshi Shiraishi received the BS degree in mathematics in 1998 and the MS and Dr degrees in mathematical science from Waseda University, Japan in 2004 and 2007, respectively. He joined the GE Edison Life Insurance Company, the Prudential Life Insurance Company of Japan and the Hannover-Re Reinsurance Company, in 1998, 2000 and 2005, respectively. His research interests are actuarial science, time series analysis, econometric theory and financial engineering. In particular, he investigates the statistical estimation of optimal dividend problems in the field of the actuarial science. He is currently an associate professor in the Department of Mathematics, Keio University, Japan. He is a fellow of the Institute of Actuary of Japan (FIAJ).

PUBLIC INTEREST STATEMENT

In this paper, we review some results for insurance risk theory. Once an insurance policy is insured against the insurance risk, the insured party pays an appropriate premium to the insurer at the start of the period of insurance cover. Then, the insured party will make a claim under the insurance policy. From an insurer's point of view, the (pure) surplus is defined by "initial surplus + (aggregated) premium income – (aggregated) claim payment". We introduce the insurance risk models proposed thus far. In classical insurance risk analysis, it is interested in the ruin probability which is the probability that the surplus is below zero. We introduce its extended quantity the so-called Gerber-Shiu function. In addition, we discuss some applications for the insurance risk theory such as the dividend, reinsurance, tax problems. Finally, we discuss the statistical estimation of the Gerber-Shiu function.

Here, there are two sources of uncertainty for the insurer as follows:

- How many claims will the insured party make?
- What will the amounts of those claims be?

This general framework is used to model "(aggregated) claim payments". The Cramér–Lundberg model introduced by Lundberg (2009), Cramér (1930), and Cramér (1955) represents the (aggregated) claim payment part of the surplus as a compound Poisson process. This model is dealt with in most (introductory) books on risk theory such as Bühlmann (1970), Gerber (1979), Bowers, Gerber, Hickman, Jones, and Nesbitt (1997), and Dickson (2005).

alVarious extensions of the Cramér–Lundberg model have been proposed. For instance, Dufresne and Gerber (1991) added a diffusion process to the original model and Huzak and Perman (2004) included a Lévy process with mean 0 and no positive jumps. Andersen (1957) generalized the claim number process to a renewal process that included a Poisson process. Albrecher and Teugels (2006) considered the possible dependence between the interclaim time and claim size.

In classical insurance risk analysis, if the above surplus falls to zero or below, we say that ruin occurs, and the probability of ultimate ruin (i.e. the probability that ruin eventually occurs) is discussed. In particular, by denoting the ruin probability as a function of the initial surplus (u), the effect of u for the ruin probability is analysed. Gerber (1987) argued that the ruin probability is a crude stability criterion and discussed the seriousness of the situation when ruin occurs (the so-called "severity of ruin"). Gerber and Shiu (1998) introduced the expected discounted penalty function (the so-called "Gerber–Shiu function") to treat both the deficit at ruin (which is an indicator of the severity of ruin) and the time of ruin simultaneously. The Gerber–Shiu function is described as a function of certain control variables such as the initial surplus, dividend barrier and reinsurance parameter.

De Finetti (1957) argued that it is unrealistic to minimize the ruin probability for an infinite length of time assuming that companies can grow their surplus without limit. He proposed an alternative formulation to avoid the infinite growth of the surplus, that is, a payment of a part of the surplus to the shareholder as a dividend (i.e. a refund). However, two fundamental questions have to be addressed:

- When should dividends be distributed?
- How much of the surplus should be distributed?

The answer to these questions is based on the dividend strategy. Avanzi (2009) reviewed a variety of dividend strategies such as the constant barrier strategy, band strategy, linear barrier strategy and non-linear barrier strategy. Bühlmann (1970), Gerber and Shiu (2004), Gerber, Shiu, and Smith (2006), and Dickson and Drekic (2006) among others derived optimal dividend barriers, which maximize the expectation of the discounted dividends under the constant barrier strategy. In addition, Gerber et al. (2006) and Dickson and Drekic (2006) discussed the dividend problem under the optimization of the expected discounted dividends excluding the present value of the deficit at ruin. Further, because the explicit form of the optimal dividend barrier is obtained only in special cases, Gerber, Shiu, and Smith (2008) approximated some formulae for this case.

In this study, we introduce related topics such as "reinsurance" and "tax". For the former, we adopt an insurer's point of view to illustrate how utility theory can be applied to determine the optimal reinsurance parameters, which correspond to the optimal retention level under both the proportional and the excess-of-loss reinsurance. For the latter, we explain the "loss-carry forward system" discussed by Albrecher and Hipp (2007) and Albrecher, Badescu, and Landriault (2008).

The statistical estimation of the ruin probability and Gerber–Shiu function has been discussed by Grandell (1978/1979), Csörgő and Teugels (1990), Deheuvels and Steinebach (1990), Croux and Veraverbeke (1990), Csörgő and Steinebach (1991), Embrechts and Mikosch (1991), Bening and Korolev (2003), Mnatsakanov, Ruymgaart, and Ruymgaart (2008), and Shimizu (2012). Grandell (1978/1979), Csörgö and Teugels (1990), Deheuvels and Steinebach (1990), and Csörgö and Steinebach (1991) all discussed the statistical estimation of the adjustment coefficient, because the ruin probability is approximately expressed as a function of the adjustment coefficient based on the Cramér–Lundberg approximation. Croux and Veraverbeke (1990), Bening and Korolev (2003), Mnatsakanov et al. (2008), and Shimizu (2012) discussed the non-parametric estimation of the ruin probability and Gerber–Shiu function.

The remainder of this paper is organized as follows. In Section 2, we introduce a variety of the insurance risk models proposed thus far. Section 3 discusses whether the Gerber–Shiu function can describe some of the risk indicators. The dividend problem is reviewed in Section 4. We first show the expected discounted dividends under some distributions of the claim amount. Then, approximation methods for the optimal dividend barrier are shown. We next consider the effect of the deficit at ruin and, finally, extended dividend strategies are introduced. In Section 5, some modified models depending on reinsurance or tax are introduced. Finally, Section 6 discusses the statistical estimation of the ruin probability and Gerber–Shiu function.

2. Risk models

A variety of risk models have been introduced thus far. Here, the risk models show the modelling of the surplus at time t. The definition is as follows:

Definition 2.1 The surplus at time t is

$$U(t) = u + p(t) - S(t), \quad t \geq 0, \tag{2.1}$$

where $u \geq 0$ is the initial surplus of the company, p(t) is the (predictable) premium income with $p(0) = 0$ and $p(t) > 0$ for $t > 0$, and $\{S(t)\}$ is the aggregate claim process with $S(t) \geq 0$ for $t > 0$.

In most cases, the premium income is assumed to be received continuously at a constant rate $c > 0$ (i.e. $p(t) = ct$), which implies that the possibility of gaining or losing customers is ignored. An insurance policy is a financial agreement between the insurance company and policyholder. The insurance company agrees to pay some benefits and policyholder agrees to pay premiums to the insurance company to secure these benefits. Suppose that the net premium (in which the insurance company's expenses are excluded from the gross premiums) for a single payment per insurance policy is denoted P. If a constant number of new policies (denoted L) is continuously included and the same number of the existing policies is withdrawn from the portfolio of insurance policies, then the insurance company can continuously receive $LP(\equiv c)$ as the premium income and the aggregated premium income is $LPt(\equiv ct)$. The insurance risk for the portfolio is maintained, because the volume of the insurance portfolio is the same over time. Premium calculation principles have many desirable properties such as non-negative loading, additivity, scale invariance, consistency and no ripoff (see Dickson, 2005). From these points of view, some premium principles have been introduced such as the expected value principle, variance principle, standard deviation principle, principle of zero utility, Esscher principle and Wan principle. In what follows, we suppose that the premium rate (c) is determined based on the expected value principle, that is, $c = (1 + \theta)E\{S(1)\}$ with $\theta > 0$.

The initial surplus u becomes a parameter of the function of the risk indicators in many cases. For example, in classical risk theory, the (ultimate) ruin probability is considered to be a function of the initial surplus, which implies that the investor can decide the initial surplus to satisfy a desirable condition.

The aggregated claim ($S(t)$) expresses the total amount of claims arising from the insurance portfolio. In classical risk theory, the individual claim amount is assumed to be independent and identically distributed (i.i.d.), while the sequence of the claim amounts ($\{X_i\}$) and the number of claims (N) are assumed to be independent. However, the number of claims ($N \equiv N(t)$) that occurs in the fixed time interval $[0, t]$ is assumed to be a Poisson distribution with parameter λt (i.e. $\{N(t)\}$ is a Poisson process) to express that the interclaim time is random but independent.

Example 2.1 [The Cramér–Lundberg model] The Cramér–Lundberg model is the classical model of risk theory introduced by Lundberg (2009), Cramér (1930, 1955). Suppose that the surplus of an insurance company at time t is

$$U(t) = u + ct - S(t), \quad S(t) = \sum_{i=1}^{N(t)} X_i, \quad t \geq 0, \tag{2.2}$$

where the premium income ($p(t) = ct$) is received continuously at a constant rate $c > 0$; the aggregate claims process $\{S(t) = \sum_{i=1}^{N(t)} X_i\}$ is a compound Poisson process; the claim frequency $\{N(t)\}$ is a Poisson process with the intensity parameter $\lambda > 0$; and $\{X_i\}$ is a sequence of i.i.d. positive-valued random variables.

According to Dickson (2005), "This model is, of course, a simplification of reality. Some of the more important simplifications are that we assume claims are settled in full as soon as they occur, there is no allowance for interest on the insurer's surplus, and there is no mention of expenses that an insurer would incur. Nevertheless, this is a useful model which can give us some insight into the characteristics of an insurance operation".

Example 2.2 [De Finetti's model] In this model, the surplus at time t is

$$U(t) = u + t - \sum_{i=1}^{t} X_i, \quad t = 1, 2, 3, \ldots$$

where $X_1, X_2, \ldots$ are i.i.d. random variables with

$$P(X_i = 0) = p, \quad P(X_i = 2) = q = 1 - p, \quad i = 1, 2, 3, \ldots$$

and $p > 1/2$ (i.e. $p(t) = t$, $S(t) = \sum_{i=1}^{t} X_i$). This model, proposed by De Finetti (1957), was later reviewed by Seal (1969), Gerber and Shiu (2004), and Avanzi (2009).

Example 2.3 [Approximation of the Cramér–Lundberg model] Dickson and Waters (1991) discussed the compound Poisson process in discrete time of which the surplus at time t is

$$U(t) = u + t - \sum_{i=1}^{t} X_i, \quad t = 1, 2, \ldots.$$

De Vylder and Goovaerts (1988), Li and Garrido (2002), Claramunt, Mármol, and Alegre (2003), and Dickson and Waters (2004) also dealt with this model.

The compound binomial process of which the surplus at time t is

$$U(t) = u + t - \sum_{i=1}^{N(t)} X_i, \quad t = 1, 2, \ldots$$

where $\{N(t)\}$ is no longer a Poisson process, but a binomial process, that is, the increment $N(t + h) - N(t)$ for $h = 1, 2, \ldots$ is a binomial random variable with parameters h (number of trials) and p (probability of success). Gerber (1988), Shiu (1989), Dickson (1994), De Vylder and Marceau (1996), Cheng, Gerber, and Shiu (2000), Tan and Yang (2006), Bao (2007), and Landriault (2008b) also dealt with this model.

Example 2.4 [The Wiener process] The continuous counterpart of de Finetti's model is the Wiener process. In this model, the surplus at time t is

$$U(t) = u + \mu t + \sigma W(t), \quad t \geq 0, \tag{2.3}$$

where $\{W(t)\}$ is a standard Wiener process. As a consequence, $U(t+h) - U(t)$ for $t, h > 0$ has mean $h\mu$ and variance $h\sigma^2$. Gerber and Shiu (2003, 2004), and Leung, Kwok, and Leung (2008) dealt with this model.

Example 2.5 [With diffusion model (Wiener–Poisson risk process)] Dufresne and Gerber (1991) introduced the diffusion model, of which the surplus at time t is

$$U(t) = u + ct - S(t) + \sigma W(t), \quad t \geq 0, \tag{2.4}$$

where $S(t)$ is the same as (Equation 2.2), $\sigma > 0$ is a constant volatility parameter and $\{W(t)\}$ is a standard Wiener process. Li (2006), Li and Wu (2006), Jang (2007), Li, Feng, and Song (2007), Wan (2007), Frostig (2008), and Yuen, Lu, and Wu (2009) also dealt with this model.

Example 2.6 [The Sparre Andersen model] The Cramér–Lundberg model assumes that $\{S(t)\}$ is a compound Poisson process, but in the Sparre Andersen model, $\{S(t)\}$ is not necessarily a compound Poisson process. The idea is taken from Andersen (1957), who gave his name to the model. Li and Garrido (2004), Albrecher, Claramunt, and Mármol (2005), Albrecher, Hartinger, and Thonhauser (2007), Meng, Zhang, and Wu (2007), Albrecher and Hartinger (2006), and Yang and Zhang (2008) among others dealt with this model. Moreover, Zhang (2014) considered the Sparre Andersen model to be perturbed by a Brownian motion.

Example 2.7 [The dynamic control model] Højgaard (2002) considered a process with the dynamic control of the premium income, in which the classical Cramér–Lundberg model with premium rates $(c \equiv c_t)$ is calculated by using the expected value principle (i.e. $c_t = (1+\theta_t)E\{S(1)\}$). The company controls dynamically the relative safety loading (θ_t) with the possibility of gaining or losing customers.

Example 2.8 [The interclaim-dependent claim size model] Albrecher and Teugels (2006), Boudreault, Cossette, Landriault, and Marceau (2006), Landriault (2008a), and Huang and Li (2011) considered an extension to the classical Cramér–Lundberg model in which a particular dependence structure among the interclaim time and subsequent claim size is introduced. When the claim number process $\{N(t)\}$ is a (homogeneous) Poisson process with the intensity parameter $\lambda > 0$, the interclaim time $\Delta_i T$ between the ith claim and $(i-1)$th claim follows an independently exponential distribution with the parameter λ. In this model, we suppose that the bivariate random vectors $(X_i, \Delta_i T)$ for each $j \in \mathbb{N}$ are mutually independent, whereas X_i and $\Delta_i T$ are dependent.

Example 2.9 [The dual model] In the classical Cramér–Lundberg model, the surplus of a company at time t is

$$U(t) = u + ct - S(t)$$

where u is the initial surplus, c is the constant premium rate and $S(t)$ are the aggregate claims at time t. In the dual model, the surplus at time t is

$$U(t) = u - ct + S(t)$$

where u is also the initial surplus, c is the company's expense rate and $S(t)$ are the aggregate gains at time t. Avanzi, Gerber, and Shiu (2007), Gerber and Smith (2008), and Ng (2009) discussed the optimal barrier strategies in this model. Moreover, Avanzi and Gerber (2008) and Avanzi, Shen, and Wong (2011) introduced the dual model with diffusion.

Example 2.10 [The spectrally positive (negative) Lévy risk model] In this model, the surplus at time t is

$$U(t) = u + ct - S(t) + Z(t), \quad t \geq 0,$$

where u is the initial surplus, c is the constant premium rate and $\{S(t)\}$ and $\{Z(t)\}$ are independent with $S(0) = Z(0) = 0$, $E\{Z(1)\} = 0$ and $E\{|Z(1)|\} < \infty$. We provide a more detailed description of the surplus by emphasizing the properties of $\{S(t)\}$ and $\{Z(t)\}$. The Laplace exponent of S is given by

$$m_S(r) = \ln E\{e^{-rS(1)}\} = \int_0^\infty (e^{-rx} - 1)\nu_S(dx), \quad r \geq 0,$$

where ν_S is a non-negative measure of $\mathbb{R}_+$ satisfying $\int_0^\infty (1 \wedge x)\nu_S(dx) < \infty$. In addition, let $\tilde{Z}$ denote the dual process of Z (i.e. the process $\{\tilde{Z}(t)\}$ has the same law as $\{-Z(t)\}$). Then, the Laplace exponent of $\tilde{Z}$ is given by

$$m_{\tilde{Z}}(r) = \ln E\{e^{-r\tilde{Z}(1)}\} = \ln E\{e^{rZ(1)}\} = \frac{\sigma^2}{2} + \int_0^\infty (e^{-rx} - 1 + rx)\nu_{\tilde{Z}}(dx), \quad r \geq 0,$$

where $\sigma \geq 0$ and $\nu_{\tilde{Z}}$ are non-negative measures of $\mathbb{R}_+$ satisfying $\int_0^\infty (1 \wedge x)\nu_{\tilde{Z}}(dx) < \infty$ and $\int_1^\infty x\nu_{\tilde{Z}}(dx) < \infty$ to ensure that Z has finite expectations. The classical compound Poisson model can be recovered by assuming that $\sigma = 0$ and $\nu_{\tilde{Z}}$ are null measures and by setting $\nu_S(dx) = \lambda f(x)dx$. Huzak and Perman (2004), Loeffen (2009), Biffis and Morales (2010), Loeffen and Renaud (2010), Yin and Wen (2013), and Feng and Shimizu (2013) among others dealt with this model.

3. Gerber–Shiu function

Suppose that the surplus of an insurance company $U(t)$ follows the Cramér–Lundberg model:

$$U(t) = u + ct - S(t), \quad S(t) = \sum_{i=1}^{N(t)} X_i, \quad t \geq 0,$$

where the claim size X_i is an independent random variable with common distribution $F(x)$ and probability density $f(x)$, and $\{N(t)\}$ is a Poisson process with the intensity parameter λ.

Let T denote the time of ruin,

$$T = \inf\{t \geq 0 | U(t) < 0\}$$

($T = \infty$ if ruin does not occur). Classical ruin theory is interested in the probability of ultimate ruin as a function of the initial surplus $u \geq 0$,

$$\psi(u) = P\{T < \infty | U(0) = u\}. \tag{3.1}$$

We impose that following assumption for the premium rate c.

Assumption 3.1 This assumption ensures that $\{U(t)\}$ has a positive drift and $\psi(u) < 1$ (see Gerber & Shiu, 1998).

$$c > \lambda\mu, \quad \mu = E(X_i) = \int_0^\infty xf(x)dx.$$

Gerber (1987) discussed not only the probability of ruin, but also the severity of the situation when ruin occurs. Let $U(T) \leq 0$ denote the surplus at ruin. They introduced the following quantity:

$$\psi(u,y_0) = P\{T < \infty,\ |U(T)| \le y_0 | U(0) = u\}. \tag{3.2}$$

This quantity is an indicator of the severity of ruin, which corresponds to the probability that ruin will occur and states that the deficit at the time of ruin will be less than y_0, for a given initial surplus (u). Dufresne and Gerber (1988) and Dickson (1992) also discussed the severity of ruin.

Furthermore, Gerber and Shiu (1997, 1999) derived the joint distribution of the time of ruin, the surplus immediately before ruin (denoted by $U(T-)$), and the deficit at ruin by using the Laplace transform, which leads to the introduction of the following generalized function by Gerber and Shiu (1998).

Definition 3.1 (The expected discounted penalty function (Gerber–Shiu Function))

$$\phi(u) = E\{w(U(T-), |U(T)|)e^{-\delta T} I(T < \infty) | U(0) = u\} \tag{3.3}$$

where w(x, y) is a non-negative function of $x > 0, y > 0$, $\delta \ge 0$ is a force of interest, and I(A) is an indicator function ($I(A) = 1$ if A is true and $I(A) = 0$ if A is false).

The Gerber–Shiu function is due at ruin, and it depends on the deficit at ruin as well as the surplus immediately prior to ruin. This function is thus useful for deriving results in connection with the joint and marginal distributions of $T, U(T-)$ and $|U(T)|$.

The following shows examples of the Gerber–Shiu function.

Example 3.1 [Ruin probability] If $w(x,y) = 1$ and $\delta = 0$, then it follows that

$$\phi(u) = E\{I(T < \infty) | U(0) = u\} = P\{T < \infty | U(0) = u\} = \psi(u),$$

which implies that the ruin probability defined by (Equation 3.1) is included in the class of the Gerber–Shiu function.

Example 3.2 [Ruin probability with the deficit at the time of ruin] If $w(x,y) = I(y \le y_0)$ and $\delta = 0$, then it follows that

$$\phi(u;y_0) = E\{I(|U(T)| \le y_0) I(T < \infty) | U(0) = u\} = P\{T < \infty,\ |U(T)| \le y_0 | U(0) = u\} = \psi(u,y_0),$$

which implies that the ruin probability defined by Equation (3.2) is included in the class of the Gerber–Shiu function .

Example 3.3 [Conditional joint distribution function] If $w(x,y) = I(x \le x_0, y \le y_0)$ and $\delta = 0$, then it follows that

$$\begin{aligned}\phi(u;x_0,y_0) &= E\{I(U(T-) \le x_0, |U(T)| \le y_0) I(T < \infty) | U(0) = u\} \\ &= P\{T < \infty,\ U(T-) \le x_0,\ |U(T)| \le y_0 | U(0) = u\},\end{aligned}$$

which implies that the joint distribution of $(U(T-), |U(T)|)$ on the event $\{T < \infty\}$ is included in the class of the Gerber–Shiu function.

In addition, when the above joint distribution has the density (i.e. it satisfies a differentiable condition), the (conditional) joint density function of $(U(T-), |U(T)|)$ on the event $\{T < \infty\}$ is included in the class of the Gerber–Shiu function under $w(x,y) = I(x = x_0, y = y_0)$ and $\delta = 0$.

Example 3.4 [Single premium for the ruin] If $\delta > 0$ and w are defined in terms of the benefit amount of an insurance (or reinsurance) payable at the time of ruin, then $\phi(u)$ is the single premium of this insurance.

Example 3.5 [Expected present value of a deferred continuous annuity] Define

$$w(x,y) = \frac{1-e^{-\rho y}}{\delta}, \quad \delta > 0,$$

where ρ is the positive solution to Lundberg's fundamental equation satisfied by

$$E\{e^{-\delta(T_0-T)}|T<\infty\} = e^{\rho U(T)}$$

and T_0 (called the time of recovery) is the first time the surplus reaches zero after ruin. Then, it follows that

$$\begin{aligned}\phi(u) &= E\left\{\frac{1-e^{-\rho|U(T)|}}{\delta}e^{-\delta T}I(T<\infty)|U(0)=u\right\}\\ &= \frac{1}{\delta}E([1-E\{e^{-\delta(T_0-T)}|T<\infty\}]e^{-\delta T}I(T<\infty)|U(0)=u)\\ &= \frac{1}{\delta}E\{E(e^{-\delta T}-e^{-\delta T_0}|T<\infty)I(T<\infty)|U(0)=u\}\\ &= E\left\{\int_T^{T_0} e^{-\delta t}dtI(T<\infty)|U(0)=u\right\}.\end{aligned}$$

In this case, the Gerber–Shiu function corresponds to the expected present value of a continuous annuity at a rate of 1 per unit time, starting at the time of ruin and ending as soon as the surplus rises to zero (see Remark (v) of Section 6 in Gerber & Shiu, 1998).

Example 3.6 [Option pricing] For $t \geq 0$, let $A(t) = e^{U(t)}$ be the price of an asset such as a stock or a stock index at time t. Gerber and Shiu (1999) considered the pay-off function

$$w(x,y) = \max(K - e^{-y}, 0),$$

where K is the exercise price of a perpetual American put option. Let $T \equiv T_b = \inf\{t|A(t) < e^b\}$ where b is a number satisfying $e^b \leq \min(e^u, K)$. Then, it follows that

$$\begin{aligned}\phi(u;b) &= E\left[\max\left(K - e^{-|U(T_b)|}, 0\right)e^{-\delta T_b}I(T_b<\infty)|U(0)=u\right]\\ &= E\left[\max\left(K - A(T_b), 0\right)e^{-\delta T_b}I(T_b<\infty)|A(0)=e^u\right],\end{aligned}$$

which implies that the value of the option-exercise strategy T_b is included in the class of the Gerber–Shiu function. Then, the optimal value of b is derived as the solution to

$$\phi(b;b) = K - e^b.$$

4. Dividend problem

Suppose that in the absence of dividends, the surplus of an insurance company at time t is defined by (Equation 2.2) for $t \geq 0$ (in this section, we assume the Cramér–Lundberg model). The insurance company refunds the part that the surplus exceeds a constant barrier $b(\geq u)$ as the dividend to the shareholder. In this model, let $D_b(t)$ denote the aggregate dividend at time t. Then, we can write

$$dD_b(t) = \begin{cases} 0 & \text{if } U_b(t) < b \\ cdt & \text{if } U_b(t) = b \end{cases}, \tag{4.1}$$

where $U_b(t)$ is the (modified) surplus with the dividend barrier b given by

$$U_b(t) = U(t) - D_b(t). \tag{4.2}$$

In addition, the time of ruin with dividends is given by $T_b = \inf\{t|U_b(t) < 0\}$.

4.1. Classical dividend problem

De Finetti (1957), Bühlmann (1970), Gerber (1979), and Gerber and Shiu (1998) dealt with the problem of the expectation of the discounted dividends, namely,

$$V(u,b) = E\left[\int_0^{T_b} e^{-\delta t} dD_b(t)\right] \tag{4.3}$$

where $\delta > 0$ is a force of interest.

Suppose each claim amount (X_i) has the probability distribution function ($f(x)$). Then, the following representations are obtained (see Dickson, 2005):

$$V(u,b) = e^{-\frac{b-u}{\alpha}} V(b,b) + \frac{\lambda}{c}\int_u^b e^{-\frac{s-u}{\alpha}} \int_0^s f(x)V(s-x,b)dxds, \tag{4.4}$$

$$V(b,b) = \alpha + \frac{\lambda}{\lambda+\delta}\int_0^b f(x)V(b-x,b)dx. \tag{4.5}$$

Equation (4.4) comes from the decomposition of the aggregate dividends as

(I) The case that the surplus attains b before the first claim occurs,

(II) The case that the first claim occurs before the surplus attains b.

In case (I), since the "future aggregated dividends immediately before the first claim occur" corresponds to $V(b, b)$, we can obtain the first term on the right-hand side of Equation (4.4). In case (II), since the "future aggregated dividends immediately after the first claim occur" corresponds to $V(s-x, b)$, where $s \in (u, b)$ is the amount of the "surplus immediately before the first claim occurs" and $x \in (0, s)$ is the amount of the first claim, we can obtain the second term on the right-hand side of Equation (4.4).

Similarly, Equation (4.5) comes from the decomposition of the aggregate dividends as

(III) The part before the first claim occurs,

(IV) The part after the first claim occurs.

By using Equations (4.4) and (4.5), the following integro-differential equation is derived:

$$\frac{\partial}{\partial u}V(u,b) = \frac{\lambda+\delta}{c}V(u,b) - \frac{\lambda}{c}\int_0^u f(x)V(u-x,b)dx. \tag{4.6}$$

The optimal dividend barrier ($b_{\text{opt}} \equiv b_{\text{opt}}(u)$) is defined by the maximizer of $V(u, b)$:

Definition 4.1

$$b_{\text{opt}} := \arg\max_{b \geq u} V(u,b) \tag{4.7}$$

for a constant value $u \geq 0$.

The explicit form of b_{opt} is obtained only if each claim amount X_i follows some special distributions such as exponential distribution and the diffusion process.

Example 4.1 [Exponential distribution (Bühlmann, 1970)] Suppose that X_i follows an exponential distribution with $E(X_i) = 1/\beta$ (i.e. $f(x) = \beta e^{-\beta x}$). Then, from Equation (4.6), the second-order differential equation

$$\frac{\partial^2}{\partial u^2}V(u,b) + \left(\beta - \frac{\lambda+\delta}{c}\right)\frac{\partial}{\partial u}V(u,b) - \frac{\beta\delta}{c}V(u,b) = 0 \tag{4.8}$$

is obtained. By solving this differential equation with the boundary condition $\frac{\partial}{\partial u}V(u,b)\Big|_{u=b} = 1$ from Equations (4.4) and (4.5), we have

$$V(u,b) = \frac{h(u)}{h'(b)}, \quad h(x) = (\beta+\rho)\exp(\rho x) - (\beta - R)\exp(-Rx), \tag{4.9}$$

where $\rho > 0$ and $-R < 0$ are the solutions to the following quadratic equation with respect to x

$$x^2 + \left(\beta - \frac{\lambda+\delta}{c}\right)x - \frac{\beta\delta}{c} = 0.$$

Finally, from Equation (4.9), we obtain by solving $h''(b_{\text{opt}}) = 0$

$$b_{\text{opt}} = \frac{1}{\rho+R}\ln\frac{R^2(\beta-R)}{\rho^2(\beta+\rho)}. \tag{4.10}$$

Example 4.2 [Diffusion process (Gerber & Shiu, 2004)] Suppose that the aggregated claim process $S(t)$ is not a compound Poisson process but a diffusion process defined by $S(t) = \sigma W(t)$, where $\{W(t)\}$ is a standard Wiener process. Then, the second-order differential equation

$$\frac{\sigma^2}{2}\frac{\partial^2}{\partial u^2}V(u,b) + c\frac{\partial}{\partial u}V(u,b) - \delta V(u,b) = 0$$

is obtained. By solving this differential equation with the boundary conditions $V(0,b) = 0$ and $\frac{\partial}{\partial u}V(u,b)\Big|_{u=b} = 1$, we have

$$V(u,b) = \frac{h(u)}{h'(b)}, \quad h(x) = \exp(\rho x) - \exp(-Rx), \tag{4.11}$$

where $\rho > 0$ and $-R < 0$ are the solutions to the following quadratic equation with respect to x

$$\frac{\sigma^2}{2}x^2 + cx - \delta = 0.$$

Finally, from Equation (4.11), we obtain by solving $h''(b_{\text{opt}}) = 0$

$$b_{\text{opt}} = \frac{2}{\rho+R}\ln\frac{R}{\rho}. \tag{4.12}$$

The factorization formula

$$V(u,b) = \frac{h(u)}{h'(b)} \tag{4.13}$$

is well known [(e.g. (7.5) of Gerber & Shiu, 1998)] and satisfied for general distributions with the p.d.f. $f(x)$, where $h(x)$ is a positive solution to the following integro-differential equation:

$$ch'(x) - (\lambda+\delta)h(x) + \lambda\int_0^u h(x-y)f(y)dy = 0, \; x > 0. \tag{4.14}$$

Let $f^*(s)$ and $h^*(s)$ denote the Laplace transform of $f(x)$ and $h(x)$, namely,

$$f^*(s) = \int_0^\infty e^{-sx}f(x)dx, \quad h^*(s) = \int_0^\infty e^{-sx}h(x)dx.$$

Then, the Laplace transform of Equation (4.14) is (see e.g. Section 7.8 of Dickson, 2005)

$$c\{-h(0) + sh^*(s)\} - (\lambda+\delta)h^*(s) + \lambda f^*(s)h^*(s) = 0, \tag{4.15}$$

which yields $h^*(s) = ch(0)/\mathcal{L}(s)$ with $\mathcal{L}(s) = cs - (\lambda + \delta) + \lambda f^*(s)$. Because $\lim_{s\to\infty} f^*(s) = 0$ and $\lim_{s\to\infty} h^*(s) = 0$, it follows from Equation (4.15),

$$\lim_{s\to\infty} sh^*(s) = h(0). \tag{4.16}$$

Example 4.3 [Mixture of exponential distribution (Gerber et al., 2006)] Suppose that X_i follows a mixture of the exponential distribution with the p.d.f.

$$f(x) = \sum_{i=1}^{h} A_i \beta_i e^{-\beta_i x},$$

where $0 < \beta_1 < \cdots < \beta_h$, $A_i > 0$ with $A_1 + \cdots + A_h = 1$. Ross (2014) and others called such a density hyperexponential. In this case, from Equations (4.15) and (4.16), we can write

$$h^*(s) = h(0)\frac{\prod_{i=1}^{h}(s+\beta_i)}{\prod_{k=0}^{h}(s-\rho_k)},$$

where $\rho_0, \rho_1, \ldots, \rho_h$ are the solutions to $\mathcal{L}(s) = 0$, which are called Lundberg's fundamental equation (Gerber & Shiu, 1998), that is,

$$s - \frac{\lambda+\delta}{c} + \frac{\lambda}{c}\sum_{i=1}^{h}\frac{A_i\beta_i}{s+\beta_i} = 0,$$

with $-\beta_h < \rho_h < \cdots < -\beta_1 < \rho_1 < 0 < \rho_0$. By the method of partial fractions, we have

$$h^*(s) = h(0)\sum_{k=0}^{h} C_k \frac{1}{s-\rho_k},$$

and it follows that

$$h(x) = h(0)\sum_{k=0}^{h} C_k \exp(\rho_k x), \quad x > 0,$$

where $C_k = \prod_{i=1}^{h}(\rho_k + \beta_i)/\prod_{l\neq k}(\rho_k - \rho_l)$. The optimal dividend barrier (b_{opt}) is obtained as the solution to

$$h''(b) = h(0)\sum_{k=0}^{h} C_k \rho_k^2 \exp(\rho_k b) = 0.$$

However, we cannot obtain the explicit form of b_{opt} when $h \geq 2$.

Example 4.4 [Gamma distribution (Gerber et al., 2008)] Suppose that X_i follows a gamma distribution with the p.d.f.

$$f(x) = \frac{\beta^m}{(m-1)!}x^{m-1}e^{-\beta x},$$

where $\beta > 0$, $m \in \mathbb{N}$. In this case, we can write

$$h^*(s) = \frac{ch(0)}{\mathcal{L}(s)} = \frac{(s+\beta)^m}{\prod_{k=0}^{m}(s-\rho_k)},$$

where $\rho_0, \rho_1, \ldots, \rho_m$ are the solutions to $\mathcal{L}(s) = cs - (\lambda+\delta) + \lambda\left(\frac{\beta}{s+\beta}\right)^m = 0$. Then, in the same way as Example (4.3), we have

$$h(x) = h(0)\sum_{k=0}^{m} C_k \exp(\rho_k x), \quad x > 0,$$

where $C_k = (\rho_k + \beta)^m / \prod_{l\neq k}(\rho_k - \rho_l)$. The optimal dividend barrier (b_{opt}) is obtained as the solution to

$$h''(b) = h(0)\sum_{k=0}^{m} C_k \rho_k^2 \exp(\rho_k b) = 0.$$

However, we cannot obtain the explicit form of b_{opt} when $m \geq 2$.

From the factorization formula (4.11), the optimal dividend barrier (b_{opt}) is defined as the solution to $h''(b) = 0$; however, the explicit form cannot be obtained except in special cases. From this point of view, some asymptotic formulae for b_{opt} have been considered by Højgaard (2002), Dickson and Drekic (2006), and Gerber et al. (2008).

Example 4.5 [De Vylder approximation (A) (Gerber et al., 2008)] This approximation method aims to approximate the original model by using a model with exponential claim amounts and the formula for the optimal dividend barrier. The De Vylder approximation has been used by Højgaard (2002), Dickson and Drekic (2006), and Gerber et al. (2008). Let

$$U(t) = u - ct + S(t), \quad S(t) = \sum_{i=1}^{N(t)} X_i,$$

denote the original model where X_i has the p.d.f. $f(x)$ with the first three moments $\mu_j = E(X_i^j)$ for $j = 1, 2, 3$, and $N(t)$ has the intensity parameter λ.

Suppose that the parameters c and λ are replaced by $\tilde{c}$ and $\tilde{\lambda}$, respectively, and $f(x)$ is replaced by $\beta e^{-\beta x}$, satisfying the first three moments per unit time of the process $\{U(t)\}$. Let

$$\tilde{U}(t) = u - \tilde{c}t + \tilde{S}(t), \quad \tilde{S}(t) = \sum_{i=1}^{\tilde{N}(t)} \tilde{X}_i,$$

denote the approximated model where $\tilde{X}_i$ has the p.d.f. $\beta e^{-\beta x}$ with the first three moments

$$\tilde{\mu}_j = E(\tilde{X}_i^j) = \frac{j!}{\beta^j} \ (i = 1, 2, 3),$$

and $N(t)$ has the intensity parameter $\tilde{\lambda}$. Then, the conditions for $E\{\tilde{U}(t)^j\} = E\{U(t)^j\}$ $(j = 1, 2, 3)$ are

$$c - \lambda\mu_1 = \tilde{c} - \frac{\tilde{\lambda}}{\beta}, \quad \lambda\mu_2 = \frac{2\tilde{\lambda}}{\beta^2}, \quad \lambda\mu_3 = \frac{6\tilde{\lambda}}{\beta^3}.$$

Therefore, β, $\tilde{\lambda}$ and $\tilde{c}$ are given by

$$\beta = 3\frac{\mu_2}{\mu_3}, \quad \tilde{\lambda} = \frac{9}{2}\lambda\frac{\mu_2^3}{\mu_3^2}, \quad \tilde{c} = c - \lambda\mu_1 + \frac{3}{2}\lambda\frac{\mu_2^2}{\mu_3}.$$

By using this result and Equation (4.10), we get the approximation

$$b_{\text{opt}} \approx \frac{1}{\tilde{\rho} + \tilde{R}} \ln \frac{\tilde{R}^2(\beta - \tilde{R})}{\tilde{\rho}^2(\beta + \tilde{\rho})}, \tag{4.17}$$

where $\tilde{\rho} > 0$ and $-\tilde{R} < 0$ are the solutions to the quadratic equation

$$x^2 + \left(\beta - \frac{\tilde{\lambda} + \delta}{\tilde{c}}\right)x - \frac{\beta\delta}{\tilde{c}} = 0.$$

Example 4.6 [De Vylder approximation (B) (Gerber et al., 2008)] Another approximation method is not to modify c, but to choose $\tilde{\tilde{\lambda}}$ and $\tilde{\tilde{\beta}}$, satisfying the first two moments per unit time of the process $\{U(t)\}$. In this case, the conditions are

$$\lambda\mu_1 = \frac{\tilde{\tilde{\lambda}}}{\tilde{\tilde{\beta}}}, \quad \lambda\mu_2 = \frac{2\tilde{\tilde{\lambda}}}{\tilde{\tilde{\beta}}^2}.$$

Therefore, $\tilde{\tilde{\beta}}$ and $\tilde{\tilde{\lambda}}$ are given by

$$\tilde{\tilde{\beta}} = 2\frac{\mu_1}{\mu_2}, \quad \tilde{\tilde{\lambda}} = 2\lambda\frac{\mu_1^2}{\mu_2}.$$

By using this result and Equation (4.10), we get the approximation

$$b_{\text{opt}} \approx \frac{1}{\tilde{\tilde{\rho}} + \tilde{\tilde{R}}} \ln \frac{\tilde{\tilde{R}}^2(\tilde{\tilde{\beta}} - \tilde{\tilde{R}})}{\tilde{\tilde{\rho}}^2(\tilde{\tilde{\beta}} + \tilde{\tilde{\rho}})}, \tag{4.18}$$

where $\tilde{\tilde{\rho}} > 0$ and $-\tilde{\tilde{R}} < 0$ are the solutions to the quadratic equation

$$x^2 + \left(\tilde{\tilde{\beta}} - \frac{\tilde{\tilde{\lambda}} + \delta}{c}\right)x - \frac{\tilde{\tilde{\beta}}\delta}{c} = 0.$$

Example 4.7 [Diffusion approximation (Gerber et al., 2008)] This approximation method aims to approximate the original model by using a model with a diffusion process and the formula for the optimal dividend barrier. The idea is to replace the original model $U(t) = u + ct - S(t)$ with $U(t) = u + \bar{c}t - \bar{\sigma}W(t)$, where the drift parameter is $\bar{c} = c - \lambda\mu_1$ and the volatility parameter is $\bar{\sigma}^2 = \lambda\mu_2$. Then, from Equation (4.12), we get the approximation

$$b_{\text{opt}} \approx \frac{2}{\bar{\rho} + \bar{R}} \ln \frac{\bar{R}}{\bar{\rho}}, \tag{4.19}$$

where $\bar{\rho} > 0$ and $-\bar{R} < 0$ are the solutions to the quadratic equation

$$\frac{\bar{\sigma}^2}{2}x^2 + \bar{c}x - \delta = 0.$$

Example 4.8 [Diffusion approximation (Order 1, Order 2) (Gerber et al., 2008)] This approximation method is considered to develop a refined approximation compared with Example (4.7). Suppose that the original model ($\{U(t) = u + ct - S(t)\}$) is a compound Poisson model with the premium rate c, intensity parameter $\lambda > 0$, and p.d.f. $f(x)$ of the claim amount X_i. In this method, rescaling the p.d.f. of the claim amount is considered. For a fixed $m > 0$, we reconsider another compound Poisson model

$$U_m(t) = u + c_m t - S_m(t), \quad S_m(t) = \sum_{i=1}^{N_m(t)} X_{i,m},$$

where $\{N_m(t)\}$ is a Poisson process with the intensity parameter $\lambda_m > 0$ and $X_{i,m}$ with the p.d.f. $f_m(x)$. Here, we assume that the p.d.f. $f_m(x)$ is defined by

$$f_m(x) = \frac{1}{m} f\left(\frac{x}{m}\right), \quad x > 0$$

and c_m, λ_m are determined by the corresponding first two moments of the process $\{U_m(t)\}$ and $\{U(t)\}$. Then, it follows that $E(X_{i,m}^j) = m^j E(X_i^j) = m^j \mu_j$. Note that if $m \to 0$, $\{U_m(t)\}$ converges to a diffusion process and if $m = 1$, $\{U_m(t)\}$ corresponds to the original process.

Next, we consider a diffusion approximation for the model $\{U_m(t)\}$. We replace the model $U_m(t) = u + c_m t - S_m(t)$ with $U_0(t) = u + c_0 t - \sigma_0 W(t)$, where the drift parameter and volatility parameter are given by

$$c_0 = c_m - \lambda_m(m\mu_1), \qquad \sigma_0^2 = \lambda_m(m^2\mu_2). \tag{4.20}$$

Then, from Example (4.7), we have the diffusion approximation for b_{opt} as

$$b_{\text{opt}}^{(0)} = \frac{2}{\rho_0 + R_0}\ln\frac{R_0}{\rho_0},$$

where $\rho_0 > 0$ and $-R_0 < 0$ are the solutions to the quadratic equation

$$\mathcal{L}_0(x) = \frac{\sigma_0^2}{2}x^2 + c_0 x - \delta = 0. \tag{4.21}$$

On the contrary, we can approximately express the optimal dividend barrier b_{opt} as [(see (4.2) of Gerber et al. (2008)]

$$b_{\text{opt}} \approx \frac{1}{\rho + R}\ln\frac{\mathcal{L}'(\rho)R^2}{-\mathcal{L}'(-R)\rho^2},$$

where $\rho, -R$ are the solutions to Lundberg's fundamental equation ($\mathcal{L}(s) = 0$). Hence, we deal with the true optimal dividend barrier for the model $\{U_m(t)\}$ as

$$b_{\text{opt}}^{(m)} = \frac{1}{\rho_m + R_m}\ln\frac{\mathcal{L}_m'(\rho_m)R_m^2}{-\mathcal{L}_m'(-R_m)\rho_m^2},$$

where $\rho_m, -R_m$ are the solutions to

$$\mathcal{L}_m(s) = c_m s - (\lambda_m + \delta) + \lambda_m f_m^*(s) = 0. \tag{4.22}$$

The Taylor expansion for the Laplace transform $f_m^*(s)$ is obtained as follows:

$$f_m^*(s) = \int_0^\infty e^{-sx} f_m(x)dx = \int_0^\infty e^{-msy} f(y)dy = f^*(ms) = \sum_{k=0}^\infty \frac{(-ms)^k}{k!}\mu_k. \tag{4.23}$$

From Equations (4.21–4.23), the Lundberg function is written as

$$\mathcal{L}_m(s) = \frac{\sigma_0^2}{2}(s - \rho_0)(s + R_0) - \frac{1}{6}m\frac{\mu_3}{\mu_2}\sigma_0^2 s^3 + o(m), \tag{4.24}$$

which leads to

$$\rho_m = \rho_0 + \frac{1}{3}m\frac{\mu_3}{\mu_2}\frac{\rho_0^3}{\rho_0 + R_0} + o(m), \quad -R_m = -R_0 + \frac{1}{3}m\frac{\mu_3}{\mu_2}\frac{(-R_0)^3}{\rho_0 + R_0} + o(m),$$

because of $\mathcal{L}_m(\rho_m) = \mathcal{L}_m(-R_m) = 0$. Furthermore, by differentiating Equation (4.24), we have

$$\mathcal{L}_m'(\rho_m) = \frac{\sigma_0^2}{2}\left\{\rho_0 + R_0 + \frac{2}{3}m\frac{\mu_3}{\mu_2}\frac{\rho_0^3}{\rho_0 + R_0} - m\frac{\mu_3}{\mu_2}\rho_0^2 + o(m)\right\},$$

$$\mathcal{L}_m'(-R_m) = \frac{\sigma_0^2}{2}\left\{-\rho_0 - R_0 + \frac{2}{3}m\frac{\mu_3}{\mu_2}\frac{(-R_0)^3}{-\rho_0 - R_0} - m\frac{\mu_3}{\mu_2}(-R_0)^2 + o(m)\right\},$$

which implies that we obtain $b_{\text{opt}} \approx b_{\text{opt}}^{(m)} \approx b_{\text{opt}}^{(0)} - m\Delta_1$ with

$$\Delta_1 = \frac{1}{3}\frac{\mu_3}{\mu_2}\frac{\rho_0^3 + (-R_0)^3}{(\rho_0 + R_0)^2}b_{\text{opt}}^{(0)} + \frac{1}{3}\frac{\mu_3}{\mu_2}\frac{3\rho_0^2 - 2\rho_0(-R_0) + 3(-R_0)^2}{(\rho_0 + R_0)^2}. \tag{4.25}$$

In a similar way, we have $b_{\text{opt}} \approx b_{\text{opt}}^{(m)} \approx b_{\text{opt}}^{(0)} - m\Delta_1 + m^2\Delta_2$ with

$$\Delta_2 = \frac{1}{6}\frac{\mu_4}{\mu_2}\frac{(\rho_0 - R_0)(2\rho_0^2 - 3\rho_0(-R_0) + 2(-R_0)^2)}{(\rho_0 + R_0)^2}$$
$$- \frac{1}{18}\frac{\mu_3^2}{\mu_2^2}\frac{(\rho_0^3 - R_0)(\rho_0^2 - 4\rho_0(-R_0) + (-R_0)^2)(5\rho_0^2 - 4\rho_0(-R_0) + 5(-R_0)^2)}{(\rho_0 + R_0)^4} \tag{4.26}$$
$$+ \left\{ \frac{1}{12}\frac{\mu_4}{\mu_2}\frac{\rho_0^4 + (-R_0)^4}{(\rho_0 + R_0)^2} - \frac{1}{9}\frac{\mu_3^2}{\mu_2^2}\frac{\rho_0^6 - 3\rho_0^5(-R_0) - 2\rho_0^3(-R_0)^3 - 3\rho_0(-R_0)^5 + (-R_0)^6}{(\rho_0 + R_0)^4} \right\} b^{(0)}_{opt}.$$

4.2. The deficit at ruin

Dickson and Waters (2004), Gerber et al. (2006), and Dickson (2005) considered that the expected present value of the deficit at ruin ($E[e^{-\delta T_b}|U_b(T_b)|]$) and the initial surplus (u) should be excluded from the expected present value of net income to shareholders. In that case, the objective function becomes

$$L(u, b) = V(u, b) - Y(u, b) - u, \tag{4.27}$$

where $Y(u, b) = E[e^{-\delta T_b}|U_b(T_b)|]$ with the deficit at ruin $|U_b(T_b)|$.

Similar to Equation (4.6), the expected present value of the deficit at ruin $Y(u, b)$ has the following integro-differential equation (see (8.34) of Dickson, 2005):

$$\frac{\partial}{\partial u}Y(u, b) = \frac{\lambda + \delta}{c}Y(u, b) - \frac{\lambda}{c}\int_0^u f(x)Y(u - x, b)dx - \frac{\lambda}{c}\int_u^\infty (x - u)f(x)dx. \tag{4.28}$$

The optimal dividend barrier ($b^L_{opt} \equiv b^L_{opt}(u)$) is defined by the maximizer of $V(u, b)$.

Definition 4.2

$$b^L_{opt} := \arg\max_{b \ge u} L(u, b) \tag{4.29}$$

for a constant value $u \ge 0$.

In this case, the explicit form of b^L_{opt} is not obtained even if each claim amount X_i follows an exponential distribution.

Example 4.9 [Exponential distribution (Dickson, 2005, Gerber et al., 2006)] Suppose that X_i follows an exponential distribution with $E(X_i) = 1/\beta$ (i.e. $f(x) = \beta e^{-\beta x}$). Then, from Equation (4.28), we obtain the second-order differential equation

$$\frac{\partial^2}{\partial u^2}Y(u, b) + \left(\beta - \frac{\lambda + \delta}{c}\right)\frac{\partial}{\partial u}Y(u, b) - \frac{\beta\delta}{c}Y(u, b) = 0, \tag{4.30}$$

which yields

$$Y(u, b) = D_0(b)e^{\rho u} + D_1(b)e^{-Ru}, \tag{4.31}$$

where $\rho > 0$ and $-R < 0$ are the same as in Example (4.1). From this result and the boundary condition $\frac{\partial}{\partial u}Y(u, b)\Big|_{u=b} = 0$, we have

$$D_0(b) = Re^{-Rb}D(b), \quad D_1(b) = \rho e^{\rho b}D(b), \quad D(b) = \frac{(\beta + \rho)(\beta - R)}{\beta^2\left\{\rho(\beta + \rho)e^{\rho b} + R(\beta - R)e^{-Rb}\right\}}.$$

In addition, since we can write from Example (4.1)

$$V(u, b) = C_0(b)e^{\rho u} + C_1(b)e^{-Ru},$$

where

$$C_0(b) = (\beta + \rho)C(b), \quad C_1(b) = -(\beta - R)C(b), \quad C(b) = \frac{1}{\rho(\beta + \rho)e^{\rho b} + R(\beta - R)e^{-Rb}},$$

it follows that

$$L(u, b) = \{C_0(b) - D_0(b)\}e^{\rho u} + \{C_1(b) - D_1(b)\}e^{-Ru}. \tag{4.32}$$

The optimal dividend barrier b^L_{opt} is obtained by solving

$$\frac{\partial L(u, b)}{\partial b} = \{C_0(b)' - D_0(b)'\}e^{\rho u} + \{C_1(b)' - D_1(b)'\}e^{-Ru} = 0$$

or

$$\rho^2\{C_0(b) - D_0(b)\}e^{\rho b} + R^2\{C_1(b) - D_1(b)\}e^{-Rb} = 0,$$

which comes from another boundary condition $\frac{\partial^2}{\partial u^2}L(u, b)\Big|_{u=b} = 0$. However, they must be solved numerically.

Example 4.10 [Mixture of exponential distributions (Gerber et al., 2006)] Suppose that X_i follows a mixture of exponential distributions with the p.d.f.

$$f(x) = \sum_{i=1}^{h} A_i\beta_i e^{-\beta_i x},$$

where $0 < \beta_1 < \cdots < \beta_h$, $A_i > 0$ with $A_1 + \cdots + A_h = 1$. By applying the operator[1] $\prod_{i=1}^{h}\left(\frac{\partial}{\partial x} + \beta_i\right)$ to the integro-differential equation (Equation 4.6) and

$$\frac{\partial}{\partial u}Y(u, b) = \frac{\lambda + \delta}{c}Y(u, b) - \frac{\lambda}{c}\int_0^u f(x)Y(u - x, b)dx, \tag{4.33}$$

we have

$$V(u, b) = \sum_{k=0}^{h} C_k(b)e^{\rho_k u}, \quad Y(u, b) = \sum_{k=0}^{h} D_k(b)e^{\rho_k u},$$

where $\rho_0, \rho_1, \ldots, \rho_h$ are the same as in Example (4.3), and $C_k(b), D_k(b)$, $k = 0, \ldots, h$ are determined by

$$\sum_{k=0}^{h} \frac{C_k(b)}{\beta_i + \rho_k} = 0, \quad \sum_{k=0}^{h} \frac{D_k(b)}{\beta_i + \rho_k} = \frac{1}{\beta_i^2}, \quad i = 1, \ldots, h$$

and

$$\sum_{k=0}^{h} C_k(b)\rho_k e^{\rho_k b} = 1, \quad \sum_{k=0}^{h} D_k(b)\rho_k e^{\rho_k b} = 0.$$

Hence, we have

$$L(u, b) = \sum_{k=0}^{h} \{C_k(b) - D_k(b)\}e^{\rho_k u}. \tag{4.34}$$

The optimal dividend barrier b^L_{opt} is obtained by solving

$$\frac{\partial L(u, b)}{\partial b} = \sum_{k=0}^{h} \{C_k(b)' - D_k(b)'\}e^{\rho_k u} = 0$$

or

$$\sum_{k=0}^{h} \rho_k^2 \{C_k(b) - D_k(b)\} e^{\rho_k b} = 0.$$

However, they must be solved numerically.

4.3. Other dividend strategies

Thus far, we have considered that when the surplus attains a constant level b, the premium income is paid to shareholders as dividends. By contrast, Avanzi (2009) introduced other dividend strategies as follows.

Example 4.11 [Band/ (Multiple) Threshold/ Constant Barrier Strategies] For a "band strategy"[2], dividends are paid according to the region in which the surplus $U_A(t)$ is located. Given a region $A = \{(0, b_0), [a_1, b_1), \dots, [a_h, b_h)\}$ for $0 < b_0 \le a_1 \le b_1 \le \cdots \le a_h \le b_h$, $h \in \mathbb{N}$, while $U_A(t)$ is located in the region A, no dividends are paid. On the contrary, while $U_A(t)$ is located in the region $\mathbb{R}_+ \setminus A$, dividends are paid. In a band strategy on the region A, the aggregate dividend $D_A(t)$ is defined by

$$dD_A(t) = \begin{cases} 0 & \text{if } U_A(t) \in A \\ \tilde{c}dt & \text{if } U_A(t) \in \mathbb{R}_+ \setminus A \end{cases} \tag{4.35}$$

with $0 \le \tilde{c} \le c$. Note that if $\tilde{c} = c$, $U_A(t) \le b_0$.

A "threshold strategy" is a special case of the band strategy. When we set $h = 1$, $b_0 = b$ under the band strategy, the definition of the aggregate dividend $D_b(t) \equiv D_A(t)$ is

$$dD_b(t) = \begin{cases} 0 & \text{if } U_b(t) < b \\ \tilde{c}dt & \text{if } U_b(t) \ge b \end{cases} \tag{4.36}$$

with $0 \le \tilde{c} \le c$. Frostig (2005), Gerber et al. (2006), Lin and Pavlova (2006), Fang and Wu (2007), Wan (2007), Frostig (2008), Cheung, Dickson, and Drekic (2008), Yuen et al. (2009), and Avanzi (2009) discussed the threshold strategy.

A "constant barrier strategy" is a special case of the threshold strategy, in which the definition of $D_b(t)$ corresponds to Equation (4.1) when we set $\tilde{c} = c$ under the threshold strategy. A "multiple threshold strategy" or "multilayer strategy" is a combination of the threshold strategy. When we set $a_i = b_i$ and $0 \le \tilde{c}_i \le c$ for $i = 1, \dots, h$ under the band strategy, the definition of the aggregate dividend $D_A(t)$ is

$$dD_A(t) = \begin{cases} 0 & \text{if } U_A(t) \in A \\ \tilde{c}_i dt & \text{if } U_A(t) \in [b_{i-1}, b_i),\ i = 1, \dots, h \end{cases} \tag{4.37}$$

where $b_h = \infty$. Albrecher and Hartinger (2007), Badescu and Landriault (2008), and Lin and Sendova (2008) discussed the multilayer strategy.

Example 4.12 [Linear barrier strategies] Gerber (1974) argued that a (constant) barrier strategy does not sufficiently take into account the safety of the company, since the probability of ruin is 1 (i.e. $P(T_b < \infty) = 1$). In a linear barrier strategy with the parameters (a, b), the aggregate dividend $D_{(a,b)}(t)$ is defined by

$$dD_{(a,b)}(t) = \begin{cases} 0 & \text{if } U_{(a,b)}(t) < b + at \\ (c - a)dt & \text{if } U_{(a,b)}(t) = b + at \end{cases} \tag{4.38}$$

Gerber (1981), Siegl and Tichy (1999) and Albrecher, Hartinger, and Tichy (2005) also discussed the linear strategy.

Example 4.13 [Non-linear barrier strategies] These are a general family of strategies allowing for a (reflecting) barrier $b(t)$ as a function of time. In this strategy, the aggregate dividend $D_b(t)$ is defined by

$$dD_b(t) = \begin{cases} 0 & \text{if } U_b(t) < b(t) \\ (c - b(t)')dt & \text{if } U_b(t) = b(t) \end{cases} \tag{4.39}$$

Gerber (1974), Alegre, Claramunt, and Marmol (1999), Albrecher and Kainhofer (2002), and Albrecher, Kainhofer, and Tichy (2003) discussed the non-linear strategy.

5. Other related topics

5.1. Reinsurance

Suppose that the surplus of an insurance company at time t is the Cramér–Lundberg model defined by (Equation 2.2) for $t \geq 0$. If the insurance company (i.e. insurer) effects reinsurance by paying a reinsurance premium continuously at a constant rate, then the process becomes a net of the reinsurance surplus process $\{U_a(t)\}_{t\geq 0}$ given by

$$U^{(a)}(t) = u + c^{(a)}t - \sum_{i=1}^{N(t)} X_i^{(a)} \tag{5.1}$$

where u is the initial surplus, $c^{(a)}$ is the insurer's net premium income per unit time, $\{N(t)\}$ is a Poisson process with the intensity parameter $\lambda > 0$ and $X_i^{(a)}$ is the amount the insurer pays on the i-th claim. Note that the insurer's net premium rate $c^{(a)}$ and payable amount $X_i^{(a)}$ depend on the reinsurance scheme and reinsurance parameter $a \geq 0$.

Example 5.1 [Proportional Reinsurance] Under a proportional reinsurance arrangement, the insurer pays a fixed proportion $a \in [0, 1]$ of each claim that occurs during the period of the reinsurance arrangement. The remaining, $1 - a$, of each claim is paid by the reinsurer, which implies that $X_i^{(a)} = aX_i$. On the contrary, the insurer has to pay the reinsurance premium corresponding to the proportion of the premium rate calculated by the reinsurer. If you consider the premium rate in the expected value principle (see 3.3.2 of Dickson, 2005), the premium rates of the insurer (c) and reinsurer (c_R) are given by

$$c = (1 + \theta)E\{S(1)\}, \quad c_R = (1 + \theta_R)E\{S(1)\},$$

where $E\{S(1)\}$ is the (re)insurer's expected claim under the risk, given by $\lambda\mu$ with $\lambda = E\{N(1)\}$, and $\mu = E(X_i)$; θ, and θ_R are the premium loading factors defined by the insurer (θ) and reinsurer (θ_R) with $\theta_R \geq \theta > 0$. As a consequence, the insurer's net premium rate $c^{(a)}$ is

$$c^{(a)} = c - (1 - a)c_R = \{1 + \theta - (1 + \theta_R)(1 - a)\}\lambda\mu.$$

Note that if $0 < \theta_R < \theta$, the insurance company can earn a profit without any risk by $a = 1$.

Example 5.2 [Excess-of-loss Reinsurance] Under an excess-of-loss reinsurance arrangement, a claim is shared between the insurer and reinsurer only if it exceeds a fixed amount called the retention level $a \geq 0$. If the claim amount is smaller than a, the insurer has to pay the entire claim amount. Hence, the amount the insurer pays on the i-th claim is $X_i^{(a)} = \min(a, X_i)$ and the remaining $X_i - \min(a, X_i) = \max(X_i - a, 0)$ is paid by the reinsurer.

Let $f(x)$ denote the p.d.f. of the claim amount X_i. Then, the reinsurance premium for the reinsurance portion $\max(X_i - a, 0)$ is written as

$$c_R^{(a)} = (1+\theta_R)\lambda E\{\max(X_i - a, 0)\} = (1+\theta_R)\lambda \int_a^{\infty} (x-a)f(x)dx := (1+\theta_R)\lambda\mu^{(a)}.$$

As a consequence, the insurer's net premium rate $c^{(a)}$ under the expected value principle is given by

$$c^{(a)} = c - c_R^{(a)} = \{(1+\theta)\mu - (1+\theta_R)\mu^{(a)}\}\lambda.$$

Note that if $0 < \theta_R < \theta$, the insurance company can earn a profit without any risk by $a = 0$.

The optimal reinsurance parameter ($a_{\text{opt}} \equiv a_{\text{opt}}(u)$) is defined by the maximizer of an objective function $Q(a)$:

Definition 5.1

$$a_{\text{opt}} := \arg\max_{a \geq 0} Q(a),$$

where $Q(a)$ is satisfied with $Q(a) < Q(a_{\text{opt}}) < \infty$ for any $a \geq 0$.

Example 5.3 [Adjustment Coefficient] Let $M_X^{(a)}(r)$ denote the moment-generating function of the insurer's payable amount $X_i^{(a)}$, that is, $M_X^{(a)}(r) = E(e^{rX_i^{(a)}})$ for $r \in \mathbb{R}$. Then, the reinsurance adjustment coefficient ($R(a)$) is defined by the positive solution to

$$\ell^{(a)}(r) := \ln E\{e^{-r(U^{(a)}(1)-u)}\} = \lambda M_X^{(a)}(r) - \lambda - c^{(a)}r = 0,$$

i.e. $R(a)$ is satisfied with

$$M_X^{(a)}(R(a)) - \frac{c^{(a)}R(a)}{\lambda} = 1.$$

Waters (1983) and Centeno (1986) among others discussed the optimality of the reinsurance parameter under $Q(a) \equiv R(a)$.

Example 5.4 [Ruin Probability] Define

$$Q(a) = -P\{T^{(a)} < \infty | U^{(a)}(0) = u\},$$

which implies that the optimal reinsurance parameter a_{opt} is the minimizer of the ruin probability given $u \geq 0$. Dickson and Waters (1996) discussed the optimal proportion retained a_{opt} based on the ruin probability. According to Dickson (2005), the a_{opt} should be the maximizer of the adjustment coefficient rather than the minimizer of the ruin probability, at least for large values. In addition, the excess-of-loss reinsurance maximizes the adjustment coefficient (and thus minimizes the probability of ruin) (see Bowers et al., 1997).

Recently, many researchers have considered the optimal reinsurance and dividend control problem of an insurance company. As for proportional reinsurance, Højgaard and Taksar (1998), Taksar (2000), David (2005), and Cao and Wan (2009) among others considered optimal investment and reinsurance strategies to maximize (or minimize) the utility functions (included the ruin probability) for a diffusion risk model or extended models. As for the excess-of-loss reinsurance, Asmussen, Højgaard, and Taksar (2000), Beveridge, Dickson, and Wu (2007), Gu, Guo, Li, and Zeng (2012), Zhao, Rong, and Zhao (2013), and Liu and Hu (2014) among others discussed the optimization (i.e. the maximization or minimization of utility functions) of the excess-of-loss reinsurance and investment problems for an insurer with diffusion or jump-diffusion risk models.

5.2. Tax

Albrecher and Hipp (2007) discussed the effect of tax payments under a loss-carry forward system in the Cramér–Lundberg model without dividends. The loss-carry forward system is motivated by its realistic feature of carried forward losses (i.e. occurred losses can be deducted from later income and hence they reduce the taxable profit of the business). Under this system, the classical Cramér–Lundberg model is modified as follows.

Example 5.5 [The Cramér–Lundberg Model under the loss-carry forward system] Suppose that the surplus of an insurance company at time t is defined by (Equation 2.2) for $t \geq 0$ (in this section, we assume the Cramér–Lundberg model):

$$U(t) = u + ct - S(t), \quad S(t) = \sum_{i=1}^{N(t)} X_i.$$

We assume that tax is paid at a fixed rate $\gamma \in (0,1)$ of the insurer's premium income whenever it is in a profitable situation. Suppose that the initial surplus is u. We now consider the risk process $\{U_\gamma(t);t \geq 0\}$, which is the surplus process with tax. Before the first claim occurs, the (aggregated) premium income at time t becomes $c(1-\gamma)t$ and the (aggregated) tax payment at time t becomes $c\gamma t$. Let $\Delta_1 T (\equiv T_1)$ be the time of the first claim. Then, the gains level (the surplus immediately before the first claim occurs) is set to $M_1 := u + c(1-\gamma)\Delta_1 T$. The surpluses immediately before and after the first claim occurs are written as

$$U_\gamma(T_1-) = u + c(1-\gamma)\Delta_1 T, \quad U_\gamma(T_1) = U_\gamma(T_1-) - X_1,$$

respectively. If $0 < U(T_1)$ and $c\Delta_2 > X_1$, the insurer does not have to pay any tax during $[T_1, T_1 + X_1/c)$, because the surplus is less than M_1 and we set the time that the surplus equals M_1 as σ_1, i.e. $\sigma_1 := T_1 + X_1/c$. Then, we can see that $U_\gamma(\sigma_1) = M_1$ and from $t = \sigma_1$ to $t = T_1 + \Delta_2 T = T_2$, the tax is paid. By denoting the term at which the tax is paid as W_1, W_2, we can write $W_1 = T_1$ and $W_2 = T_2 - \sigma_1$. Then, the new gains level is set to $M_2 = M_1 + c(1-\gamma)W_2$. Let $N(t)$ be the underlying claims arrival process. Albrecher and Hipp (2007) formally defined a sequence of gains levels (M_n), waiting times (term at which tax is paid) (W_n) and starting times of periods with profit (σ_n) as

$$M_n = M_{n-1} + c(1-\gamma)W_n,$$
$$W_n = \inf\{t > 0; N(\sigma_{n-1} + t) > N(\sigma_{n-1})\},$$
$$\sigma_n = \inf\{t > \sigma_{n-1} + W_n; U_\gamma(t) = M_n\},$$

with $\sigma_0 = 0$ and $M_0 = u$ for $n = 1, 2, \ldots$, where the surplus process $U_\gamma(t)$ at time t is

$$U_\gamma(t) = M_{\Lambda(t)} + c(1-\gamma)\min\left\{W_{\Lambda(t)}, t - \sigma_{\Lambda(t)}\right\} - \sum_{i=N(\sigma_{\Lambda(t)})+1}^{N(t)} X_i,$$

where $\Lambda(t) = \sup\{n \in \mathbb{N}; \sigma_n \leq t\}$.

Albrecher et al. (2008) introduced the following modified dual risk model, which is reasonable to model the surplus process for pharmaceutical and petroleum companies.

Example 5.6 [The Dual Risk Model under the Loss-carry Forward System] Let $\{Y_j\}$ denote a sequence of i.i.d. random variables, where Y_j corresponds to the revenue associated with the j-th innovation, and let $\{\Delta_j T\}$ denote a sequence of the inter-innovation times (i.i.d. random variables). Then, the sequence of innovation times $\{T_j\}$ is defined by $T_0 = 0$ and $T_j = \sum_{i=1}^{j} \Delta_i T$. Suppose that the innovation number process $\{N(t), t \geq 0\}$ is a renewal process with the inter-innovation times $\Delta_1 T, \Delta_2 T, \ldots$. The following dual risk model is thus reasonable to model the surplus process (without tax) for pharmaceutical and petroleum companies:

$$R(t) = u - ct + S(t), \quad S(t) = \sum_{i=1}^{N(t)} Y_i,$$

where u is the company's initial surplus, c is now the constant expense rate and $\{S(t), t \geq 0\}$ is the aggregate revenue process. Let $\gamma \in (0, 1)$ be the tax rate at which the insurance company has to pay the profit times γ as the tax. In this system, the profit is defined as the excess of each new record high of the surplus over the previous one. The value of the nth record high (J_n) is defined as follows:

$$J_0 = u, \quad J_n = J_{n-1} + (1-\gamma) \sum_{j=k_{n-1}+1}^{k_n} (Y_j - c\Delta_j T), \quad n = 1, 2, \ldots$$

where $k_n = \inf_{k \in \mathbb{N}} \left\{ k > k_{n-1} \mid \sum_{j=k_{n-1}+1}^{k} (Y_j - c\Delta_j T) > 0 \right\}$ with $k_0 = 0$ is the number of innovations up to the time of the n-th record high. The resulting surplus with tax is given by

$$\begin{aligned} R_\gamma(t) &= J_{\Lambda(t)} - \sum_{j=k_{\Lambda(t)}+1}^{N(t)} (c\Delta_j T - Y_j) - c(t - T_{N(t)}) \\ &= J_{\Lambda(t)} - c(t - T_{k_{\Lambda(t)}}) + \sum_{j=k_{\Lambda(t)}+1}^{N(t)} Y_j, \end{aligned}$$

where $\Lambda(t) = \sup\{n \in \mathbb{N}; T_{k_n} \leq t\}$ and $\sum_{j=k_{\Lambda(t)}+1}^{N(t)} Y_j = 0$ if $k_{\Lambda(t)} = N(t)$.

Note that $Y_j - cT_j$ is the "profit and loss" from the $(i-1)$-th innovation to i-th innovation, and thus $k_n - k_{n-1}$ corresponds to the minimum term that the "aggregated profit and loss" is positive. Under the loss-carry forward system, the loss is carried forward, which implies that while the aggregated profit and loss is negative, the company does not have to pay the tax. That is the reason why k_n is calculated. In addition, $\{J_n\}$ corresponds to the surplus at the time that the n-th tax payment is made.

Table 1 shows an example of the dual risk model under the loss-carry forward system.

Table 1. Example of the loss-carry forward system with $u = 2$, $c = 2$ and $\gamma = 30\%$

j	**1**	**2**	**3**	**4**	**5**	**6**	**7**	**8**	**9**	**10**	**…**
$\Delta_j T$	1.0	2.0	1.0	1.0	3.0	1.0	1.0	2.0	1.0	1.0	…
Y_j	1.0	10.0	1.0	2.0	10.0	1.0	2.0	1.0	10.0	11.0	…
T_j	1.0	3.0	4.0	5.0	8.0	9.0	10.0	12.0	13.0	14.0	…
cT_j	2.0	6.0	8.0	10.0	16.0	18.0	20.0	24.0	26.0	28.0	…
$S(T_j)$	1.0	11.0	12.0	14.0	24.0	25.0	27.0	28.0	38.0	49.0	…
$R(T_j)$	**9.0**	**15.0**	**14.0**	**14.0**	**18.0**	**17.0**	**17.0**	**14.0**	**22.0**	**31.0**	…
$Y_j - c\Delta_j T$	-1.0	6.0	-1.0	0	4.0	-1.0	0	-3.0	8.0	9.0	…
$\sum_{i=1}^{j}(Y_i - c\Delta_i T)$	-1.0	**5.0**	-								$k_1 = 2$
$\sum_{i=3}^{j}(Y_i - c\Delta_i T)$	-		-1.0	-1.0	**3.0**	-					$k_2 = 5$
$\sum_{i=6}^{j}(Y_i - c\Delta_i T)$	-					-1.0	-1.0	-4.0	**4.0**		$k_3 = 9$
$\sum_{i=9}^{j}(Y_i - c\Delta_i T)$	-									**9.0**	$k_4 = 10$
$\Lambda(T_j)$	0	1	1	1	2	2	2	2	3	4	…
$J_{\Lambda(T_j)}$	10.0	14.0	14.0	14.0	16.4	16.4	16.4	16.4	19.6	26.8	…
$c(T_j - T_{k_{\Lambda(T_j)}})$	2.0	0	2.0	4.0	0	2.0	4.0	8.0	0	0	…
$\sum_{j=k_{\Lambda(T_j)}+1}^{N(T_j)} Y_j$	1.0	0	1.0	3.0	0	1.0	3.0	4.0	0	0	…
$R_\gamma(T_j)$	**9.0**	**14.0**	**13.0**	**13.0**	**16.4**	**15.4**	**15.4**	**12.4**	**19.6**	**26.8**	…

Albrecher, Renaud, and Zhou (2008) discussed the same problem for the general Lévy risk model case, while Albrecher, Borst, Boxma, and Resing (2009) derived a tax identity valid to arbitrary surplus-dependent tax rates.

6. Statistical estimation

In this section, we discuss the statistical estimation of the ruin probability and Gerber–Shiu function. The statistical inference for the ultimate ruin probability has been studied by many authors. Grandell (1978/1979), Csörgő and Teugels (1990), Deheuvels and Steinebach (1990), Csörgő and Steinebach (1991), and Embrechts and Mikosch (1991) discussed the statistical estimation of the adjustment coefficient, because the ruin probability is approximately expressed as a function of the adjustment coefficient by using the Cramér–Lundberg approximation. On the contrary, Frees (1986), Croux and Veraverbeke (1990), Bening and Korolev (2003), and Mnatsakanov et al. (2008) discussed the non-parametric estimation of the ruin probability. Shimizu (2012) proposed a non-parametric estimator of the Gerber–Shiu function when the surplus process $\{U(t)\}$ is a compound Poisson claim process plus a diffusion perturbation (called the Wiener–Poisson risk model, defined by Equation (2.4)).

6.1. Statistical estimation of the adjustment coefficient

Suppose that the surplus of an insurance company at time t follows the Cramér–Lundberg model defined by Equation (2.2), where we assume that the claim size X_i has the probability density $f(x)$ for $x \geq 0$. Let $\psi(u)$ denote the ruin probability defined by Equation (3.1). We introduce the adjustment coefficient (R), which is the positive solution to

$$-rc + \lambda M_X(r) - \lambda = 0 \tag{6.1}$$

with respect to r, where $M_X(r) = E(e^{rX}) = \int_0^\infty e^{rx} f(x)dx$. Hence, the adjustment coefficient satisfies

$$M_X(R) - \frac{cR}{\lambda} = 1.$$

Here, the left-hand side of Equation (6.1) comes from the Laplace exponent of $U(1) - u$, that is,

$$\ell(r) := \ln E\{e^{-r(U(1)-u)}\} = -rc + \ln E\{M_X(r)^{N(1)}\} = -rc + \ln e^{\lambda(M_X(r)-1)}$$

which corresponds to Lundberg's fundamental equation with $\delta = 0$. Then, the Cramér–Lundberg approximation

$$\psi(u) \approx Ce^{-Ru}, \quad C = \frac{c/\lambda - E(X)}{E(Xe^{RX}) - c/\lambda}$$

is derived (see Minkova, 2010). A natural problem is then to consider the estimation of the adjustment coefficient R. Grandell (1978/1979) proposed the following estimator for R.

Definition 6.1 Suppose that $X_1, \ldots, X_{N(T)}, N(T)$ and T are observed, where X_i is the i-th claim amount, T is the ruin time and $N(T)$ is the claim frequency up to the ruin. Then, an estimator (R_T) of R is defined as the solution to

$$\widehat{M}_X(r) - \frac{cr}{\hat{\lambda}} = 1,$$

with respect to r, where $\widehat{M}_X(r) = \frac{1}{N(T)} \sum_{k=1}^{N(T)} e^{rX_k}$ and $\hat{\lambda} = \frac{N(T)}{T}$.

Furthermore, Grandell (1978/1979) showed its asymptotic normality as follows:

Theorem 6.1 *If $E(X_i) - c/\lambda < 0$ and $\lim_{r\to\infty} \{M_X(r) - cr/\lambda\} = \infty$ are satisfied, then*

$$\sqrt{T}(R_T - R) \xrightarrow{d} N\left(0, \frac{y(2R)}{\lambda(y'(R))^2}\right)$$

where $y(r) = M_X(r) - cr/\lambda$.

Csörgő and Teugels (1990) derived an estimator of R based on the sample of size n. Deheuvels and Steinebach (1990) and Csörgő and Steinebach (1991) derived an estimator of R by using intermediate order statistics. Embrechts and Mikosch (1991) discussed a bootstrap procedure to estimate the adjustment coefficient R. Furthermore, Shimizu (2009) considered the estimation of the adjustment coefficient R when the underlying process consists of a Brownian motion plus negative jumps where each jump size of the process does not necessarily correspond to each claim size.

6.2. Non-parametric estimation of the ruin probability

Suppose that the surplus $U(t)$ is defined by (Equation 2.2), where the claim size X_i has the mean $E(X_i) = \mu$, the distribution function $F(x)$, and the p.d.f. $f(x)$. Let $\psi(u)$ denote the ruin probability defined by (Equation 3.1). Define $\bar{\psi}(u) = 1 - \psi(u)$ as the non-ruin probability. Then, it follows the following the integro-differential equation (see (7.6) of Dickson, 2005):

$$\frac{d}{du}\bar{\psi}(u) = \frac{\lambda}{c}\bar{\psi}(u) - \frac{\lambda}{c}\int_0^u \bar{\psi}(u-x)f(x)dx, \quad u \geq 0. \tag{6.2}$$

Integrating Equation (6.2) from 0 to t leads to (see e.g. Minkova, 2010.)

$$\bar{\psi}(t) = \bar{\psi}(0) + \frac{\lambda}{c}\int_0^t \bar{\psi}(t-x)(1-F(x))dx, \quad t \geq 0. \tag{6.3}$$

Let $\bar{\psi}^*(s), \bar{\psi}^{**}(s)$ denote the Laplace transforms

$$\bar{\psi}^*(s) = \int_0^\infty e^{-sz}\bar{\psi}(z)dz, \quad \bar{\psi}^{**}(s) = \int_0^\infty e^{-sz}d\bar{\psi}(z).$$

Taking the Laplace transform in both sides of Equation (6.3) gives

$$\bar{\psi}^*(s) = \frac{\psi(0)}{s\left\{1 - \frac{\lambda}{c}f^*(s)\right\}} = \frac{\psi(0)}{s}\sum_{r=0}^{\infty}\left(\frac{\lambda}{c}f^*(s)\right)^r,$$

where $f^*(s)$ is the Laplace transform of $f(x)$. Since it follows that $\bar{\psi}^{**}(s) = s\bar{\psi}^*(s)$, we have

$$\bar{\psi}^{**}(s) = \psi(0)\sum_{r=0}^{\infty}\left(\frac{\lambda\mu}{c}\right)^r\left(\frac{f^*(s)}{\mu}\right)^r. \tag{6.4}$$

The inverse Laplace transform of Equation (6.4) leads to the following formula (the so-called "Pollaczek–Khinchin formula"):

$$\psi(u) = \left(1 - \frac{\lambda\mu}{c}\right)\sum_{r=1}^{\infty}\left(\frac{\lambda\mu}{c}\right)^r\{1 - G^{(r)}(u)\}, \tag{6.5}$$

where $G(u) = \frac{1}{\mu}\int_0^u(1-F(y))dy$ and $G^{(r)}$ denote the r-fold convolution of the distribution function G itself[3]. The following formula is equivalent to Equation (6.5):

$$\psi(u) = \frac{\lambda\mu}{c} - \left(1 - \frac{\lambda\mu}{c}\right)\sum_{r=1}^{\infty}\left(\frac{\lambda\mu}{c}\right)^r G^{(r)}(u)$$

Assuming that μ and λ are known, Croux and Veraverbeke (1990) proposed its estimator as

$$\psi_n(u) = \frac{\lambda\mu}{c} - \left(1 - \frac{\lambda\mu}{c}\right)\sum_{r=1}^{m_n}\left(\frac{\lambda\mu}{c}\right)^r U_{nr}(u), \tag{6.6}$$

where $\{m_n\}$ is a sequence of natural numbers that depends on the sample size n satisfied with $m_n \le n$, $\lim_{n\to\infty} m_n = +\infty$, and $U_{nr}(u)$ is a U-statistic given as follows.

Definition 6.2 Suppose that $X_1, \ldots, X_n$ are observed, where X_i is the ith claim amount. Then, the estimator $(U_{nr}(u))$ of $G^{(r)}(u)$ for a fixed r, u is defined by

$$U_{nr}(u) = \binom{n}{r}^{-1} \sum_{1\le i_1<\cdots<i_r\le n} h_r(X_{i_1}, \ldots, X_{i_r}; u),$$

where $\sum_{1\le i_1<\cdots<i_r\le n}$ is the sum over all $\binom{n}{r}$ distinct combinations of $\{i_1, i_2, \ldots, i_r\}$ satisfied with $1 \le i_1 < \cdots < i_r \le n$, and $h_r(x_1, \ldots, x_r; u)$ is a symmetric kernel

$$h_r(x_1, \ldots, x_r; u) = \frac{1}{\mu^r}\int_0^\infty \cdots \int_0^\infty I(y_1 + \cdots + y_r \le u)\prod_{j=1}^r I(x_j > y_j)dy_1 \cdots dy_r.$$

Note that for $r = 1, 2, \ldots$, we can interpret $G^{(r)}(u) = P(Y_1 + \cdots + Y_r \le x)$, where $Y_1, \ldots, Y_r$ are independent with a common density function $(1/\mu)(1 - F(y))$ for $y \ge 0$. Hence, we can write

$$G^{(r)}(u) = \frac{1}{\mu^r}\int_0^\infty \cdots \int_0^\infty I(y_1 + \cdots + y_r \le u)\prod_{j=1}^r (1 - F(y_j))dy_1 \cdots dy_r.$$

Therefore, we can confirm that $U_{nr}(u)$ is an unbiased estimator of $G^{(r)}(u)$, since

$$\begin{aligned} E(U_{nr}(u)) &= E(h_r(X_1, \ldots, X_r; u)) \\ &= \frac{1}{\mu^r}\int_0^\infty \cdots \int_0^\infty I(y_1 + \cdots + y_r \le u)\prod_{j=1}^r E(I(X_j > y_j))dy_1 \cdots dy_r \\ &= \frac{1}{\mu^r}\int_0^\infty \cdots \int_0^\infty I(y_1 + \cdots + y_r \le u)\prod_{j=1}^r (1 - F(y_j))dy_1 \cdots dy_r = G^{(r)}(u). \end{aligned}$$

Frees (1986) showed the consistency for the linear combination of U-statistics, i.e.

$$\sum_{r=1}^{m_n}\left(\frac{\lambda\mu}{c}\right)^r U_{nr}(u) \overset{a.s.}{\to} \sum_{r=1}^{\infty}\left(\frac{\lambda\mu}{c}\right)^r G^{(r)}(u).$$

Furthermore, Croux and Veraverbeke (1990) showed the asymptotic normality for $\psi_n(u)$ as follows.

THEOREM 6.2 *If* $\log n = o(m_n)$, $n \to \infty$, *then for each* $u > 0$

$$\sqrt{n}(\psi_n(u) - \psi(u)) \overset{d}{\to} N\left(0, \left(1 - \frac{\lambda\mu}{c}\right)^2 \sum_{r,s=1}^{\infty} rs\left(\frac{\lambda\mu}{c}\right)^{r+s} \xi_{rs}(1; u)\right),$$

where

$$\xi_{rs}(1; u) = E\{h_{r1}(X_1; u)h_{s1}(X_1; u)\} - G^{(r)}(u)G^{(s)}(u),$$

$$h_{r1}(x; u) = \frac{1}{\mu}\int_0^\infty G^{(r-1)}(u - y)I(x < y)dy.$$

Bening and Korolev (2003) mentioned that the above results can rarely be applied in practice since

- The Croux–Veraverbeke estimator Equation (6.6) is constructed from a sample with a non-random size,

- The property of the asymptotic normality of the proposed estimator cannot be directly used to construct the (asymptotic) confidence intervals since the limit distribution of the estimator (more precisely, its variance) depends on the unknown distribution of claims.

For these problems, they considered the following model. Let $\Lambda(t)$, $t \geq 0$ be a random process independent of $N(t)$ and possessing non-decreasing almost surely finite right-continuous sample paths starting from the origin. A Cox process, also called a doubly stochastic Poisson process, guided by the process $\Lambda(t)$, is defined as

$$\tilde{N}(t) = N(\Lambda(t)), \quad t \geq 0.$$

Consider the (new) process

$$U(t) = u + c\Lambda(t) - \tilde{S}(t), \quad \tilde{S}(t) = \sum_{i=1}^{\tilde{N}(t)} X_i, \quad t \geq 0.$$

Under this model, we suppose that the claims $X_1, \ldots, X_{N(t)}$ are obtained at time t. Then, the following estimator for $\psi(u)$ is introduced:

$$\psi_{N(t)}(u) = \frac{\lambda\mu}{c} - \left(1 - \frac{\lambda\mu}{c}\right) \sum_{r=1}^{m(N(t))} \left(\frac{\lambda\mu}{c}\right)^r U_{N(t),r}(u),$$

where $m(n)$ is an integer with $m(n) \leq n$, $\lim_{n\to\infty} m(n) = +\infty$ and $U_{N(t),r}(u)$ is the same as $U_{nr}(u)$ when $N(t) = n$. Moreover, a consistent estimator of asymptotic variance is introduced as follows:

$$\sigma^2_{N(t)} = \left(1 - \frac{\lambda\mu}{c}\right)^2 \sum_{r,s=1}^{k(N(t))} rs\left(\frac{\lambda\mu}{c}\right)^{r+s} \bar{\sigma}_{N(t),r,s}(u),$$

where $k(n)$ is an integer with $k(n) \leq n$, $\lim_{n\to\infty} k(n) = +\infty$ and

$$\bar{\sigma}_{n,r,s}(u) = \frac{1}{n} \sum_{i=1}^{n} \bar{h}_{n,r}(X_i;u)\bar{h}_{n,s}(X_i;u) - U_{n,r}(u)U_{n,s}(u),$$

$$\bar{h}_{n,r}(x;u) = \frac{1}{\mu} \int_0^{\infty} U_{n,r-1}(u-y)I(x > y)dy.$$

Then, Bening and Korolev (2003) showed the asymptotic normality for $\psi_{N(t)}(u)$ as follows.

THEOREM 6.3 *Suppose* $\log n = o(m(n))$ *and*

$$\frac{1}{\sqrt{n}} \sum_{r=1}^{k(n)} \left(\frac{u}{c}\right)^r r^{3/2} \cdot \max\left\{1, \sum_{l=1}^{k(n)} l\left(\frac{u}{c}\right)^l\right\} \to 0$$

as $n \to \infty$ *are satisfied. Assume* $\Lambda(t) \xrightarrow{p} \infty$ *as* $t \to \infty$. *Then,*

$$\sqrt{N(t)}\frac{\psi_{N(t)}(u) - \psi(u)}{\sigma_{N(t)}} \xrightarrow{d} N(0,1).$$

By contrast, Mnatsakanov et al. (2008) discussed the estimation of the ruin probability by using the inverse of the Laplace transform. In the same way to Equation (4.15), the Laplace transform of Equation (6.2) is

$$-\psi(0) + s\psi^*(s) = \frac{\lambda}{c}\psi^*(s) - \frac{\lambda}{c}\psi^*(s)f^*(s),$$

where $\psi^*(s), f^*(s)$ are the Laplace transforms of ψ, f, respectively. By stating $\psi(0) = 1 - \rho \leq \psi(u) \leq 1$ for $u \geq 0$, we can write

$$\psi^*(s) = \frac{1-\rho}{D(s)}, \quad D(s) = s - \frac{\lambda}{c}\{1 - f^*(s)\}, \quad s \geq 0. \tag{6.7}$$

Assuming that λ is known, Mnatsakanov et al. (2008) proposed its estimator as follows.

Definition 6.3 Suppose that $X_1, \ldots, X_n$ are observed, where X_i is the ith claim amount. Then, the estimator ($\widehat{\psi^*}(s)$) of $\psi^*(s)$ for a fixed s is defined by

$$\widehat{\psi^*}(s) = \frac{1-\hat{\rho}}{\hat{D}(s)}, \quad \hat{D}(s) = s - \frac{\lambda}{c}\left\{1 - \widehat{f^*}(s)\right\},$$

where

$$\widehat{f^*}(s) = \frac{1}{n}\sum_{i=1}^{n} e^{-sX_i}, \quad \hat{\rho} = \frac{\lambda}{c}\frac{1}{n}\sum_{i=1}^{n} X_i.$$

Although $\widehat{\psi^*}$ is close to ψ^*, there is no guarantee that the inverse of the Laplace transform $\mathcal{L}^{-1}\hat{\psi}^*$ will be close to $\mathcal{L}^{-1}\psi^* \equiv \psi$, because $\widehat{\mathcal{L}^{-1}}$ is not continuous. Hence, Mnatsakanov et al. (2008) considered slightly modifying the functions $\widehat{\psi^*}$ and ψ^*, that is, for an arbitrary fixed $\theta > 0$,

$$\widehat{\psi^*_\theta}(s) = \widehat{\psi^*}(s+\theta), \quad \psi^*_\theta(s) = \psi^*(s+\theta), \quad s \geq 0$$

are defined. Then, $\widehat{\psi^*_\theta}(s), \psi^*_\theta(s) \in L^2(0,\infty)$ a.s. are satisfied because the non-continuity for $\mathcal{L}^{-1}\hat{\psi}^*$ and $\mathcal{L}^{-1}\psi^*$ is due to the behaviour of $D(s)$ and $\hat{D}(s)$ near 0. For $\widehat{\psi^*_\theta}(s), \psi^*_\theta(s)$, the inverse of the Laplace transform is introduced as follows:

$$(\mathcal{L}^{-1}_\alpha \widehat{\psi^*_\theta})(u) = \frac{1}{\pi^2}\int_0^\infty \int_0^\infty R_\alpha(s)\frac{1}{\sqrt{s}} e^{-uts}\widehat{\psi^*_\theta}(t)dtds,$$

$$(\mathcal{L}^{-1}_\alpha \psi^*_\theta)(u) = \frac{1}{\pi^2}\int_0^\infty \int_0^\infty R_\alpha(s)\frac{1}{\sqrt{s}} e^{-uts}\psi^*_\theta(t)dtds$$

for a fixed $\alpha > 0$, where

$$R_\alpha(s) = \frac{1}{2}\int_{\{\rho(x)\geq\alpha\}} (\cosh \pi x)\cos(x \log s)dx, \quad \rho(x) = \frac{\pi}{\cosh \pi x}, \quad s \in (0,\infty), \quad x \in \mathbb{R}.$$

As a result, the following estimators are introduced.

Definition 6.4

$$\widehat{\psi^\theta_\alpha}(u) = e^{\theta u}\widehat{\psi_{\theta,\alpha}}(u) = (\mathcal{L}^{-1}_\alpha \widehat{\psi^*_\theta})(u), \quad \widetilde{\psi^\theta_\alpha}(u) = e^{\theta u}\psi_{\theta,\alpha}(u) = (\mathcal{L}^{-1}_\alpha \psi^*_\theta)(u).$$

Then, the following consistency in the ISE (integrated squared error) sense is shown.

THEOREM 6.4 *Suppose that $0 < d\psi(u)/du \leq M$ is satisfied for a fixed M and $\forall u \geq 0$, and Assumption 3.1 is made. Then, if f is unknown,*

$$\|\widehat{\psi^\theta_\alpha} - \psi\|_B^2 = O_p\left(\frac{1}{\log n}\right), \quad \text{as } n \to \infty,$$

where $\alpha = \alpha(n) = \sqrt{(\log n)/n}$ and if f is known

$$\|\widetilde{\psi^\theta_\alpha} - \psi\|_B^2 = O\left(\frac{1}{\log(1/\alpha)}\right), \quad \text{as } \alpha \downarrow 0.$$

Here, $\|\cdot\|_B$ is the norm in $L^2(0,B)$ for an arbitrary fixed $0 < B < \infty$, which implies that the comparison between $\psi_\alpha^\theta(\widehat{\psi_\alpha^\theta})$ and ψ_α is carried out in $L^2(0,B)$ by restricting the functions to $[0,B]$.

6.3. Statistical estimation of the Gerber–Shiu function

Shimizu (2012) proposed an estimator of the Gerber-Shiu function for the Wiener–Poisson risk model (corresponding to the model in Example (2.5)) in a similar way to Mnatsakanov et al. (2008), and showed the consistency in the ISE sense.

The Laplace transform of the Gerber–Shiu function for the Wiener–Poisson risk process (Equation 2.4) was obtained by Tsai and Willmot (2002), Morales (2007), and Biffis and Morales (2010). The following result was given by Morales (2007).

THEOREM 6.5 *Suppose that the penalty function w is uniformly bounded, or it satisfies* $|w(x,y) - w_0| \le k(x+y)^\alpha$ *for a universal constant* $k > 0$ *and* $\alpha > 1$. *Then,*

$$\phi^*(s) = \frac{K^*(\rho) - K^*(s) + w_0(s-\rho)D}{\{c + (s+\rho)D\}(s-\rho) - \lambda\{f^*(\rho) - f^*(s)\}}, \quad s \ge 0, \tag{6.8}$$

where f is a p.d.f. of X_i, f^* and K^* are the Laplace transforms of f and K, $D = \sigma^2/2$, the function K is given by

$$K(x) = \lambda \int_x^\infty w(x, y-x) f(y) dy,$$

ρ is the unique non-negative solution to the following Lundberg's fundamental equation with respect to s:

$$\ell(s) = cs - \lambda(1 - f^*(s)) + Ds^2 - \delta = 0.$$

In particular, $\rho = 0$ as $\delta = 0$.

Shimizu (2012) proposed its estimator as follows:

Definition 6.5 Suppose that $\{U(t) | t \equiv t_i^n = i\Delta_n,\ i = 0,1,2,\dots,n\}$ for some $\Delta_n > 0$ and the claim data $X_1, \dots, X_{N(T_n)}$ where $T_n = n\Delta_n$ are observed. Then, the estimator $(\widehat{\phi^*}(s))$ of $\phi^*(s)$ for a fixed s is defined by

$$\widehat{\phi^*}(s) = \frac{\widehat{K^*}(\hat\rho) - \widehat{K^*}(s) + w_0(s-\hat\rho)\hat D}{\left\{c + (s+\hat\rho)\hat D\right\}(s-\hat\rho) - \lambda\left\{\widehat{f^*}(\hat\rho) - \widehat{f^*}(s)\right\}},$$

where

$$\widehat{f^*}(s) = \frac{1}{N_{T_n}} \sum_{i=1}^{N_{T_n}} e^{-sX_i}, \quad \widehat{K^*}(s) = \frac{1}{T_n} \sum_{i=1}^{N_{T_n}} \int_0^{X_i} e^{-sy} w(y, X_i - x) dy, \quad \hat\lambda = \frac{N_{T_n}}{T_n}$$

and

$$\hat D = \frac{\widehat{\sigma^2}}{2}, \quad \widehat{\sigma^2} = \frac{1}{T_n} \sum_{i=1}^{n} (U(t_i^n) - U(t_{i-1}^n))^2 I\{\tau_j \notin (t_{i-1}^n, t_i^n]; j = 1, \dots, N_{T_n}\}.$$

Here, τ_j is the j-th claim occurring time and $\hat\rho$ is defined as the following M estimator:

$$\hat\rho = \arg\min_{s \in I} |\hat\ell(s)|, \quad \hat\ell(s) = cs - \hat\lambda(1 - \widehat{f^*}(s)) + \hat D s^2 - \delta,$$

where $I \subset (0,\infty)$ is a compact set in which ρ exists, and we set $\hat\rho = 0$ for $\delta = 0$.

In the same way to Mnatsakanov et al. (2008), Shimizu (2012) considered slightly modifying the functions $\widehat{\phi}^*$ for an arbitrary fixed $\theta > 0$,

$$\widehat{\phi_\theta^*}(s) = \widehat{\phi}^*(s+\theta), \quad s \geq 0.$$

For $\widehat{\phi_\theta^*}(s)$, the inverse of the Laplace transform is introduced as follows:

$$(\mathcal{L}_m^{-1}\widehat{\phi_\theta^*})(u) = \frac{1}{\pi^2}\int_0^\infty \int_0^\infty \Phi_m(s)\frac{1}{\sqrt{s}}e^{-uts}\widehat{\phi_\theta^*}(t)dtds,$$

where $m > 0$ is a constant and

$$\Phi_m(s) = \int_0^{a_m} (\cosh \pi x)\cos(x \log s)dx, \quad a_m = \frac{1}{\pi \cosh \pi m} > 0, \quad s \in (0,\infty).$$

As a result, the following estimator is introduced.

Definition 6.6

$$\widehat{\phi_m^\theta}(u) = e^{\theta u}\widehat{\phi_{\theta,m}}(u) = (\mathcal{L}_m^{-1}\widehat{\phi_\theta^*})(u)$$

THEOREM 6.6 *Suppose that*

(i) $E(|X_i|^p) < \infty$ for $p > 0$,

(ii) There exists a compact $I \subset (0,\infty)$ such that ρ belongs to the interior of I,

(iii) $c > \lambda E(X_i)$ (i.e. Assumption 3.1),

(iv) $T_n\Delta_n \to 0$ as $n \to \infty$,

(v) The penalty function w is uniformly bounded, or it satisfies $|w(x,y) - w_0| \leq k(x+y)^\alpha$ for a universal constant $k > 0$ and $\alpha > 1$,

(vi) There exists a constant $k' > 0, \alpha' > 0$ such that $|\int_0^\infty w(x,dy)| \leq k'(1+|x|^{\alpha'})$ for any $x \in \mathbb{R}^+$,

(vii) The Gerber–Shiu function ϕ and first derivative ϕ' are of polynomial growth.

Then, for numbers $m \equiv m(n)$ such that $m(n) = (T_n/\log T_n)^{1/2}$ as $n \to \infty$ and a number $\theta > 0$ such that $(c - \lambda E(X_i))\theta > \delta$, the following hold true:

(I) For any constant $B > 0$

$$\|\widehat{\phi_m^\theta} - \phi\|_B = O_p\left((\log T_n)^{-1/2}\right), \quad \text{as } n \to \infty,$$

(II) If we take $\theta \in (0,1/2)$, it follows for a sequence $B_n = \log\log T_n$ that,

$$\|\widehat{\phi_m^\theta} - \phi\|_{B_n} = O_p\left((\log T_n)^{\theta-1/2}\right), \quad \text{as } n \to \infty.$$

Funding

This work was supported by JSPS KAKENHI Grant Number 24730193.

Author details

Hiroshi Shiraishi[1]

E-mail: shiraishi@math.keio.ac.jp

[1] Department of Mathematics, Keio University, Yokohama, Kanagawa, Japan.

Notes

1. Applying the operator $\prod_{i=1}^{h}\left(\frac{\partial}{\partial x}+\beta_i\right)$ to an equation $A = B$ corresponds to the calculation of $\prod_{i=1}^{h}\left(\frac{\partial}{\partial x}+\beta_i\right)A = \prod_{i=1}^{h}\left(\frac{\partial}{\partial x}+\beta_i\right)B$.
2. Note that we modified Avanzi's (2009) definition.
3. i.e. $G^{(r)}(u) = \frac{1}{\mu^r}\int\cdots\int_{y_1+\cdots+y_r\leq u}\prod_{s=1}^{r}(1-F(y_s))dy_1\cdots dy_r$.

References

Albrecher, H., Badescu, A., & Landriault, D. (2008). On the dual risk model with tax payments. *Insurance: Mathematics and Economics, 42*, 1086–1094.

Albrecher, H., Borst, S., Boxma, O., & Resing, J. (2009). The tax identity in risk theory: A simple proof and an extension. *Insurance: Mathematics and Economics, 44*, 304–306.

Albrecher, H., Claramunt, M. M., & Mármol, M. (2005). On the distribution of dividend payments in a sparre andersen model with generalized erlang (n) interclaim times. *Insurance: Mathematics and Economics, 37*, 324–334.

Albrecher, H., & Hartinger, J. (2006). On the non-optimality of horizontal barrier strategies in the sparre andersen model. *HERMES International Journal of Computer Mathematics and Its Applications, 7*, 1–14.

Albrecher, H., & Hartinger, J. (2007). A risk model with multilayer dividend strategy. *North American Actuarial Journal, 11*, 43–64.

Albrecher, H., Hartinger, J., & Thonhauser, S. (2007). On exact solutions for dividend strategies of threshold and linear barrier type in a sparre andersen model. *Astin Bulletin, 37*, 203–233.

Albrecher, H., Hartinger, J., & Tichy, R. F. (2005). On the distribution of dividend payments and the discounted penalty function in a risk model with linear dividend barrier. *Scandinavian Actuarial Journal, 2005*, 103–126.

Albrecher, H., & Hipp, C. (2007). Lundberg's risk process with tax. *Blätter der DGVFM, 28*, 13–28.

Albrecher, H., & Kainhofer, R. (2002). Risk theory with a nonlinear dividend barrier. *Computing, 68*, 289–311.

Albrecher, H., Kainhofer, R., & Tichy, R. F. (2003). Simulation methods in ruin models with non-linear dividend barriers. *Mathematics and Computers in Simulation, 62*, 277–287.

Albrecher, H., Renaud, J-.F., & Zhou, X. (2008). A L{\'e} vy insurance risk process with tax. *Journal of Applied Probability, 45*, 363–375.

Albrecher, H., & Teugels, J. L. (2006). Exponential behavior in the presence of dependence in risk theory. *Journal of Applied Probability, 43*, 257–273.

Alegre, A., Claramunt, M., & Marmol, M. (1999). Dividend policy and ruin probability. In *Proceedings of the Third International Congress on Insurance, Mathematics & Economics*. London.

Andersen, E. S. (1957). On the collective theory of risk in case of contagion between claims. *Bulletin of the Institute of Mathematics and its Applications, 12*, 275–279.

Asmussen, S., Højgaard, B., & Taksar, M. (2000). Optimal risk control and dividend distribution policies. Example of excess-of loss reinsurance for an insurance corporation. *Finance and Stochastics, 4*, 299–324.

Avanzi, B. (2009). Strategies for dividend distribution: A review. *North American Actuarial Journal, 13*, 217–251.

Avanzi, B., & Gerber, H. U. (2008). Optimal dividends in the dual model with diffusion. *Astin Bulletin, 38*, 653–667.

Avanzi, B., Gerber, H. U., & Shiu, E. S. (2007). Optimal dividends in the dual model. *Insurance: Mathematics and Economics, 41*, 111–123.

Avanzi, B., Shen, J., & Wong, B. (2011). Optimal dividends and capital injections in the dual model with diffusion. *Astin Bulletin, 41*, 611–644.

Badescu, A., & Landriault, D. (2008). Recursive calculation of the dividend moments in a multi-threshold risk model. *North American Actuarial Journal, 12*, 74–88.

Bao, Z.-H. (2007). A note on the compound binomial model with randomized dividend strategy. *Applied Mathematics and Computation, 194*, 276–286.

Bening, V. E., & Korolev, V. Y. (2003). Nonparametric estimation of the ruin probability for generalized risk processes. *Theory of Probability & its Applications, 47*, 1–16.

Beveridge, C. J., Dickson, D. C., & Wu, X. (2007). *Optimal dividends under reinsurance*. Melbourne: Centre for Actuarial Studies, Department of Economics, University of Melbourne.

Biffis, E., & Morales, M. (2010). On a generalization of the Gerber--Shiu function to path-dependent penalties. *Insurance: Mathematics and Economics, 46*, 92–97.

Boudreault, M., Cossette, H., Landriault, D., & Marceau, E. (2006). On a risk model with dependence between interclaim arrivals and claim sizes. *Scandinavian Actuarial Journal, 2006*, 265–285.

Bowers, N. L., Gerber, H., Hickman, J., Jones, D., & Nesbitt, C. (1997). *Actuarial mathematics*. Itasca, MN: Society of Actuaries.

Bühlmann, H. (1970). *Mathematical methods in risk theory* (Die Grundlehren der mathematischen Wissenschaften, Band 172). New York, NY: Springer-Verlag.

Cao, Y., & Wan, N. (2009). Optimal proportional reinsurance and investment based on Hamilton--Jacobi--Bellman equation. *Insurance: Mathematics and Economics, 45*, 157–162.

Centeno, L. (1986). Measuring the effects of reinsurance by the adjustment coefficient. *Insurance: Mathematics and Economics, 5*, 169–182.

Cheng, S., Gerber, H. U., & Shiu, E. S. (2000). Discounted probabilities and ruin theory in the compound binomial model. *Insurance: Mathematics and Economics, 26*, 239–250.

Cheung, E. C., Dickson, D. C., & Drekic, S. (2008). Moments of discounted dividends for a threshold strategy in the compound Poisson risk model. *North American Actuarial Journal, 12*, 299–318.

Claramunt, M. M., Mármol, M., & Alegre, A. (2003). A note on the expected present value of dividends with a constant barrier in the discrete time model. *Bulletin of the Swiss Association of Actuaries, 2*, 149–159.

Cramér, H. (1930). *On the mathematical theory of risk*. Skandia Jubilee Volume, Stockholm.

Cramér, H. (1955). *Collective risk theory: A survey of the theory from the point of view of the theory of stochastic processes*. Stockholm: Nordiska bokhandeln

Croux, K., & Veraverbeke, N. (1990). Nonparametric estimators for the probability of ruin. *Insurance: Mathematics and Economics, 9*, 127–130.

Csörgő, M., & Steinebach, J. (1991). On the estimation of the adjustment coefficient in risk theory via intermediate order statistics. *Insurance: Mathematics and Economics, 10*, 37–50.

Csörgő, S., & Teugels, J. L. (1990). Empirical Laplace transform and approximation of compound distributions. *Journal of Applied Probability, 27*, 88–101.

David Promislow, S., & Young, V. R. (2005). Minimizing the probability of ruin when claims follow Brownian motion with drift. *North American Actuarial Journal, 9*, 110–128.

De Finetti, B. (1957). Su un' impostazione alternativa della teoria collettiva del rischio [On a setting of the alternative theory of collective risk]. *Transactions of the XVth International Congress of Actuaries, 2*, 433–443.

De Vylder, F., & Goovaerts, M. J. (1988). Recursive calculation of finite-time ruin probabilities. *Insurance: Mathematics and Economics, 7*(1), 1–7.

De Vylder, F., & Marceau, E. (1996). Classical numerical ruin probabilities. *Scandinavian Actuarial Journal, 1996*, 109–123.

Deheuvels, P., & Steinebach, J. (1990). On some alternative estimates of the adjustment coefficient in risk theory. *Scandinavian Actuarial Journal*, 135–159. doi:10.1080/034 61238.1990.10413878

Dickson, D. (1994). Some comments on the compound binomial model. *Astin Bulletin, 24*, 33–45.

Dickson, D., & Waters, H. R. (1991). Recursive calculation of survival probabilities. *Astin Bulletin, 21*, 199–221.

Dickson, D. C. (1992). On the distribution of the surplus prior to ruin. *Insurance: Mathematics and Economics, 11*, 191–207.

Dickson, D. C., & Drekic, S. (2006). Optimal dividends under a ruin probability constraint. *Annals of Actuarial Science, 1*, 291–306.

Dickson, D. C., & Waters, H. R. (1996). Reinsurance and ruin. *Insurance: Mathematics and Economics, 19*, 61–80.

Dickson, D. C. M. (2005). *Insurance risk and ruin. International series on actuarial science*. Cambridge: Cambridge University Press. doi:10.1017/CBO9780511624155

Dickson, D. C. M., & Waters, H. R. (2004). Some optimal dividends problems. *Astin Bulletin, 34*, 49–74. doi:10.2143/AST.34.1.504954

Dufresne, F., & Gerber, H. U. (1988). The surpluses immediately before and at ruin, and the amount of the claim causing ruin. *Insurance: Mathematics and Economics, 7*, 193–199.

Dufresne, F., & Gerber, H. U. (1991). Risk theory for the compound Poisson process that is perturbed by diffusion. *Insurance: Mathematics and Economics, 10*, 51–59.

Embrechts, P., & Mikosch, T. (1991). A bootstrap procedure for estimating the adjustment coefficient. *Insurance: Mathematics and Economics, 10*, 181–190.

Fang, Y., & Wu, R. (2007). Optimal dividend strategy in the compound poisson model with constant interest. *Stochastic Models, 23*, 149–166.

Feng, R., & Shimizu, Y. (2013). On a generalization from ruin to default in a Lévy insurance risk model. *Methodology and Computing in Applied Probability, 15*, 773–802.

Frees, E. W. Nonparametric renewal function estimation. *Annals of Statistics, 14*, 1366–1378. doi:10.1214/aos/1176350163

Frostig, E. (2005). On the expected time to ruin and the expected dividends when dividends are paid while the surplus is above a constant barrier. *Journal of Applied Probability, 42*, 595–607.

Frostig, E. (2008). On risk model with dividends payments perturbed by a Brownian motion—An algorithmic approach. *Astin Bulletin, 38*, 183–206.

Gerber, H. (1974). The dilemma between dividends and safety and a generalization of the Lundberg-Cramér formulas. *Scandinavian Actuarial Journal, 1974*, 46–57.

Gerber, H. U. (1979). *An introduction to mathematical risk theory*, volume 8 of S.S. Heubner Foundation Monograph Series. Philadelphia: University of Pennsylvania, Wharton School, S.S. Huebner Foundation for Insurance Education. Distributed by Richard D. Irwin Inc, Homewood, Ill, With a foreword by James C. Hickman.

Gerber, H. U. (1981). On the probability of ruin in the presence of a linear dividend barrier. *Scandinavian Actuarial Journal, 1981*, 105–115.

Gerber, H. U. (1988). Mathematical fun with the compound binomial process. *Astin Bulletin, 18*, 161–168.

Gerber, H. U., Goovaerts, M. J., & Kaas, R. (1987). On the probability and severity of ruin. *Astin Bulletin, 17*, 151–163.

Gerber, H. U., & Shiu, E. S. (1998). On the time value of ruin. *North American Actuarial Journal, 2*, 48–72.

Gerber, H. U., & Shiu, E. S. (1999). From ruin theory to pricing reset guarantees and perpetual put options. *Insurance: Mathematics and Economics, 24*, 3–14.

Gerber, H. U., & Shiu, E. S. (2003). Geometric Brownian motion models for assets and liabilities: From pension funding to optimal dividends. *North American Actuarial Journal, 7*, 37–51.

Gerber, H. U., & Shiu, E. S. W. (1998). On the time value of ruin (With discussion and a reply by the authors). *North American Actuarial Journal, 2*, 48–78. doi:10.1080/10920277.1998.10595671

Gerber, H. U., & Shiu, E. S. W. (2004). Optimal dividends: Analysis with Brownian motion. *North American Actuarial Journal, 8*(1), 1–20. doi:10.1080/10920277.2004.10596125

Gerber, H. U., Shiu, E. S. W., & Smith, N. (2006). Maximizing dividends without bankruptcy. *Astin Bulletin, 36*, 5–23. doi:10.2143/AST.36.1.2014143

Gerber, H. U., Shiu, E. S. W., & Smith, N. (2008). Methods for estimating the optimal dividend barrier and the probability of ruin. *Insurance: Mathematics and Economics, 42*, 243–254. doi:10.1016/j.insmatheco.2007.02.002

Gerber, H. U., & Smith, N. (2008). Optimal dividends with incomplete information in the dual model. *Insurance: Mathematics and Economics, 43*, 227–233.

Grandell, J. (1978/1979). Empirical bounds for ruin probabilities. *Stochastic Processes and their Applications, 8*, 243–255. doi:10.1016/0304-4149(79)90001-2

Gu, A., Guo, X., Li, Z., & Zeng, Y. (2012). Optimal control of excess-of-loss reinsurance and investment for insurers under a CEV model. *Insurance: Mathematics and Economics, 51*, 674–684.

Højgaard, B. (2002). Optimal dynamic premium control in non-life insurance. Maximizing dividend pay-outs. *Scandinavian Actuarial Journal, 2002*, 225–245.

Højgaard, B., & Taksar, M. (1998). Optimal proportional reinsurance policies for diffusion models. *Scandinavian Actuarial Journal, 1998*, 166–180.

Huang, Y. J., & Li, A. Q. (2011). On a risk model with dependence between interclaim arrivals and claim sizes with multiple thresholds under stochastic interest. *Advanced Materials Research* (Vol. 179, pp. 1086–1090). Pfaffikon: Trans Tech Publications.

Huzak, M., Perman, M., Šikić, H., & Vondraček, Z. (2004). Ruin probabilities for competing claim processes. *Journal of Applied Probability, 41*, 679–690.

Jang, J. (2007). Jump diffusion processes and their applications in insurance and finance. *Insurance: Mathematics and Economics, 41*, 62–70.

Landriault, D. (2008). Constant dividend barrier in a risk model with interclaim-dependent claim sizes. *Insurance: Mathematics and Economics, 42*, 31–38. doi:10.1016/j.insmatheco.2006.12.002

Landriault, D. (2008). Randomized dividends in the compound binomial model with a general premium rate. *Scandinavian Actuarial Journal, 2008*(1), 1–15.

Leung, K. S., Kwok, Y. K., & Leung, S. Y. (2008). Finite-time dividend-ruin models. *Insurance: Mathematics and Economics, 42*, 154–162.

Li, L.-L., Feng, J., & Song, L. (2007). The dividend payments for general claim size distributions under interest rate. *International Journal of Mathematics Sciences, 1*, 135–138.

Li, S. (2006). The distribution of the dividend payments in the compound poisson risk model perturbed by diffusion. *Scandinavian Actuarial Journal, 2006*, 73–85.

Li, S., & Garrido, J. (2002). *On the time value of ruin in the discrete time risk model.* Workings Paper. Bussiness Economics. Retrieved from Name website: http://hdl.handle.net/10016/67

Li, S., & Garrido, J. (2004). On a class of renewal risk models with a constant dividend barrier. *Insurance: Mathematics and Economics, 35*, 691–701.

Li, S., & Wu, B. (2006). *The diffusion perturbed compound Poisson risk model with a dividend barrier.* Preprint. Retrieved from: http://hdl.handle.net/11343/34313

Lin, X. S., & Pavlova, K. P. (2006). The compound Poisson risk model with a threshold dividend strategy. *Insurance: Mathematics and Economics, 38*, 57–80.

Lin, X. S., & Sendova, K. P. (2008). The compound Poisson risk model with multiple thresholds. *Insurance: Mathematics and Economics, 42*, 617–627.

Liu, W., & Hu, Y. (2014). Optimal financing and dividend control of the insurance company with excess-of-loss reinsurance policy. *Statistics & Probability Letters, 84*, 121–130.

Loeffen, R. (2009). An optimal dividends problem with a terminal value for spectrally negative L{\'e}vy processes with a completely monotone jump density. *Journal of Applied Probability, 46*, 85–98.

Loeffen, R. L., & Renaud, J.-F. (2010). De Finetti' s optimal dividends problem with an affine penalty function at ruin. *Insurance: Mathematics and Economics, 46*, 98–108.

Lundberg, F. (1909). Über die theorie der rückversicherung [On the theory of reinsurance]. *Transactions VI International Congress of Actuaries, 1*, 877–948.

Meng, H., Zhang, C., & Wu, R. (2007). The expectation of aggregate discounted dividends for a Sparre Anderson risk process perturbed by diffusion. *Applied Stochastic Models in Business and Industry, 23*, 273–291.

Minkova, L. D. (2010). *Insurance risk theory*. TEMPUS Project SEE doctoral studies in mathematical sciences: Lecture notes.

Mnatsakanov, R., Ruymgaart, L. L., & Ruymgaart, F. H. (2008). Nonparametric estimation of ruin probabilities given a random sample of claims. *Mathematical Methods of Statistics, 17*, 35–43. doi:10.3103/S1066530708010031

Morales, M. (2007). On the expected discounted penalty function for a perturbed risk process driven by a subordinator. *Insurance: Mathematics and Economics, 40*, 293–301.

Ng, A. C. (2009). On a dual model with a dividend threshold. *Insurance: Mathematics and Economics, 44*, 315–324.

Ross, S. M. (2014). *Introduction to probability models*. Amsterdam: Academic Press.

Seal, H. L. (1969). *Stochastic theory of a risk business*. New York, NY: Wiley.

Shimizu, Y. (2009). A new aspect of a risk process and its statistical inference. *Insurance: Mathematics and Economics, 44*, 70–77.

Shimizu, Y. (2012). Non-parametric estimation of the Gerber--Shiu function for the Wiener--Poisson risk model. *Scandinavian Actuarial Journal, 2012*, 56–69.

Shiu, E. S. (1989). The probability of eventual ruin in the compound binomial model. *Astin Bulletin, 19*, 179–190.

Siegl, T., & Tichy, R. F. (1999). A process with stochastic claim frequency and a linear dividend barrier. *Insurance: Mathematics and Economics, 24*, 51–65.

Taksar, M. I. (2000). Optimal risk and dividend distribution control models for an insurance company. *Mathematical Methods of Operations Research, 51*(1), 1–42.

Tan, J., & Yang, X. (2006). The compound binomial model with randomized decisions on paying dividends. *Insurance: Mathematics and Economics, 39*(1), 1–18.

Tsai, C. C.-L., & Willmot, G. E. (2002). A generalized defective renewal equation for the surplus process perturbed by diffusion. *Insurance: Mathematics and Economics, 30*, 51–66.

Wan, N. (2007). Dividend payments with a threshold strategy in the compound Poisson risk model perturbed by diffusion. *Insurance: Mathematics and Economics, 40*, 509–523.

Waters, H. R. (1983). Some mathematical aspects of reinsurance. *Insurance: Mathematics and Economics, 2*, 17–26.

Yang, H., & Zhang, Z. (2008). Gerber--Shiu discounted penalty function in a sparre andersen model with multi-layer dividend strategy. *Insurance: Mathematics and Economics, 42*, 984–991.

Yin, C., & Wen, Y. (2013). Optimal dividend problem with a terminal value for spectrally positive Lévy processes. *Insurance: Mathematics and Economics, 53*, 769–773.

Yuen, K. C., Lu, Y., & Wu, R. (2009). The compound Poisson process perturbed by a diffusion with a threshold dividend strategy. *Applied Stochastic Models in Business and Industry, 25*, 73–93.

Zhang, Z. (2014). On a perturbed Sparre Andersen risk model with threshold dividend strategy and dependence. *Journal of Computational and Applied Mathematics, 255*, 248–269.

Zhao, H., Rong, X., & Zhao, Y. (2013). Optimal excess-of-loss reinsurance and investment problem for an insurer with jump-diffusion risk process under the Heston model. *Insurance: Mathematics and Economics, 53*, 504–514.

8

Knapsack constraint reformulation: A new approach that significantly reduces the number of sub-problems in the branch and bound algorithm

Elias Munapo[1] and Santosh Kumar[2*]

*Corresponding author: Santosh Kumar, Department of Mathematics and Statistics, University of Melbourne, Parkville, Australia
E-mails: skumar@ms.unimelb.edu.au, Santosh.Kumarau@gmail.com.

Reviewing editor: Akiko Yoshise, University of Tsukuba, Japan

Abstract: The paper presents a new approach to significantly reduce the number of sub-problems required to verify optimality in the branch and bound algorithm. The branch and bound algorithm is used to solve linear integer models and these models have application in areas such as scheduling, resource allocation, transportation, facility allocation and capital budgeting. The single constraint of the knapsack linear integer problem (KLIP) is reformulated in such a way that the number of standard branch and bound sub-problems required to verify optimality is significantly reduced. Computational results of the proposed approach on randomly generated KLIPs are also presented.

Subjects: Advanced Mathematics; Discrete Mathematics; Finite Mathematics; Mathematics & Statistics; Operational Mathematics; Science

Keywords: knapsack integer problem; reformulation; branch and bound; computational complexity

Subject classification codes: 90C10; 90C11; 90C27

ABOUT THE AUTHORS

Elias Munapo and Santosh Kumar have interest in discrete optimization methods and they have put forward several ideas that may provide further insight into discrete optimization and linear programming methodology. Their approaches need an appropriate software support, which is a challenge not only for the authors but for the field of optimization. They have made direct contributions to the field of integer programming and special models like quadratic assignment, travelling salesman, network routing by link weight modification. The current paper on knapsack constraint reformulation opens up many interesting questions that can only be answered provided this paper reaches researchers in discrete optimization. We believe, our work challenges software developers to raise funds, support university research for higher degrees. We can provide supervision, if required and hope universities will welcome such a move from the software companies.

PUBLIC INTEREST STATEMENT

Resource constraints often arise in all studies. Mathematical classification is generally based on intrinsic property of the situation. One such problem in optimization has been described by a generic name, "Knapsack Problem", meaning that the knapsack has a finite limited capacity and our interest is to fit in that knapsack as much as possible from the available opportunities. Its applications arise in various areas. A typical knapsack constraint, which deals with only positive numbers, can be written as:

$$a_1x_1 + a_2x_2 + \dots + a_nx_n \leq b$$

Here, $a_1, a_2 + \dots + a_n \geq 0$ are the known amounts of resource that will be used per unit and $x_1, x_2 + \dots + x_n \geq 0$ represent the unknown integers to be determined with some objective to be achieved. This paper reformulates the above constraint and reduces the required computational effort. This reformulation may have applications in other areas, which requires further attention.

1. Introduction

The general linear integer programming (LIP) problem has many applications in real life, for example, they arise in set covering, travelling salesman, assignment, transportation, knapsack, capital budgeting, facility location, timetabling and airline scheduling. Many of these applications have been presented in Taha (2004) and Winston (2004), where these problems have been formulated as a linear integer programming (LIP) model. Because of real-life applications, see Chinneck (2004), the LIP model has attracted so much attention from researchers, yet a consistent and efficient general purpose method has not been developed. We are not aware of any polynomial time algorithm for the general LIP model. In fact, the LIP model is NP complete and a polynomial algorithm for the LIP is believed not to exist. This paper presents a new approach, which takes advantage of the presence of a single constraint to solve a knapsack linear integer problem (KLIP). The problem is reformulated in such a way that the number of standard branch and bound sub-problems required to verify optimality is significantly reduced. Computational results of the proposed approach on randomly generated KLIPs are also presented.

A mathematical statement of the knapsack problem is given in Section 2 and in Section 3, a few problems are discussed where the branch and bound (B & B) performance is poor. We have categorized these problems according to the possible source leading to poor performance. These problems have been subdivided in seven classes. Knapsack constraint reformulation is presented in Section . A summary of the computational experiments on randomly generated problems is presented in Section 5 and finally, the paper has been concluded in Section 6.

2. The knapsack linear integer problem

Maximize or Minimize $Z = c_1x_1 + c_2x_2 + \ldots + c_nx_n$,

such that:

$$a_1x_1 + a_2x_2 + \ldots + a_nx_n \leq \text{ or } \geq b, \tag{1}$$

where a_j, b and c_j are given nonnegative constants, and and $x_j \geq 0$ are integer valued quantities and $j = 1, 2, \ldots, n$.

3. A collection of LIP models where the B & B method performs poorly

3.1. Standard branch and bound algorithm

The B & B algorithm was first proposed by Land and Doig (1960) for solving integer programs. The algorithm was further modified by Dakin (1965) to solve both pure and mixed integer programs. The B & B algorithm in general relies on the usual strategy of first relaxing the integer problem into a linear programming (LP) model. If the linear programming optimal solution is an integer then, the optimal solution to the integer problem has been obtained and the search concludes. If the LP optimal solution is not an integer, then a variable with a fractional value is selected to create two sub-problems such that part of the feasible region is discarded without eliminating any of the feasible integer solutions. The process is repeated on all variables with fractional values until an integer solution is found, see Bealie (1979), Beasley (1996), Mitchell and Lee (2001), Taha (2004) and Winston (2004). In this paper, the standard B & B refers to that version of the B & B algorithm proposed by Dakin (1965) and we assume that there are no state-of-art branching rules, cuts or pricing.

In the following, several problems have been presented where B & B performance is poor.

3.2. Complexity of the standard B & B method with numerical illustrations

The worst case complexity of the B & B algorithm is discussed for the LIP, which is NP Complete. The number of sub-problems can easily reach unmanageable levels even for very small problems. In this section, we present some classes of the LIP models that cause serious challenges for the standard B & B algorithm.

Class 1: Knapsack binary linear problem, taken from Kumar, Munapo, and Jones (2007) .

$$\text{Maximize } Z = \sum_{i=1}^{n-1} x_i, \text{ or Minimize } Z = x_n$$

such that:

$$2\sum_{i=1}^{n-1} x_i \pm x_n = n - 1$$

where $x_j = 0$ or $1 \forall j$ and n is even.

The behaviour of the standard branch and bound method for n = 4, 6, 8, 16 and 40 is given in Table 1.

Class 2: Step pattern formed by negative signs.

The second class is the integer problems that have constraints with negative signs forming a step pattern. Also, the coefficients in the objective function as well as the constants on the right-hand side of the constraints are significantly different. The standard B & B algorithm on this class of problems can behave in a very bizarre way if the branching is not properly managed. The following three numerical illustrations 3.2.1–3.2.2 were taken from Kumar et al. (2007). Numerical illustration 3.2.3 was slightly modified.

Numerical illustration 3.2.1

$$\text{Maximize } Z = 3x_1 + 5x_2 + 7x_3$$

such that:

$$\begin{aligned} &4x_1 + 9x_2 - 8x_3 \leq 81, \\ &5x_1 - 7x_2 + x_3 \leq 42, \\ &-2x_1 + x_2 + 7x_3 \leq 10000, \end{aligned}$$

where $x_1, x_2, x_3 \geq 0$ and integer.

Table 1. Complexity of the problem as n value increases

Value of n in the model	Number of sub-problems created by the standard branch and bound approach to reach the optimum solution
4	11
6	39
8	139
16	25,739
40	Computer was stopped after the sub-problems exceeded 30,000

Note the step formed by the minus signs and the fluctuating values of constants in the right-hand side. Using the standard B & B algorithm, it generated 209 sub-problems to verify optimality. The optimal solution is: $x_1 = 590, x_2 = 544, x_3 = 897$ and $Z = 10,769$.

Numerical illustration 3.2.2

Maximize $Z = 11x_1 + 21x_2 + 17x_3 + 25x_4 + 15x_5$

such that:

$$10x_1 + 20x_2 + 15x_3 + 12x_4 - 3x_5 \leq 789,$$
$$5x_1 + 18x_2 + 21x_3 - 7x_4 + 25x_5 \leq 678,$$
$$12x_1 + 24x_2 - 10x_3 + 19x_4 + 13x_5 \leq 290,$$
$$24x_1 - 8x_2 + 18x_3 + 19x_4 + 13x_5 \leq 1568,$$
$$-15x_1 + 22x_2 + 28x_3 + 16x_4 + 17x_5 \leq 230,$$

where $x_1, x_2, x_3, x_4, x_5 \geq 0$ and integer.

Using the standard B & B algorithm, it generated 605 sub-problems to verify optimality. The optimal solution is: $x_1 = 36, x_2 = 0, x_3 = 24, x_4 = 5, x_5 = 0$ and $Z = 929$.

Numerical illustration 3.2.3

Maximize $Z = 5x_1 + 90x_2 + 12x_3 + 27x_4 + 56x_5 + 56x_6 + 23x_7 + 36x_8 + 8x_9 + 178x_{10}$,

such that:

$$-x_1 + x_2 + x_3 + x_4 + x_5 + x_6 + x_7 + x_8 + x_9 + x_{10} \leq 95,$$
$$x_1 - 6x_2 + x_3 + x_4 + 56x_5 + x_6 + x_7 + x_8 + x_9 + x_{10} \leq 5679,$$
$$x_1 + x_2 - 5x_3 + x_4 + x_5 + x_6 + 16x_7 + 20x_8 + x_9 + x_{10} \leq 1990,$$
$$5x_1 + x_2 + x_3 - 8x_4 + x_5 + x_6 + x_7 + x_8 + x_9 + x_{10} \leq 450,$$
$$x_1 + 19x_2 + x_3 + x_4 - 166x_5 + 3x_6 + x_7 + x_8 + x_9 + x_{10} \leq 670,$$
$$x_1 + x_2 + x_3 + x_4 + x_5 - 12x_6 + x_7 + x_8 + x_9 + x_{10} \leq 80,$$
$$x_1 + x_2 + x_3 + x_4 + 90x_5 + 8x_6 - x_7 + x_8 + 7x_9 + x_{10} \leq 8887,$$
$$x_1 + x_2 + 34x_3 + 5x_4 + x_5 + x_6 + x_7 - 9x_8 + x_9 + x_{10} \leq 68,$$
$$x_1 + x_2 + x_3 + 0x_4 + x_5 + 81x_6 + x_7 + x_8 - x_9 + 25x_{10} \leq 350,$$
$$23x_1 + x_2 + x_3 + x_4 + x_5 + 5x_6 + x_7 + x_8 + x_9 - 10x_{10} \leq 523,$$

where $x_1, x_2, \ldots, x_{10} \geq 0$ and integer.

The standard B & B requires 2101 sub-problems to verify optimality, which is given by: $x_1 = 6, x_2 = 87, x_5 = 7, x_6 = 2, x_8 = 1, x_{10} = 3, x_3 = x_4 = x_7 = x_9 = 0$ and $Z = 8934$.

The other classes of linear integer problems that make the B & B an unreliable method are presented in Classes 3 to 7.

Class 3: Complementary constraints.

Minimize $Z = x_1$

such that:

$$a_1x_1 + a_2x_2 + \ldots + a_jx_j + \ldots + a_nx_n \geq b \text{ and}$$
$$-(a_1x_1 + a_2x_2 + \ldots + a_jx_j + \ldots + a_nx_n) \geq b,$$

where $x_j \geq 0$ and integer, a_j and b are integers and constants, $j = 1, 2, \ldots, n$ and $-\bar{a}_j < a_j \; \forall j \neq 1$. Also note that the first constraint in this class of problem contains both positive and negative coefficients. The two constraints are complementary in the sense that when added give a zero on the left-hand side. More examples are given as Class 3.1 and Class 3.2.

Class 3.1

Minimize $Z = x_1$,

such that:

$$3x_1 + 6x_2 - 8x_3 \geq 100,$$
$$-3x_1 - 6x_2 + 8x_3 \geq 100$$

where $x_j \geq 0 \; \forall j$ and integer.

Class 3.2

Minimize $Z = x_1$

such that:

$$4x_1 + 2x_2 - 9x_3 + 7x_4 - 5x_5 + 12x_6 \geq 210$$
$$-4x_1 - 2x_2 + 9x_3 - 7x_4 + 5x_5 - 12x_6 \geq 210$$

where $x_j \geq 0 \; \forall j$ and integer.

Class 4: Binary knapsack problem.

This is an extension of class 1. The problem becomes more difficult for the standard B & B algorithm if it is slightly modified as given in below.

Maximize $Z = \sum_{j=1}^{n-1} x_j$

such that:

$$2\sum_{j=1}^{n-1} x_j \pm \kappa x_n = n - 1$$

where $x_j = 0$ or $1 \; \forall j, \kappa \leq n - 1, \kappa$ and n are even.

Class 4 has an alternate form also as was the case in Class 1. This alternate form will have the objective function as given below and constraints remain unchanged. This alternative objective is given by: Minimize $Z = x_n$.

For the Class 4 problem, the behaviour of the B & B method for n = 4, 6, 8 and 16 becomes worse as given in Table 2.

Class 5: Knapsack problem: A pure integer case.

Mere changing of variables from binary to pure integer makes any LIP worse for the standard B & B algorithm. For example, consider:

Table 2. Complexity of the problem as n value increases	
Value of n in the model	**Number of sub-problems created by the branch and bound approach to reach the optimum**
4	$\kappa = 3$, Sub-problems =17
6	$\kappa = 3$, Sub-problems =59
	$\kappa = 5$, Sub-problems =59
8	$\kappa = 3$, Sub-problems =209
	$\kappa = 5$, Sub-problems =209
	$\kappa = 7$, Sub-problems =209
16	Computer was stopped after the number of sub-problems exceeded 30,000

$$\text{Maximize } Z = \sum_{j=1}^{n-1} x_j$$

such that:

$$2\sum_{j=1}^{n-1} x_j + \kappa x_n = n - 1$$

where $x_j \geq 0$ and integer $\forall j, 1 \leq \kappa \leq n-1, \kappa$ is odd and n is even.

Once again, an alternate problem for class 5 is when objective function is changed but constraints remain unchanged. This alternative objective function is given by:

$$\text{Minimize } Z = x_n.$$

The computational behaviour of the standard branch and bound method for n = 4, 6, 8 and 16 is given in Table 3.

Class 6: Hard Knapsack Problems.

$$\text{Maximize } Z = \sum_{j=1}^{n-1} x_j$$

Table 3. Complexity of the problem as n value increases	
Value of n in the model	**Number of sub-problems created by the branch and bound approach to reach the optimum solution**
4	$\kappa = 3$, Sub-problems =23
6	$\kappa = 3$, Sub-problems =129
	$\kappa = 5$, Sub-problems =129
8	$\kappa = 3$, Sub-problems =755
	$\kappa = 5$, Sub-problems =755
	$\kappa = 7$, Sub-problems =755
16	Computer was stopped when the number of sub-problems exceeded 30,000

such that:

$$2\sum_{j=1}^{n-1} x_j + \kappa x_n = \lambda,$$

where $x_j \geq 0$ and integer $\forall j, 2(n-1) + \kappa = \lambda$, and $\kappa, \lambda \geq 0$ are odd and n even. Once again, an alternate problem can be for a minimizing objective function given by: Minimize: $Z = x_n$.

The standard B & B method cannot solve most of these problems for large values of κ. For example, a knapsack problem with the parameters: $n = 4, \kappa = 91$, and $\lambda = 97$ becomes

Minimize $Z = x_4$,

such that:

$$2x_1 + 2x_2 + 2x_3 + 91x_4 = 97$$

where $x_j \geq 0$ and integer $\forall j$.

The standard B & B method requires 7,449 sub-problems to verify the optimal solution. For large values of λ, the knapsack problems cannot be solved by standard B & B algorithm on its own.

Class 7: Hard pure integer models.

The following general integer model is also very difficult to solve by the standard B & B algorithm on its own.

Minimize $Z = \omega_1 x_1 + \omega_2 x_2 + \ldots + \omega_n x_n$

such that :

$$\begin{aligned} &\alpha_{11}x_1 + \alpha_{12}x_2 + \ldots + \alpha_{1n}x_n \geq \beta_1 \\ &\alpha_{21}x_1 + \alpha_{22}x_2 + \ldots + \alpha_{2n}x_n \geq \beta_2 \\ &\vdots \\ &\alpha_{1m}x_1 + \alpha_{m2}x_2 + \ldots + \alpha_{mn}x_n \geq \beta_m \end{aligned}$$

where $\forall j, \omega_1 = 1$ or some large number. There must be one large number in the objective row and $\forall ij, a_{ij} = 2$ or some large odd number. There must be one large odd number in every row and in every column. Further, it must satisfy $\alpha_{i1} + \alpha_{i2} + \ldots + \alpha_{in} < \beta_i \forall i$.

A numerical illustration of the above model

Minimize $Z = x_1 + 141x_2 + x_3$

such that:

$$\begin{aligned} &55x_1 + 2x_2 + 2x_3 \geq 73 \\ &2x_1 + 91x_2 + 2x_3 \geq 97 \\ &2x_1 + 2x_2 + 85x_3 \geq 99 \end{aligned}$$

where $x_1, x_2, x_3 \geq 0$ and integer.

A total of 653 standard B & B sub-problems are required to verify optimality. The optimal solution is: $x_1 = 45, x_2 = 0, x_3 = 4$ and $Z = 49$.

There are so many other classes of difficult integer models that occur in real life. These include the travelling salesman problem (TSP) and the generalized assignment problem (GAP). At the moment, it may be very difficult to come up with an efficient general purpose algorithm. Instead, it makes sense to study one class and then propose an efficient way of solving it. In this paper, the KLIP is targeted because of its special features.

4. Reformulation of the knapsack problem

The knapsack problem has special features that we can take advantage of for reformulation. This problem has only one constraint and all coefficients are nonnegative. The coefficients in the constraint of any KLIP can be arranged in ascending order i.e.

$$a_1x_1 + a_2x_2 + \dots + a_nx_n \text{ where } a_1 \le a_2 \le \dots \le a_n. \tag{2}$$

This is possible for all KLIP models since all coefficients are nonnegative, and since a_1 is the smallest one can rewrite the expression (2) as follows:

$$a_1x_1 + a_1x_2 + \dots + a_1x_n + (a_2 - a_1)x_2 + (a_3 - a_1)x_3 + \dots + (a_n - a_1)x_n$$

Let $a_j^1 = a_j - a_1$ for $j = 2, 3, \dots, n$ then (2) can be expressed as given by (3).

$$(a_1x_1 + a_1x_2 + \dots + a_1x_n) + a_2^1x_2 + a_2^1x_3 \dots + a_2^1x_n, \text{ where } a_2^1 \le a_3^1 \dots \le a_n^1 \tag{3}$$

Repeating the process given in (3), where smallest coefficient is a_2^1, we obtain

$$\begin{aligned}(a_1x_1 + a_1x_2 + \dots + a_1x_n) + a_2^1x_2 + a_2^1x_3 \dots + a_2^1x_n \\ + (a_3^1 - a_2^1)x_3 + (a_4^1 - a_2^1)x_4 + \dots + (a_n^1 - a_2^1)x_n, \\ \text{where } a_2^1 \le a_3^1 \dots \le a_n^1\end{aligned} \tag{4}$$

If the process repeated until there is no term left, the transformed constraint will become as shown in (5).

$$\begin{aligned}(a_1x_1 + a_1x_2 + \dots + a_1x_n) + a_2^1x_2 + a_2^1x_3 \dots + a_2^1x_n \\ + a_3^2x_3 + a_3^2x_4 + \dots + a_3^2x_n + \dots + a_n^kx_n,\end{aligned} \tag{5}$$

$$\begin{aligned}a_1x_1 + a_1x_2 + \dots + a_1x_n &= a_1y_1 \\ a_2^1x_2 + a_2^1x_3 \dots + a_2^1x_n &= a_2^1y_2 \\ a_3^2x_3 + a_3^2x_4 + \dots + a_3^2x_n &= a_3^2y_3 \\ \dots \\ a_n^kx_n &= a_n^ky_k\end{aligned} \tag{6}$$

Note, y_j are integer valued quantities and $j = 1, 2, \dots, k$

The KLIP becomes

$$\text{Maximize or Minimize } Z = c_1x_1 + c_2x_2 + \dots + c_nx_n$$

such that:

$$\begin{aligned}a_1x + a_1x_2 + \dots + a_1x_n &= a_1y_1 \\ a_2^1x_2 + a_2^1x_3 \dots + a_2^1x_n &= a_2^1y_2 \\ a_3^2x_3 + a_3^2x_4 + \dots + a_3^2x_n &= a_3^2y_3 \\ \dots \\ a_n^kx_n &= a_n^ky_k \\ a_1y_1 + a_2^1y_2 + \dots + a_n^ky_n &\le \text{or} \ge b\end{aligned} \tag{7}$$

Once again, y_j are integer value quantities and $j = 1, 2, \ldots, k$.

Phase 1: Solve reformulated problem (7) using the standard B & B method with the integral restriction on y_i only. If solution is an integer, then it is also optimal to the original KLIP, else go to Phase 2.

Phase 2: Continue with the standard B & B method used in Phase 1 but this time with the integral restriction extended to x_j also.

This reformulated model is easier to solve than the original KLIP. The standard B & B method takes a significantly smaller number of sub-problems to verify optimality on the reformulated model than the original KLIP. The proposed approach works better if $a_1, a_2^1, a_3^2, \ldots a_n^k$ are different.

Numerical Illustrations 4.1

Minimize $Z = 20x_1 + 8x_2 + 3x_3 + 5x_4 + 33x_5$,

such that:

$$29x_1 + 20x_2 + 18x_3 + 24x_4 + 12x_5 \geq 679 \tag{8}$$

where $x_j \geq 0$ and integer $\forall j$.

The standard B & B method takes 231 sub-problems to verify the optimal solution of (8), which is given by: $x_1 = x_2 = 0, x_3 = 38, x_4 = x_5 = 0$ and $Z = 114$.

Note the coefficients, when arranged in an increasing order are given by 12, 18, 20, 24 and 29. Thus, we have $a_1 = 12, a_2^1 = 18 - 12 = 6, a_3^2 = 20 - 12 - 6 = 2, a_4^3 = 24 - 12 - 6 - 2 = 4$ and $a_5^4 = 29 - 12 - 6 - 2 - 4 = 5$. After the reformulation, the KILP (8) becomes (9).

Minimize $Z = 20x_1 + 8x_2 + 3x_3 + 5x_4 + 33x_5$,

Such that :

$$\begin{aligned}
&12y_1 + 6y_2 + 2y_3 + 4y_4 + 5y_5 \geq 679\\
&12(x_1 + x_2 + x_3 + x_4 + x_5) = 12y_1\\
&6(x_1 + x_2 + x_3 + x_4) = 6y_2\\
&2(x_1 + x_2 + x_4) = 2y_3\\
&4(x_1 + x_4) = 4y_4\\
&5x_1 = 5y_5
\end{aligned} \tag{9}$$

Note that the coefficients 12, 6, 2, 4 and 5 are different.

Phase 1: Where $y_j \geq 0$ and integer $\forall j$. In phase 1, there is no integral restriction on the variables x_j. The number of sub-problems necessary to verify the optimal solution reduces to 37, resulting once again in the solution given by: $x_3 = y_1 = y_2 = 38, x_1 = x_2 = x_4 = x_5 = y_3 = y_4 = y_5 = 0$ and $Z = 114$.

Note that Phase 2 was not necessary since Phase 1 solution was optimal and the number of sub-problems required to reach the optimal solution dropped from 231 to 37, which is a significant reduction. This shows that the reformulation is effective for the KLIP. Here are two more illustrations before we present a summary of computational experiments.

Numerical illustration 4.2

Table 4. Computational experiments on randomly generated Knapsack problems

KLIP S No.	No. of Variables	No. of sub-problems before	No. of sub problems after	% Reduction in sub-problems
1	5	25	4	84.0
2	10	41	7	82.9
3	20	158	29	81.6
4	30	132	12	90.9
5	40	89	9	80.0
6	50	*	32	-
7	60	289	18	93.7
8	70	6590	47	99.3
9	80	2198	29	98.7
10	90	1789	15	99.2
11	100	691	106	84.7
12	150	11209	89	99.2
13	200	721	19	97.4
14	250	*	51	-
15	300	2187	71	96.8
16	350	6842	29	99.6
17	400	895	67	92.5
18	450	17866	27	99.8
19	500	*	105	-
20	550	96	11	88.5
21	600	14	3	78.6
22	650	871	69	92.1
23	700	956	17	98.2
24	750	4008	189	95.3
25	800	38	2	94.7
26	850	2983	63	97.9
27	900	345	22	93.6
28	950	27689	58	99.8
29	1000	116	26	77.6
30	1500	19	4	78.9
31	2000	4578	23	99.5
32	2500	89	13	85.4
33	3000	8104	269	96.7
34	3500	978	47	95.2
35	4000	67	15	77.6
36	4500	*	44	-
37	5000	24672	56	99.8
38	6000	*	391	-
39	6500	897	35	96.1
40	7000	1946	219	88.7

*Number of sub-problems exceeded 30,000.

Table 5. More computational experiments on randomly generated Knapsack problems

KLIP S No.	No. of variables	No. of sub-problems before	No. of sub problems after	% Reduction in sub-problems
41	7500	*	456	-
42	8000	15008	375	97.5
43	8500	6895	132	98.1
44	9000	*	249	-
45	9500	1193	83	93.1
46	10000	5829	683	88.3
47	10500	*	43	-
48	12000	*	102	-
49	13000	28972	219	99.2
50	15000	*	147	-

*Number of sub-problems exceeded 30,000.

Minimize $Z = x_4$,

Such that :

$$2x_1 + 2x_2 + 2x_3 + 91x_4 = 97 \tag{10}$$

where $x_j \geq 0$ and integer $\forall j$.

This KLIP is a special case from Class 6. The number of sub-problems reduces from 7449 to only 5 after reformulation. Optimal solution was obtained in Phase 1.

Numerical illustration 4.3

Minimize $Z = x_n$,

such that:

$$2\sum_{i=1}^{n-1} x_i \pm x_n = n - 1, \tag{11}$$

where $x_j \geq 0$ and integer $\forall j$ and $n = 16$.

This KLIP comes from Class 1 but in this case, variables are integer and not necessarily binary. The number of sub-problems reduces from over 30, 000 to only 11 after reformulation and the optimal solution was obtained in Phase 1.

5. Computational experiments

Fifty randomly generated knapsack problems of different sizes were used in the analysis. The objective of the computational experiments was to determine whether the number of sub-problems decrease after the reformulation. The computational results were tabulated as given in Tables 4 and 5. MATLAB R2013 (version 8.2) running on an Intel Pentium Dual desktop (Dual core G2020 2.9 GHz CPU, 2GB DDR3 1333 RAM) was used for the computational experiments. In all the fifty cases, it was observed that the number of standard B & B sub-problems decreased significantly after reformulation, as can be seen from the results in Tables 4 and 5.

6. Conclusions

What has emerged from the computational experiments is that the number of standard B & B sub-problems required to verify optimality can be significantly reduced by reformulation. In all the fifty cases analysed, the number of sub-problems were significantly reduced after reformulation. So many improvements have been done on the branch and bound algorithm in terms of addition of cuts to get the branch and cut algorithm (Brunetta, Conforti, & Rinaldi, 1997; Mitchell, 2001; Padberg & Rinaldi, 1991), pricing to get the branch and price algorithm (Barnhart, Johnson, Nemhauser, Savelsbergh, & Vance, 1998; Salvelsbergh, 1997) and also combining the two improved versions to get the branch cut and price hybrid algorithm (Barnhart, Hane, & Vance, 2010; Fukasawa et al., 2006; Ladanyi, Ralphs, & Trotter, 2001). Here, in this paper, we have we have achieved improvement in the standard B & B algorithm on its own before combining it with other approaches. In addition to using cuts and pricing within the context of a branch and bound algorithm, preprocessing given in Savelsbergh (1994) can reduce the number of sub-problems needed to verify optimality. Reformulation proposed in this paper significantly reduces the number of B & B sub-problems required to verify optimality.

In subsequent publications, attempt will be made to:

(1) Extend the proposed constraint reformulation to the general LIP model, when applicable.

(2) Search for the mathematical reasons that give rise to the efficiency observed.

(3) Extend the reformaulation concept to other situations.

Acknowledgements
Authors are grateful to the referees for their constructive suggestions for improving this paper.

Funding
The authors received no direct funding for this research.

Author details
Elias Munapo[1]
E-mail: emunapo@gmail.com
Santosh Kumar[2]
E-mails: skumar@ms.unimelb.edu.au, Santosh.Kumarau@gmail.com
ORCID ID: http://orcid.org/0000-0001-9321-5622
[1] Graduate School of Business and Leadership, University of KwaZulu-Natal, Westville Campus, Durban, South Africa.
[2] Department of Mathematics and Statistics, University of Melbourne, Parkville, Australia.

References

Barnhart, C., Johnson, E. L., Nemhauser, G. L., Savelsbergh, M. W. P., & Vance, P. H. (1998). Branch and price column generation for solving huge integer programs. *Operations Research, 46*, 316–329.

Barnhart, C., Hane, C. A., & Vance, P. H. (2010). Using branch-and-price-and-cut to solve origin--destination integer multicommodity flow problems. *Operations Research, 48*, 318–326.

Bealie, E. M. L. (1979). Branch and bound methods for mathematical programming systems. *Annals of Discrete Mathematics, 5*, 201–219.

Beasley, J. E. (Ed.). (1996). *Advances in linear and integer programming*. New York, NY: Oxford University Press.

Benders, J. F. (1962). Partitioning procedures for solving mixed variables programming problems. *Numerische Mathematik, 4*, 238–252.

Brunetta, L., Conforti, M., & Rinaldi, G. (1997). A branch and cut algorithm for the equicut problem. *Mathematical Programming, 78*, 243–263.

Chinneck, J. W. (2004). *Practical optimization: A gentle introduction lecture notes*. Retrieved from http://www.sce.carleton.ca/faculty/chinneck/po/Chapter13.pdf

Dakin, R. J. (1965). A tree search algorithm for mixed integer programming problems. *The Computer Journal, 8*, 250–255.

Fukasawa, R., Longo, H., Lysgaard, J., Poggi de Aragao, M., Uchoa, E., & Werneck, R. F. (2006). Robust branch-and-cut-price for the capacitated vehicle routing problem. *Mathematical Programming Series A, 106*, 491–511.

Kumar, S., Munapo, E., & Jones, B. C. (2007). An integer equation controlled descending path to a protean pure integer program. *Indian Journal of Mathematics, 49*, 211–237.

Land, A. H., & Doig, A. G. (1960). An automatic method for solving discrete programming problems. *Econometrica, 28*, 497–520.

Ladanyi, L., Ralphs, T. K., & Trotter, L. E. (2001). Branch, cut and price: Sequential and parallel. In N. Naddef & M. Jenger (Eds.), *Computational combinatorial optimization* (pp. 223–260). Berlin: Springer.

Mitchell, J. E., & Lee, E. K. (2001). Branch and bound methods for integer programming. In C. A. Floudas & P. M. Pardalos (Eds.), *Encyclopedia of optimization* (pp. 329–347). Dordrecht: Kluwer Academic.

Mitchell, J. E. (2001). Branch and cut algorithms for integer programming. In C. A. Floudas & P. M. Pardalos (Eds.), *Encyclopedia of optimization* (pp. 295–309). Dordrecht: Kluwer Academic.

Padberg, M., & Rinaldi, G. (1991). A branch and cut algorithm for the resolution of large-scale symmetric traveling salesman problems. *SIAM Review, 33*, 60–100.

Savelsbergh, M. W. P. (1994). Preprocessing and probing techniques for mixed integer programming problems. *ORSA Journal on Computing, 6*, 445–454.

Salvelsbergh, M. W. P. (1997). A branch and price algorithm to solve the generalized assignment problem. *Operations Research, 45*, 381–841.

Taha, H. A. (2004). *Operations research: An introduction*. (7th ed.). Upper Saddle River, NJ: Pearson Educators.

Winston, W. L. (2004). *Operations research applications and algorithms*. (4th ed.). Belmont, CA: Duxbury Press.

9

Some new Hermite–Hadamard type inequalities for differentiable co-ordinated convex functions

Xu-Yang Guo[1], Feng Qi[2,3]* and Bo-Yan Xi[1]

*Corresponding author: Feng Qi, Department of Mathematics, College of Science, Tianjin Polytechnic University, Tianjin City 300387, China; Institute of Mathematics, Henan Polytechnic University, Jiaozuo City, Henan Province 454010, China
E-mail: qifeng618@gmail.com

Reviewing editor: Igor Boglaev, Massey University, New Zealand

Abstract: In the paper, the authors establish some new Hermite–Hadamard type inequalities for differentiable co-ordinated convex functions of two variables.

Subjects: Advanced Mathematics; Analysis - Mathematics; Mathematical Analysis; Mathematics & Statistics; Real Functions; Science; Special Functions

Keywords: Hermite–Hadamard type inequality; differentiable co-ordinated convex functions

1. Introduction

The following definitions are well known in the literature.

Definition 1.1. A function $f{:}I \subseteq \mathbb{R} = (-\infty, +\infty) \to \mathbb{R}$, if

$$f(\lambda x + (1-\lambda)y) \le \lambda f(x) + (1-\lambda) f(y)$$

is valid for all $x, y \in I$ and $\lambda \in [0,1]$, then we say that f is a convex function on I.

Many important inequalities have been established for the class of convex functions, but the most famous is the Hermite–Hadamard inequality (see for instance Pečarić, Proschan, & Tong, 1991). This double inequality is stated as

$$f\left(\frac{a+b}{2}\right) \le \frac{1}{b-a}\int_a^b f(x)\mathrm{d}x \le \frac{f(a)+f(b)}{2},$$

where $f{:}I \subseteq \mathbb{R} \to \mathbb{R}$ a convex function, $a, b \in I$ with $a < b$.

A modification for convex functions on Δ, which are also known as co-ordinated convex functions, was introduced by Dragomir (2001) and Dragomir and Pearce (2000) as follows.

Definition 1.2. A function $f{:}\ \Delta \to \mathbb{R}$ is said to be convex on the co-ordinates on $\Delta = [a,b] \times [c,d] \subseteq \mathbb{R}^2$ with $a < b$ and $c < d$ if the partial mappings

ABOUT THE AUTHOR

Xu-Yang Guo is being a graduate for master degree of science in applied mathematics at Inner Mongolia University for Nationalities. Her supervisor is the third author, Professor Bo-Yan Xi. Currently, her research interests are in the areas of mathematical inequalities and convex analysis.

PUBLIC INTEREST STATEMENT

In the paper, the authors establish some new Hermite–Hadamard type inequalities for differentiable co-ordinated convex functions of two variables.

$f_y: [a,b] \to \mathbb{R}, f_y(u) = f_y(u,y)$ and $f_x: [c,d] \to \mathbb{R}, f_x(v) = f_x(x,v)$

are convex where defined for all $(x, y) \in \Delta$.

A formal definition for co-ordinated convex functions may be stated as follows.

Definition 1.3. A function $f: \Delta \to \mathbb{R}$ is said to be convex on the co-ordinates on $\Delta = [a, b] \times [c, d] \subseteq \mathbb{R}^2$ with $a < b$ and $c < d$ if

$$f(tx + (1-t)z, \lambda y + (1-\lambda)w) \le t\lambda f(x,y) + t(1-\lambda)f(x,w) + (1-t)\lambda f(z,y) + (1-t)(1-\lambda)f(z,w)$$

for all $t, \lambda \in [0, 1], (x,y), (z, w) \in \Delta$.

The following Hermite-Hadamard type inequality for co-ordinated convex functions on the rectangle form the plane $\mathbb{R}^2$ was also proved in Dragomir (2001).

THEOREM 1.1. (Dragomir, 2001) *Let $f: \Delta = [a,b] \times [c,d] \subseteq \mathbb{R}^2 \to \mathbb{R}$ be convex on the co-ordinates on Δ. Then*

$$\begin{aligned} f\left(\frac{a+b}{2}, \frac{c+d}{2}\right) &\le \frac{1}{2}\left[\frac{1}{b-a}\int_a^b f\left(x, \frac{c+d}{2}\right)dx + \frac{1}{d-c}\int_c^d f\left(\frac{a+b}{2}, y\right)dy\right] \\ &\le \frac{1}{(b-a)(d-c)}\int_a^b\int_c^d f(x,y)dydx \le \frac{1}{4}\left[\frac{1}{b-a}\left(\int_a^b f(x,c)dx + \int_a^b f(x,d)dx\right)\right. \\ &\left. + \frac{1}{d-c}\left(\int_c^d f(a,y)dy + \int_c^d f(b,y)dy\right)\right] \le \frac{1}{4}[f(a,c) + f(b,c) + f(a,d) + f(b,d)]. \end{aligned}$$

THEOREM 1.2. (Ozdemir, Akdemir, Kavurmaci, & Avci, 2011) *Let $f:\Delta = [a,b] \times [c,d] \subseteq \mathbb{R}^2 \to \mathbb{R}$ be a partial differentiable function on Δ. If $\left|\frac{\partial^2 f}{\partial x \partial y}\right|$ is convex on the co-ordinates on Δ, then*

$$\begin{aligned} &\left|\frac{1}{9}\left[f\left(a, \frac{c+d}{2}\right) + f\left(b, \frac{c+d}{2}\right) + 4f\left(\frac{a+b}{2}, \frac{c+d}{2}\right) + f\left(\frac{a+b}{2}, c\right) + f\left(\frac{a+b}{2}, d\right)\right]\right. \\ &\left. + \frac{1}{36}\{f(a,c) + f(a,d) + f(b,c) + f(b,d)\} + \frac{1}{(b-a)(d-c)}\int_a^b\int_c^d f(x,y)dxdy - A\right| \\ &\le \left(\frac{5}{72}\right)^2 (b-a)(d-c)\left\{\left|\frac{\partial^2}{\partial t \partial \lambda}f(a,c)\right| + \left|\frac{\partial^2}{\partial t \partial \lambda}f(a,d)\right| + \left|\frac{\partial^2}{\partial t \partial \lambda}f(b,c)\right| + \left|\frac{\partial^2}{\partial t \partial \lambda}f(b,d)\right|\right\}, \end{aligned}$$

where

$$A = \frac{1}{b-a}\int_a^b\left[\frac{f(x,c) + 4f\left(x, \frac{c+d}{2}\right) + f(x,d)}{6}\right]dx + \frac{1}{d-c}\int_c^d\left[\frac{f(a,y) + 4f\left(\frac{a+b}{2}, y\right) + f(b,y)}{6}\right]dy.$$

THEOREM 1.3. (Latif & Dragomir, 2012) *Let $f:\Delta = [a,b] \times [c,d] \subseteq \mathbb{R}^2 \to \mathbb{R}$ be a partial differentiable function on Δ. If $\left|\frac{\partial^2 f}{\partial x \partial y}\right|$ is convex on the co-ordinates on Δ, then*

$$\begin{aligned} &\left|\frac{1}{(b-a)(d-c)}\int_a^b\int_c^d f(x,y)dxdy + f\left(\frac{a+b}{2}, \frac{c+d}{2}\right) - A\right| \\ &\le \frac{(b-a)(d-c)}{64}\left\{\left|\frac{\partial^2}{\partial x \partial y}f(a,c)\right| + \left|\frac{\partial^2}{\partial x \partial y}f(a,d)\right| + \left|\frac{\partial^2}{\partial x \partial y}f(b,c)\right| + \left|\frac{\partial^2}{\partial x \partial y}f(b,d)\right|\right\}, \end{aligned}$$

where

$$A = \frac{1}{b-a}\int_a^b f\left(x, \frac{c+d}{2}\right)dx + \frac{1}{d-c}\int_c^d f\left(\frac{a+b}{2}, y\right)dy.$$

2. Main results

The following lemma is necessary and plays an important role in establishing our main results:

LEMMA 2.1. *Let $f:\Omega \subseteq \mathbb{R}^2 \to \mathbb{R}$ be a twice partial differentiable mapping on Ω° (the interior of Ω) and let $\Delta := [a, b] \times [c, d] \subseteq \Omega^\circ$ with $a < b$ and $c < d$. If $\frac{\partial^2 f}{\partial x \partial y} \in L_1(\Delta)$, where $L_1(\Delta)$ denotes the set of all Lebesgue integrable functions on Δ, then*

$$
\begin{aligned}
I(f) :&= \frac{16}{(d-c)(b-a)}\left[f\left(\frac{a+b}{2},\frac{c+d}{2}\right) - \frac{1}{d-c}\int_c^d f\left(\frac{a+b}{2},y\right)dy - \frac{1}{b-a}\int_a^b f\left(x,\frac{c+d}{2}\right)dx\right.\\
&\left.+\frac{1}{(d-c)(b-a)}\int_c^d\int_a^b f(x,y)dxdy\right] = \int_0^1\int_0^1 t\lambda\frac{\partial^2}{\partial x\partial y}f\left(\frac{t}{2}a+\left(1-\frac{t}{2}\right)b,\frac{\lambda}{2}c+\left(1-\frac{\lambda}{2}\right)d\right)dtd\lambda\\
&+\int_0^1\int_0^1 t\lambda\frac{\partial^2}{\partial x\partial y}f\left(\left(1-\frac{t}{2}\right)a+\frac{t}{2}b,\left(1-\frac{\lambda}{2}\right)c+\frac{\lambda}{2}d\right)dtd\lambda\\
&-\int_0^1\int_0^1 t\lambda\frac{\partial^2}{\partial x\partial y}f\left(\frac{t}{2}a+\left(1-\frac{t}{2}\right)b,\left(1-\frac{\lambda}{2}\right)c+\frac{\lambda}{2}d\right)dtd\lambda - \int_0^1\int_0^1 t\lambda\frac{\partial^2}{\partial x\partial y}f\\
&\left(\left(1-\frac{t}{2}\right)a+\frac{t}{2}b,\frac{\lambda}{2}c+\left(1-\frac{\lambda}{2}\right)d\right)dtd\lambda.
\end{aligned}
$$

Proof By integration by parts, we have

$$
\begin{aligned}
&\int_0^1\int_0^1 t\lambda\frac{\partial^2}{\partial x\partial y}f\left(\frac{t}{2}a+\left(1-\frac{t}{2}\right)b,\frac{\lambda}{2}c+\left(1-\frac{\lambda}{2}\right)d\right)dtd\lambda = \frac{4}{(b-a)(d-c)}\left[f\left(\frac{a+b}{2},\frac{c+d}{2}\right)\right.\\
&-\int_0^1 f\left(\frac{a+b}{2},\frac{\lambda}{2}c+\left(1-\frac{\lambda}{2}\right)d\right)d\lambda - \int_0^1 f\left(\frac{t}{2}a+\left(1-\frac{t}{2}\right)b,\frac{c+d}{2}\right)dt\\
&\left.+\int_0^1\int_0^1 f\left(\frac{t}{2}a+\left(1-\frac{t}{2}\right)b,\frac{\lambda}{2}c+\left(1-\frac{\lambda}{2}\right)d\right)dtd\lambda\right]\\
&=\frac{4}{(b-a)(d-c)}\left[f\left(\frac{a+b}{2},\frac{c+d}{2}\right) - \frac{2}{d-c}\int_{\frac{c+d}{2}}^d f\left(\frac{a+b}{2},y\right)dy\right.\\
&\left.-\frac{2}{b-a}\int_{\frac{a+b}{2}}^b f\left(x,\frac{c+d}{2}\right)dx + \frac{4}{(b-a)(d-c)}\int_{\frac{c+d}{2}}^d\int_{\frac{a+b}{2}}^b f(x,y)dxdy\right].
\end{aligned}
$$

Similarly, we have

$$
\begin{aligned}
&\int_0^1\int_0^1 t\lambda\frac{\partial^2}{\partial x\partial y}f\left(\left(1-\frac{t}{2}\right)a+\frac{t}{2}b,\left(1-\frac{\lambda}{2}\right)c+\frac{\lambda}{2}d\right)dtd\lambda = \frac{4}{(b-a)(d-c)}\left[f\left(\frac{a+b}{2},\frac{c+d}{2}\right)\right.\\
&\left.-\frac{2}{d-c}\int_c^{\frac{c+d}{2}} f\left(\frac{a+b}{2},y\right)dy - \frac{2}{b-a}\int_a^{\frac{a+b}{2}} f\left(x,\frac{c+d}{2}\right)dx + \frac{4}{(b-a)(d-c)}\int_c^{\frac{c+d}{2}}\int_a^{\frac{a+b}{2}} f(x,y)dxdy\right],
\end{aligned}
$$

$$
\begin{aligned}
&\int_0^1\int_0^1 t\lambda\frac{\partial^2}{\partial x\partial y}f\left(\frac{t}{2}a+\left(1-\frac{t}{2}\right)b,\left(1-\frac{\lambda}{2}\right)c+\frac{\lambda}{2}d\right)dtd\lambda = -\frac{4}{(b-a)(d-c)}\left[f\left(\frac{a+b}{2},\frac{c+d}{2}\right)\right.\\
&\left.-\frac{2}{d-c}\int_c^{\frac{c+d}{2}} f\left(\frac{a+b}{2},y\right)dy - \frac{2}{b-a}\int_{\frac{a+b}{2}}^b f\left(x,\frac{c+d}{2}\right)dx + \frac{4}{(b-a)(d-c)}\int_c^{\frac{c+d}{2}}\int_{\frac{a+b}{2}}^b f(x,y)dxdy\right],
\end{aligned}
$$

and

$$
\begin{aligned}
&\int_0^1\int_0^1 t\lambda\frac{\partial^2}{\partial x\partial y}f\left(\left(1-\frac{t}{2}\right)a+\frac{t}{2}b,\frac{\lambda}{2}c+\left(1-\frac{\lambda}{2}\right)d\right)dtd\lambda = -\frac{4}{(b-a)(d-c)}\left[f\left(\frac{a+b}{2},\frac{c+d}{2}\right)\right.\\
&\left.-\frac{2}{d-c}\int_{\frac{c+d}{2}}^d f\left(\frac{a+b}{2},y\right)dy - \frac{2}{b-a}\int_a^{\frac{a+b}{2}} f\left(x,\frac{c+d}{2}\right)dx + \frac{4}{(b-a)(d-c)}\int_{\frac{c+d}{2}}^d\int_a^{\frac{a+b}{2}} f(x,y)dxdy\right].
\end{aligned}
$$

The proof of Lemma 2.1 is complete.

THEOREM 2.1. *Let $f:\Omega \subseteq \mathbb{R}^2 \to \mathbb{R}$ be a twice partial differentiable mapping on Ω° (the interior of Ω) and let $\Delta := [a, b] \times [c, d] \subseteq \Omega^\circ$ with $a < b$, $c < d$ and $\frac{\partial^2 f}{\partial x \partial y} \in L_1(\Delta)$, where $L_1(\Delta)$ denotes the set of all Lebesgue* *integrable functions on Δ. If* $\left|\frac{\partial^2 f}{\partial x \partial y}\right|^q$ *is convex on the co-ordinates on Δ and* $q \ge 1$*, then the following inequality holds:*

$$|I(f)| \le \frac{1}{4}\left(\frac{1}{9}\right)^{1/q}\left\{g_q(1,2,2,4) + g_q(4,2,2,1) + g_q(2,1,4,2) + g_q(2,4,1,2)\right\},$$

where $f_{xy}(x,y) = \frac{\partial^2 f(x,y)}{\partial x \partial y}$ and

$$g_q(r_1,r_2,r_3,r_4) = \left[r_1\left|f_{xy}(a,c)\right|^q + r_2\left|f_{xy}(a,d)\right|^q + r_3\left|f_{xy}(b,c)\right|^q + r_4\left|f_{xy}(b,d)\right|^q\right]^{1/q}.$$

Proof. Using Lemma 2.1, since $\left|\frac{\partial^2 f}{\partial x \partial y}\right|^q$ is convex on the co-ordinates on Δ and Hölder inequality, then

$$\begin{aligned}
|I(f)| &\le \int_0^1\int_0^1 t\lambda\left|\frac{\partial^2 f}{\partial x \partial y}\left(\frac{t}{2}a + \left(1-\frac{t}{2}\right)b, \frac{\lambda}{2}c + \left(1-\frac{\lambda}{2}\right)d\right)\right|dtd\lambda \\
&+ \int_0^1\int_0^1 t\lambda\left|\frac{\partial^2 f}{\partial x \partial y}\left(\left(1-\frac{t}{2}\right)a + \frac{t}{2}b, \left(1-\frac{\lambda}{2}\right)c + \frac{\lambda}{2}d\right)\right|dtd\lambda \\
&+ \int_0^1\int_0^1 t\lambda\left|\frac{\partial^2 f}{\partial x \partial y}\left(\frac{t}{2}a + \left(1-\frac{t}{2}\right)b, \left(1-\frac{\lambda}{2}\right)c + \frac{\lambda}{2}d\right)\right|dtd\lambda \\
&+ \int_0^1\int_0^1 t\lambda\left|\frac{\partial^2 f}{\partial x \partial y}\left(\left(1-\frac{t}{2}\right)a + \frac{t}{2}b, \frac{\lambda}{2}c + \left(1-\frac{\lambda}{2}\right)d\right)\right|dtd\lambda \\
&\le \left(\int_0^1\int_0^1 t\lambda dtd\lambda\right)^{1-1/q}\left\{\left[\int_0^1\int_0^1 t\lambda\left(\frac{t\lambda}{4}\left|f_{xy}(a,c)\right|^q + \frac{t}{2}\left(1-\frac{\lambda}{2}\right)\left|f_{xy}(a,d)\right|^q\right.\right.\right. \\
&+ \left.\left.\left(1-\frac{t}{2}\right)\frac{\lambda}{2}\left|f_{xy}(b,c)\right|^q + \frac{(1-2/t)(1-2/\lambda)}{4}\left|f_{xy}(b,d)\right|^q\right)dtd\lambda\right]^{1/q} \\
&+ \left[\int_0^1\int_0^1 t\lambda\left(\left(1-\frac{t}{2}\right)\left(1-\frac{\lambda}{2}\right)\left|f_{xy}(a,c)\right|^q + \left(1-\frac{t}{2}\right)\frac{\lambda}{2}\left|f_{xy}(a,d)\right|^q\right.\right. \\
&+ \left.\left.\frac{t}{2}\left(1-\frac{\lambda}{2}\right)\left|f_{xy}(b,c)\right|^q + \frac{t\lambda}{4}\left|f_{xy}(b,d)\right|^q\right)dtd\lambda\right]^{1/q} \\
&+ \left[\int_0^1\int_0^1 t\lambda\left(\frac{t}{2}\left(1-\frac{\lambda}{2}\right)\left|f_{xy}(a,c)\right|^q + \frac{t\lambda}{4}\left|f_{xy}(a,d)\right|^q\right.\right. \\
&+ \left.\left.\left(1-\frac{t}{2}\right)\left(1-\frac{\lambda}{2}\right)\left|f_{xy}(b,c)\right|^q + \left(1-\frac{2}{t}\right)\left(1-\frac{2}{\lambda}\right)\left|f_{xy}(b,d)\right|^q\right)dtd\lambda\right]^{1/q} \\
&+ \left[\int_0^1\int_0^1 t\lambda\left(\left(1-\frac{t}{2}\right)\frac{\lambda}{2}\left|f_{xy}(a,c)\right|^q + \left(1-\frac{t}{2}\right)\left(1-\frac{\lambda}{2}\right)\left|f_{xy}(a,d)\right|^q\right.\right. \\
&+ \left.\left.\left.\frac{t\lambda}{4}\left|f_{xy}(b,c)\right|^q + \frac{t}{2}\left(1-\frac{\lambda}{2}\right)\left|f_{xy}(b,d)\right|^q\right)dtd\lambda\right]^{1/q}\right\} \\
&\le \frac{1}{4}\left(\frac{1}{9}\right)^{1/q}\left\{g_q(1,2,2,4) + g_q(4,2,2,1) + g_q(2,1,4,2) + g_q(2,4,1,2)\right\}
\end{aligned}$$

Theorem 2.1 is proved.

If taking $q = 1$ in Theorem 2.1, we can derive the following corollary.

Corollary 2.1.1. *Under the conditions of Theorem 2.1, when* $q = 1$*, we have*

$$|I(f)| \le \frac{1}{4}\left[\left|f_{xy}(a,c)\right| + \left|f_{xy}(a,d)\right| + \left|f_{xy}(b,c)\right| + \left|f_{xy}(b,d)\right|\right].$$

Theorem 2.2. *Let* $f:\Omega \subseteq \mathbb{R}^2 \to \mathbb{R}$ *be a twice partial differentiable mapping on* Ω° *(the interior of Ω) and let* $\Delta := [a,b] \times [c,d] \subseteq \Omega^\circ$ *with* $a < b$*,* $c < d$ *and* $\frac{\partial^2 f}{\partial x \partial y} \in L_1(\Delta)$*, where* $L_1(\Delta)$ *denotes the set of all Lebesgue integrable functions on Δ. If* $\left|\frac{\partial^2 f}{\partial x \partial y}\right|^q$ *is convex on the co-ordinates on Δ and* $q > 1$*, then the following inequality holds:*

$$|I(f)| \le \frac{1}{16}\left(\frac{4(q-1)}{2q-1}\right)^{2(1-1/q)}\left\{g_q(1,3,3,9) + g_q(9,3,3,1) + g_q(3,1,9,3) + g_q(3,9,1,3)\right\},$$

where $g(r_1, r_2, r_3, r_4)$ is defined in Theorem 2.1.

Proof. Using Lemma 2.1, and $\left|\frac{\partial^2 f}{\partial x \partial y}\right|^q$ is convex on the co-ordinates on Δ and Hölder's inequality, we have

$$\begin{aligned}
|I(f)| \le & \left(\int_0^1\int_0^1 (t\lambda)^{q/(q-1)} dt d\lambda\right)^{1-1/q} \left\{\left[\int_0^1\int_0^1 \left(\frac{t\lambda}{4}|f_{xy}(a,c)|^q + \frac{t}{2}\left(1-\frac{\lambda}{2}\right)|f_{xy}(a,d)|^q\right.\right.\right. \\
& \left.\left. +\left(1-\frac{t}{2}\right)\frac{\lambda}{2}|f_{xy}(b,c)|^q + \left(1-\frac{t}{2}\right)\left(1-\frac{\lambda}{2}\right)|f_{xy}(b,d)|^q\right) dt d\lambda\right]^{1/q} \\
& + \left[\int_0^1\int_0^1 \left(\left(1-\frac{t}{2}\right)\left(1-\frac{\lambda}{2}\right)|f_{xy}(a,c)|^q + \left(1-\frac{t}{2}\right)\frac{\lambda}{2}|f_{xy}(a,d)|^q\right.\right. \\
& \left.\left. +\frac{t}{2}\left(1-\frac{\lambda}{2}\right)|f_{xy}(b,c)|^q + \frac{t\lambda}{4}|f_{xy}(b,d)|^q\right) dt d\lambda\right]^{1/q} \\
& + \left[\int_0^1\int_0^1 \left(\frac{t}{2}\left(1-\frac{\lambda}{2}\right)|f_{xy}(a,c)|^q + \frac{t\lambda}{4}|f_{xy}(a,d)|^q\right.\right. \\
& \left.\left. +\left(1-\frac{t}{2}\right)\left(1-\frac{\lambda}{2}\right)|f_{xy}(b,c)|^q + \left(1-\frac{t}{2}\right)\frac{\lambda}{2}|f_{xy}(b,d)|^q\right) dt d\lambda\right]^{1/q} \\
& + \left[\int_0^1\int_0^1 \left(\left(1-\frac{t}{2}\right)\frac{\lambda}{2}|f_{xy}(a,c)|^q + \left(1-\frac{t}{2}\right)\left(1-\frac{\lambda}{2}\right)|f_{xy}(a,d)|^q\right.\right. \\
& \left.\left.\left. +\frac{t\lambda}{4}|f_{xy}(b,c)|^q + \frac{t}{2}\left(1-\frac{\lambda}{2}\right)|f_{xy}(b,d)|^q\right) dt d\lambda\right]^{1/q}\right\} \\
= & \frac{1}{16}\left(\frac{4(q-1)}{2q-1}\right)^{2(1-1/q)} \left\{g_q(1,3,3,9) + g_q(9,3,3,1) + g_q(3,1,9,3) + g_q(3,9,1,3)\right\}.
\end{aligned}$$

Theorem 2.2 is proved.

THEOREM 2.3. *Let $f:\Omega \subseteq \mathbb{R}^2 \to \mathbb{R}$ be a twice partial differentiable mapping on Ω° (the interior of Ω) and let $\Delta := [a, b] \times [c, d] \subseteq \Omega^\circ$ with $a < b$, $c < d$ and $\frac{\partial^2 f}{\partial x \partial y} \in L_1(\Delta)$, where $L_1(\Delta)$ denotes the set of all Lebesgue integrable functions on Δ. If $\left|\frac{\partial^2 f}{\partial x \partial y}\right|^q$ is convex on the co-ordinates on Δ and $q > 1$, then the following inequality holds:*

$$|I(f)| \le \frac{q-1}{2(2q-1)}\left(\frac{2q-1}{12(q-1)}\right)^{1/q} \left\{g_q(1,3,2,6) + g_q(6,2,3,1) + g_q(3,1,6,2) + g_q(2,6,1,3)\right\},$$

where $g(r_1, r_2, r_3, r_4)$ is defined in Theorem 2.1.

Proof By Lemma 2.1, since $\left|\frac{\partial^2 f}{\partial x \partial y}\right|^q$ is convex on the co-ordinates on Δ and Hölder's inequality, we get

$$\begin{aligned}
|I(f)| \le & \left(\int_0^1\int_0^1 t\lambda^{q/(q-1)} dt d\lambda\right)^{1-1/q} \left\{\left[\int_0^1\int_0^1 t\left(\frac{t\lambda}{4}|f_{xy}(a,c)|^q + \frac{t}{2}\left(1-\frac{\lambda}{2}\right)|f_{xy}(a,d)|^q\right.\right.\right. \\
& \left.\left. +\left(1-\frac{t}{2}\right)\frac{\lambda}{2}|f_{xy}(b,c)|^q + \left(1-\frac{t}{2}\right)\left(1-\frac{\lambda}{2}\right)|f_{xy}(b,d)|^q\right) dt d\lambda\right]^{1/q} \\
& + \left[\int_0^1\int_0^1 t\left(\left(1-\frac{t}{2}\right)\left(1-\frac{\lambda}{2}\right)|f_{xy}(a,c)|^q + \left(1-\frac{t}{2}\right)\frac{\lambda}{2}|f_{xy}(a,d)|^q\right.\right. \\
& \left.\left. +\frac{t}{2}\left(1-\frac{\lambda}{2}\right)|f_{yy}(b,c)|^q + \frac{t\lambda}{4}|f_{xy}(b,d)|^q\right) dt d\lambda\right]^{1/q} \\
& + \left[\int_0^1\int_0^1 t\left(\frac{t}{2}\left(1-\frac{\lambda}{2}\right)|f_{xy}(a,c)|^q + \frac{t\lambda}{4}|f_{xy}(a,d)|^q + \left(1-\frac{t}{2}\right)\left(1-\frac{\lambda}{2}\right)|f_{xy}(b,c)|^q\right.\right. \\
& \left.\left. +\left(1-\frac{t}{2}\right)\frac{\lambda}{2}|f_{xy}(b,d)|^q\right) dt d\lambda\right]^{1/q} \\
& + \left[\int_0^1\int_0^1 t\left(\left(1-\frac{t}{2}\right)\frac{\lambda}{2}|f_{xy}(a,c)|^q + \left(1-\frac{t}{2}\right)\left(1-\frac{\lambda}{2}\right)|f_{xy}(a,d)|^q\right.\right. \\
& \left.\left.\left. +\frac{t\lambda}{4}|f_{xy}(b,c)|^q + \frac{t}{2}\left(1-\frac{\lambda}{2}\right)|f_{xy}(b,d)|^q\right) dt d\lambda\right]^{1/q}\right\} \\
= & \frac{q-1}{2(2q-1)}\left(\frac{2q-1}{12(q-1)}\right)^{1/q} \left\{g_q(1,3,2,6) + g_q(6,2,3,1) + g_q(3,1,6,2) + g_q(2,6,1,3)\right\}.
\end{aligned}$$

Theorem 2.3 is proved.

THEOREM 2.4. *Let $f:\Omega \subseteq \mathbb{R}^2 \to \mathbb{R}$ be a twice partial differentiable mapping on Ω° (the interior of Ω) and let $\Delta := [a, b] \times [c, d] \subseteq \Omega^\circ$ with $a < b$, $c < d$ and $\frac{\partial^2 f}{\partial x \partial y} \in L_1(\Delta)$, where $L_1(\Delta)$ denotes the set of all Lebesgue integrable functions on Δ. If $\left|\frac{\partial^2 f}{\partial x \partial y}\right|^q$ is convex on the co-ordinates on Δ and $q \geq 1$ and $q \geq r, s > 0$ then the following inequality holds:*

$$
\begin{aligned}
|I(f)| \leq & \left(\frac{1}{4(r+2)(s+2)}\right)^{1/q}\left(\frac{(q-1)^2}{(2q-r-1)(2q-s-1)}\right)^{1-1/q} \\
& \times\left\{g_q\left(1, \frac{s+3}{s+1}, \frac{r+3}{r+1}, \frac{(r+3)(s+3)}{(r+1)(s+1)}\right)+g_q\left(\frac{(r+3)(s+3)}{(r+1)(s+1)}, \frac{r+3}{r+1}, \frac{s+3}{s+1}, 1\right)\right. \\
& \left.+g_q\left(\frac{s+3}{s+1}, 1, \frac{(r+3)(s+3)}{(r+1)(s+1)}, \frac{r+3}{r+1}\right)+g_q\left(\frac{r+3}{r+1}, \frac{(r+3)(s+3)}{(r+1)(s+1)}, 1, \frac{s+3}{s+1}\right)\right\}.
\end{aligned}
$$

where $g(r_1, r_2, r_3, r_4)$ is defined in Theorem 2.1.

Proof Using Lemma 2.1, and $\left|\frac{\partial^2 f}{\partial x \partial y}\right|^q$ is convex on the co-ordinates on Δ and Hölder inequality, we have

$$
\begin{aligned}
|I(f)| \leq & \left(\int_0^1\int_0^1 t^{(q-r)/(q-1)}\lambda^{(q-s)/(q-1)}dtd\lambda\right)^{1-1/q} \\
& \times\left\{\left[\int_0^1\int_0^1 t^r\lambda^s\left(\frac{t\lambda}{4}|f_{xy}(a,c)|^q+\frac{t}{2}\left(1-\frac{\lambda}{2}\right)|f_{xy}(a,d)|^q\right.\right.\right. \\
& \left.\left.+\left(1-\frac{t}{2}\right)\frac{\lambda}{2}|f_{xy}(b,c)|^q+\left(1-\frac{t}{2}\right)\left(1-\frac{\lambda}{2}\right)|f_{xy}(b,d)|^q\right)dtd\lambda\right]^{1/q} \\
& +\left[\int_0^1\int_0^1 t^r\lambda^s\left(\left(1-\frac{t}{2}\right)\left(1-\frac{\lambda}{2}\right)|f_{xy}(a,c)|^q+\left(1-\frac{t}{2}\right)\frac{\lambda}{2}|f_{xy}(a,d)|^q\right.\right. \\
& \left.\left.+\frac{t}{2}\left(1-\frac{\lambda}{2}\right)|f_{xy}(b,c)|^q+\frac{t\lambda}{4}|f_{xy}(b,d)|^q\right)dtd\lambda\right]^{1/q} \\
& +\left[\int_0^1\int_0^1 t^r\lambda^s\left(\frac{t}{2}\left(1-\frac{\lambda}{2}\right)|f_{xy}(a,c)|^q+\frac{t\lambda}{4}|f_{xy}(a,d)|^q\right.\right. \\
& \left.\left.+\left(1-\frac{t}{2}\right)\left(1-\frac{\lambda}{2}\right)|f_{xy}(b,c)|^q+\left(1-\frac{t}{2}\right)\frac{\lambda}{2}|f_{xy}(b,d)|^q\right)dtd\lambda\right]^{1/q} \\
& +\left[\int_0^1\int_0^1 t^r\lambda^s\left(\left(1-\frac{t}{2}\right)\frac{\lambda}{2}|f_{xy}(a,c)|^q+\left(1-\frac{t}{2}\right)\left(1-\frac{\lambda}{2}\right)|f_{xy}(a,d)|^q\right.\right. \\
& \left.\left.\left.+\frac{t\lambda}{4}|f_{xy}(b,c)|^q+\frac{t}{2}\left(1-\frac{\lambda}{2}\right)|f_{xy}(b,d)|^q\right)dtd\lambda\right]^{1/q}\right\} \\
\leq & \left(\frac{1}{4(r+2)(s+2)}\right)^{1/q}\left(\frac{(q-1)^2}{(2q-r-1)(2q-s-1)}\right)^{1-1/q} \\
& \times\left\{g_q\left(1, \frac{s+3}{s+1}, \frac{r+3}{r+1}, \frac{(r+3)(s+3)}{(r+1)(s+1)}\right)+g_q\left(\frac{(r+3)(s+3)}{(r+1)(s+1)}, \frac{r+3}{r+1}, \frac{s+3}{s+1}, 1\right)\right. \\
& \left.+g_q\left(\frac{s+3}{s+1}, 1, \frac{(r+3)(s+3)}{(r+1)(s+1)}, \frac{r+3}{r+1}\right)+g_q\left(\frac{r+3}{r+1}, \frac{(r+3)(s+3)}{(r+1)(s+1)}, 1, \frac{s+3}{s+1}\right)\right\}.
\end{aligned}
$$

Theorem 2.4 is proved.

If taking $r = s = q$ in Theorem 2.4, we can derive the following corollary.

Corollary 2.4.1. *Under the conditions of Theorem2.4, when $r = s = q$, we have*

$$|I(f)| \le \left(\frac{1}{4(q+2)(q+2)}\right)^{1/q}\left\{g_q\left(1,\frac{q+3}{q+1},\frac{q+3}{q+1},\frac{(q+3)^2}{(q+1)^2}\right)+g_q\left(\frac{(q+3)^2}{(q+1)^2},\frac{q+3}{q+1},\frac{q+3}{q+1},1\right)\right.$$
$$\left.+g_q\left(\frac{q+3}{q+1},1,\frac{(q+3)^2}{(q+1)^2},\frac{q+3}{q+1}\right)+g_q\left(\frac{q+3}{q+1},\frac{(q+3)^2}{(q+1)^2},1,\frac{q+3}{q+1}\right)\right\}.$$

where $g(r_1, r_2, r_3, r_4)$ is defined in Theorem 2.1.

Acknowledgements
The authors appreciate the anonymous referees for their careful corrections to and valuable comments on the original version of this paper.

Funding
This work was partially supported by the National Natural Science Foundation (NNSF) [grant number 11361038] of China; the Foundation of the Research Program of Science and Technology at Universities of Inner Mongolia Autonomous Region [grant number NJZY14192]; NSF of Inner Mongolia Autonomous Region [grant number 2014BS0106], [grant number 2015BS0123]; the Scientific Innovation Project for Graduates at the Inner Mongolia University for Nationalities [grant number NMDSS1419] China.

Author details
Xu-Yang Guo[1]
E-mail: guoxuyang1991@qq.com
Feng Qi[2,3]
E-mail: qifeng618@gmail.com
ORCID ID: http://orcid.org/0000-0001-6239-2968
Bo-Yan Xi[1]
E-mail: baoyintu78@qq.com
ORCID ID: http://orcid.org/0000-0003-4528-2331
[1] College of Mathematics, Inner Mongolia University for Nationalities, Tongliao City 028043, China.
[2] Department of Mathematics, College of Science, Tianjin Polytechnic University, Tianjin City 300387, China.
[3] Institute of Mathematics, Henan Polytechnic University, Jiaozuo City, Henan Province 454010, China.

References

Dragomir, S. S. (2001). On Hadamard inequality for convex functions on the co-ordinates in a rectangle from the plane. *Taiwanese Journal of Mathematics,5*, 775–788.

Dragomir, S. S., & Pearce, C. E. M. (2000). *Selected topics on Hermite-Hadamard inequalities and applications* (RGMIA Monographs). Victoria University. Retrieved from http://rgmia.org/monographs/hermite_hadamard.html

Latif, M. A., & Dragomir, S. S. (2012). On some new inequalities for differentiable co-ordinated convex functions. *Journal of Inequalities and Applications, 2012*, 28. doi:10.1186/1029-242X-2012-28

Ozdemir, M. E., Akdemir, A. O., Kavurmaci, H., & Avci, M. (2011). *On the Simpson's inequality for co-ordinated convex functions*. arXiv:1101.0075.

Pečarić, J. E., Proschan, F., & Tong, Y. L. (1991). *Convex functions, partial ordering and statistical applications*. New York, NY: Academic Press.

10

Development and implementation of algorithms for vehicle routing during a no-notice evacuation

Xin Chen[1]*, Zhu Zhang[1] and Ryan Fries[2]

*Corresponding author: Xin Chen, Department of Industrial Engineering, Southern Illinois University Edwardsville, Edwardsville, IL 62026-1805, USA

E-mail: xchen@siue.edu

Reviewing editor: Ryan Loxton, Curtin University, Australia

Abstract: This article develops and implements time-dependent shortest path and assignment algorithms for vehicle routing and assignment during a no-notice evacuation. A time-dependent shortest path algorithm with arc labeling is designed to improve computer storage space efficiency. Visual Basic for Application (VBA) is used to improve time efficiency of both shortest path and assignment algorithms. Performance of the algorithms is analyzed and compared through implementation in VBA and the General Algebraic Modeling System (GAMS). Since VBA seamlessly integrates with Microsoft Excel and enables efficient data manipulation, the algorithms implemented in VBA are more efficient and obtain optimal solutions faster than those implemented in the GAMS. The shortest path algorithm with arc labeling implemented in VBA may be used for real-time vehicle routing for large road networks during no-notice evacuations.

Subjects: Engineering & Technology; Industrial Engineering & Manufacturing; Operations Research; Technology

Keywords: arc labeling; assignment problem; evacuation; time-dependent shortest path; VBA; vehicle routing

1. Introduction

In a generalized vehicle routing problem (VRP), multiple vehicles travel from origins (depots; Montoya-Torres, Franco, Isaza, Jiménez, & Herazo-Padilla, 2015) to destinations (demand points). One of the objectives of the VRP is to minimize total travel time or the latest arrival time for a fleet of vehicles while meeting demand at destinations. The VRP has a wide range of applications,

ABOUT THE AUTHORS

Xin Chen, PhD, is Associate Professor of Industrial Engineering at Southern Illinois University Edwardsville. His research focuses on network- and knowledge-centric collaborative control with applications in critical infrastructure protection and management, energy distribution, social networks, supply chains, and transportation. This article is part of his research on applying optimization tools in large, complex networks.

Zhu Zhang was a graduate student in Industrial Engineering at Southern Illinois University Edwardsville. He graduated in 2014 with a Master of Science in Industrial Engineering.

Ryan Fries is an associate professor in the Department of Civil Engineering at Southern Illinois University Edwardsville. His research focuses on design and planning of transportation systems.

PUBLIC INTEREST STATEMENT

During emergencies, evacuations, and many other situations, it is important for emergency vehicles from various locations to reach their respective destinations quickly. This article develops an efficient algorithm that calculates the best routes for multiple vehicles. When road conditions change, the algorithm computes and updates the routes within a few minutes, and can be used for real-time route planning in large road networks.

e.g. bank or postal deliveries and school bus routing. During a no-notice evacuation of a metropolitan area, emergency response vehicles must be swiftly deployed to multiple demand points to manage traffic, including blocking ramps and lanes and diverting traffic. Once an emergency response vehicle arrives at an intended demand point, it is stationed at the point until the evacuation completes, which may take from a few hours to one day. Certain demand points may need multiple emergency response vehicles whereas others may need only one vehicle. Before a no-notice evacuation begins, emergency response vehicles are usually stationed at depots controlled by state or local government agencies, e.g. the state department of transportation. Since road conditions change rapidly, there is a strong need for real-time emergency vehicle routing during a no-notice evacuation.

The objective of this article is to investigate effective algorithms which may be implemented to identify optimal vehicle assignments and routes for the VRP in real time. The main contributions of this article are as follows:

(a) Implementing and integrating time-dependent shortest path (TDSP) and assignment algorithms to minimize total travel time for emergency response vehicles during a no-notice evacuation of a metropolitan area or other emergency response and evacuation situations; and

(b) Improving time and space efficiency of the algorithms for real-time vehicle assignment and routing in large road networks.

(i) The TDSP algorithm with arc labeling requires less computer memory compared to node labeling, and improves space efficiency. The TDSP with arc labeling and assignment algorithms are applied to road networks with up to 2,000 nodes, and are able to identify the optimal vehicle routes and assignments within eight minutes. (The largest road network studied previously had 1,000 nodes; Almoustafa, Hanafi, & Mladenović, 2013); and

(ii) The algorithms are implemented in Visual Basic for Applications (VBA®). Input and output of the algorithms are stored and processed using Microsoft Excel®. The seamless integration of VBA® and Excel® greatly improves time efficiency and enables real-time vehicle assignment and routing.

2. Background

Research on the VRP stemmed from the traveling salesman problem (Dantzig, Fulkerson, & Johnson, 1954) and was further developed in some early applications (Balas & Toth, 1985; Clarke & Wright, 1964; Haimovich, Rinnooy Kan, & Stougie, 1988; Laporte, Nobert, & Taillefer, 1987). Recent research focused on heuristic and meta-heuristic approaches (e.g. Bräysy & Gendreau, 2005; Zare-Reisabadi & Mirmohammadi, 2015). Most early studies assumed vehicle travel time between two nodes was static and did not change regardless of traffic. During the last decade, time-dependent vehicle routing problems (TDVRPs) have gained increasing attention (Almoustafa et al., 2013; Ando & Taniguchi, 2006; Andres Figliozzi, 2012; Chen, Hsueh, & Chang, 2006; Chen et al., 2013; Haghani & Jung, 2005; Ichoua, Gendreau, & Potvin, 2003; Kritzinger et al., 2012; Lecluyse, Van Woensel, & Peremans, 2009; Maden, Eglese, & Black, 2010; Spliet & Gabor, 2012; Vidal, Crainic, Gendreau, & Prins, 2014; Van Woensel, Kerbache, Peremans, & Vandaele, 2008). In the TDVRP, travel time between two directly connected nodes is dynamic and depends on many factors such as traffic. In many cases, travel time between two nodes may be described as a function of the node and time at which a vehicle begins to travel and the node at which the vehicle plans to arrive.

The VRP has many input parameters, including the number and origins of vehicles, the size and structure of a road network, the number and location of destinations, and vehicle travel times (dynamic or static). Numerous methods were introduced in recent years to solve the VRP; these methods stipulated various conditions for one or more parameters. One of the most common conditions was the upper limit for the size of road networks. Other conditions include static travel time and upper limit for the number of destinations. These methods become ineffective (solutions far from optimal) or inefficient (cannot identify a good or optimal solution within an acceptable amount of time) when stipulated conditions do not hold. To develop effective and efficient algorithms to solve VRPs with many parameter values remain a considerable challenge (Vidal et al., 2014).

Since exact methods that identify optimal vehicle assignments and routes are either ineffective or inefficient for generalized VRPs, heuristic methods including the genetic algorithm (Haghani & Jung, 2005), tabu search (Ichoua et al., 2003), branch and price (Almoustafa et al., 2013), and column generation algorithm (Spliet & Gabor, 2012) were studied. Haghani and Jung (2005) presented a genetic algorithm to solve a pick-up or delivery VRP with soft time windows. The study considered multiple vehicles with different capacities, real-time service requests, and dynamic travel times between destinations. Ichoua et al. (2003) conducted experiments to solve the VRP with time-dependent travel speeds, which satisfy the first-in-first-out (FIFO) property, using a parallel tabu search heuristic. Almoustafa et al. (2013) improved a branch-and-bound method to solve the asymmetric distance-constrained VRP suggested by Laporte et al. (1987). Chen et al. (2006) formulated a real-time TDVRP with time windows as a series of mixed integer programming models and developed a heuristic algorithm, which included route construction and improvement. Spliet and Gabor (2012) proposed a formulation of a time window asymmetric VRP and developed two variants of a column generation algorithm to solve the linear programming relaxation of this formulation. Kritzinger et al. (2012) applied variable neighborhood search algorithm to solve the TDVRP with time windows. Maden et al. (2010) proposed a heuristic algorithm for the VRP to minimize total travel time.

Road networks with different sizes were also studied. Laporte, Nobert, and Taillefer (1988) examined a class of asymmetrical multi-depot VRPs and location-routing problems for a network of 80 nodes. Haghani and Jung (2005) solved the TDVRP for networks with 30 demand nodes over 30 time intervals. Almoustafa et al. (2013) solved an asymmetric distance-constrained VRP for a network of 1,000 demand nodes. In summary, previous research predominately focused on developing heuristic methods for subsets of VRPs or TDVRPs. Effective and efficient algorithms which may be applied to generalized TDVRPs to obtain optimal vehicle assignments and routes were not available. Most algorithms and methods developed in previous research were not tested using real-world road networks and could not be validated for effectiveness or efficiency. The objective of this article is to provide effective and efficient algorithms for generalized TDVRPs in a no-notice evacuation. The rest of this article is organized as follows: Section 3 presents the problem and methodology. Section 4 validates the algorithms using two real-world and five simulated road networks. Section 5 concludes the article with future research directions.

3. Methodology

3.1. Problem definition

Let A, B, Γ represent sets of vehicles, depots, and demand points, respectively, in a time-dependent road network. There are total $|A|$ vehicles stationed at $|B|$ depots at the beginning of a planning period, which is a time period during which vehicles must be dispatched to demand points to meet the demand. Some or all of the $|A|$ vehicles need to be dispatched to $|\Gamma|$ demand points, each of which requires d_γ vehicles, where γ represents a demand point, $\gamma \in \Gamma$. Let $c^{\alpha}_{i,j,t}$ be the cost (time) it requires for a vehicle α, $\alpha \in A$, to travel from node i at time t and to node j. $i,j \in V$, where V is the node set in the road network. $B, \Gamma \subset V$. Let (i,j) represent an arc that originates from node i and points at node j, $(i,j) \in E$, where E is the set of arcs in the road network. $c^{\alpha}_{i,j,t} = \infty$ if $(i,j) \notin E$. When $i = \beta$, β is a depot and $\beta \in B$, $c^{\alpha}_{\beta,j,t} = \infty$, $\forall j$, if α is not ready to travel from β at time t.

In the TDVRP, the first objective is to identify the earliest time for α to arrive at a demand point γ from its depot β. α may travel from β at any time after α is ready for travel. Suppose α may travel from β at time t_1 and arrive at γ at time t_2. Alternatively, α may travel from β at time t_3 and arrive at γ at time t_4. According to the FIFO property, $t_2 < t_4$ if $t_1 < t_3$. Therefore, α should travel from β as soon as α is ready for travel. Let t_β be the time at which α becomes ready for travel at β. $c^{\alpha}_{\beta,j,t} = \infty$ when $t < t_\beta$,

$\forall j.\ c^{\alpha}_{\beta,j,t} << \infty$ when $t \geq t_{\beta}$ and $(\beta, j) \in E$. The first objective of the TDVRP is therefore to identify the TDSP for α to travel from β at t_{β} to γ. Equation (1) is a model whose optimal solution is the TDSP between β and γ when α travels from β at t_{β}.

$$\text{Minimize} \sum_{i}\sum_{j} c^{\alpha}_{i,j,t} X_{i,j}$$

Subject to:

$$\sum_{j} X_{i,j} - \sum_{j} X_{j,i} = 1 \quad \text{when } i = \beta$$

$$\sum_{j} X_{i,j} - \sum_{j} X_{j,i} = -1 \quad \text{when } i = \gamma$$

$$\sum_{j} X_{i,j} - \sum_{j} X_{j,i} = 0 \quad \text{when } i \neq \beta \text{ and } i \neq \gamma$$

$$X_{i,j} = \begin{pmatrix} 1 & (i,j) \text{ is on the path for } \alpha \text{ to travel from } \beta \text{ to } \gamma \\ 0 & \text{otherwise} \end{pmatrix} \quad \forall i, j \in V \tag{1}$$

The second objective of the TDVRP is to minimize total travel time. Let $s^{\alpha}_{\beta,\gamma}$ represent the optimal value of Equation (1), i.e. the minimum time for α to travel from β and arrive at γ. Equation (2) models an assignment problem (AP) that determines which vehicles are dispatched to a demand point to meet the demand. Note that the objective of Equation (2) is not to minimize the summation of arrival times or the latest arrival time. Since Equation (1) identifies the earliest arrival times, Equation (2) intends to minimize total travel time. In many VRPs, less travel time implies less uncertainty and more reliable vehicle routing and assignment (Chen et al., 2013). Both Equations (1) and (2) are pure integer programming problems. If $\sum_{\gamma=1}^{|\Gamma|} d_{\gamma} = |A|$, Equation (2) is a balanced transportation problem and both constrains may be changed to equality constraints. If $\sum_{\gamma=1}^{|\Gamma|} d_{\gamma} > |A|$, Equation (2) is infeasible.

$$\text{Minimize} \sum_{\alpha=1}^{|A|}\sum_{\gamma=1}^{|\Gamma|} s^{\alpha}_{\beta,\gamma} Y_{\alpha,\gamma}$$

Subject to:

$$\sum_{\gamma=1}^{|\Gamma|} Y_{\alpha,\gamma} \leq 1, \forall \alpha$$

$$\sum_{\alpha=1}^{|A|} Y_{\alpha,\gamma} \geq d_{\gamma}, \forall \gamma$$

$$Y_{\alpha,\gamma} = \begin{pmatrix} 1 & \alpha \text{ travels to } \gamma \\ 0 & \text{otherwise} \end{pmatrix} \tag{2}$$

3.2. Solution steps

Optimization software packages may be used to solve models in Equations (1) and (2). For example, the General Algebraic Modeling System (GAMS; GAMS Development Corporation, 2013) is a high-level modeling system for mathematical programming and optimization. Equations (1) and (2) may be described in algebraic statements in GAMS input files and solvers may be used to find optimal solutions and values. For medium to large road networks, however, this approach is ineffective or inefficient due to the limitation of computer memory and the inconvenience of extremely long

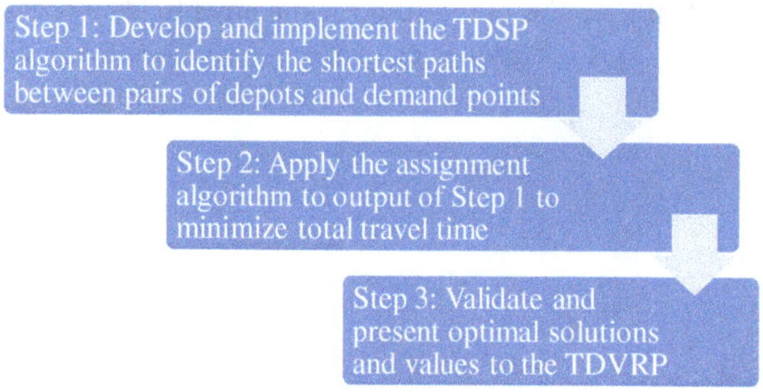

Step 1: Develop and implement the TDSP algorithm to identify the shortest paths between pairs of depots and demand points
Step 2: Apply the assignment algorithm to output of Step 1 to minimize total travel time
Step 3: Validate and present optimal solutions and values to the TDVRP

Figure 1. Three steps to solve the TDVRP for real-time transportation planning.

computation time. For real-time transportation planning, time- and space-efficient algorithms must be developed to solve the models in Equations (1) and (2). Figure 1 shows three steps to identify optimal solutions and values to the TDVRP.

3.3. TDSP algorithms

Real-time TDVRPs remain a great challenge due to time and space complexities. In the TDVRP, characteristics of the road network change with time; a shortest path computed from a snapshot of the road network may not be the shortest path at a different time. The TDSP algorithm is developed to find the shortest paths between depots and demand points in real-time. Three assumptions related to the TDSP algorithm are:

(a) The road network satisfies the FIFO principle, which specifies that if two vehicles take the same route from the same depot to the same demand point, the vehicle leaving the depot early always arrives at the demand point early. According to the FIFO principle, a vehicle should leave its depot or other nodes whenever it is ready. Waiting at any node is never beneficial because a vehicle that leaves later always arrives later;

(b) The planning period is "discretized" into sufficiently small time intervals, δ's, $\delta = 1, 2, 3, \ldots$ and $\delta \in \Delta$ where Δ is the set of time intervals over the planning period; and

(c) Travel time between two nodes connected by an arc depends on the time at which a vehicle leaves the beginning node of the arc.

The TDSP algorithm with node labeling was first developed by Dreyfus (1969) and further validated by Kaufman and Smith (1993). The TDSP algorithm with node labeling listed below finds the earliest arrival time of a vehicle at a demand point given that the vehicle is stationed at a depot when travel begins. The algorithm is executed for each pair of demand point and depot for each vehicle. Vehicles stationed at the same depot may be ready to travel from the depot at different times because different vehicles may require different preparation times, e.g. time for a driver to arrive at the depot and get ready may vary. The earliest arrival times of vehicles, which are stationed at the same depot, at the same demand point may be different and need to be calculated separately using the TDSP algorithm.

3.3.1. TDSP algorithm with node labeling

(1) Assign to every node $i, i \in V$, in a transportation network a value representing the arrival time of a vehicle $\alpha, \alpha \in A$, at the node. For a depot node β where α is stationed, set the value to a finite positive integer number representing the time at which α is ready to travel from β. Set the value to infinity for all other nodes i's, $i \in V$ and $i \neq \beta$;

(2) Mark β visited. Mark all other i's unvisited. Set β as the current node;

(3) Calculate the arrival time of α at each unvisited neighbor $j, j \in V$, of the current node $i, i \in V$. A node j is a neighbor of i if there is an arc begins at i and points at j, i.e. $(i,j) \in E$. The arrival

time at j is the summation of the value set for i and the travel time $c^{\alpha}_{i,j,\delta}$ between i and j. The travel time $c^{\alpha}_{i,j,\delta}$ is obtained from a three-dimensional matrix, $|V| \times |V| \times |\Delta|$, which stores time-dependent travel times. For example, if $i = 1, j = 2$, and the value set for node 1 is 15, the travel time between nodes 1 and 2, $c^{\alpha}_{1,2,15}$, is a component in the matrix identified by node 1, node 2, and $\delta = 15$. $\delta = 15$ indicates α begins traveling from node 1 at time 15;

(4) For each unvisited neighbor j of the current node i, compare the arrival time at j calculated in Step 3 and the value set for j. Set the value for j as the smaller one between the two;

(5) Identify the unvisited node with the smallest value. Mark the node visited. Set the node as the current node. If the node is the desired demand point $\gamma, \gamma \in \Gamma$, stop. The value set for γ is the earliest arrival time of α traveling from β at γ. Otherwise go to Step 3.

The TDSP algorithm with node labeling is implemented in VBA® and the GAMS. The algorithm can only solve TDVRPs for small road networks with less than 500 nodes when there are more than 1,000 time intervals using a Windows 7 ×64 PC, Intel i7–3770 CPU @3.40 GHZ, and 16.0 GB RAM. The main reason for size limitation on the road network is the large storage space required by node labeling. The algorithm needs to manipulate a three-dimensional matrix, $|V| \times |V| \times |\Delta|$, for node labeling. Each component in the matrix is travel time from one node to the other during a time interval. These travel times are often obtained through field observations (Rakha, El-Shawarby, Arafeh, & Dion, 2006) and real-time monitoring of road network conditions. $|V|$ is the size of the road network, i.e. the number of nodes. $|\Delta|$ is the number of time intervals. For example, if the road network size increases by tenfold, the storage space for the three-dimensional matrix increases by 100 times.

The degree of a node in a network is the number of arcs connected to the node. Most real-world road networks have a mean degree between two and four (Barabasi, 2002; Jeong, 2003). On average, each node is connected to two to four arcs. If arc labeling is used, the TDSP algorithm manipulates a two-dimensional matrix, $|E| \times |\Delta|$, where $|E|$ is the number of arcs in the road network. Each component in the two-dimensional matrix is travel time along an arc during a time interval. Travel times in the two-dimensional matrix are the same as those in the three-dimensional matrix, but are organized in a different format that reduces storage space requirement. Since $|E| \leq 4 \times |V|$ according to the mean degree of a road network, storage space requirement for arc labeling is at most $4 \times |V| \times |\Delta|$, which is much less than $|V| \times |V| \times |\Delta|$ required for node labeling for large road networks. The TDSP algorithm with arc labeling described below is used to identify the earliest arrival times of vehicles at demand points. This is a significant improvement for the TDSP algorithm and greatly increases space efficiency of the algorithm.

3.3.2. TDSP algorithm with arc labeling

(1) Assign to every arc $(i,j), (i,j) \in E$, in a transportation network a value representing the arrival time of a vehicle $\alpha, \alpha \in A$, at $j, i,j \in V$. For (i,j) in which $i = \beta$, a depot node where α is stationed, set the value to a finite positive integer number representing the time at which α arrives at j. The arrival time at j is the summation of time at which α is ready to travel from β and travel time from β to j. The travel time from β to j, $c^{\alpha}_{\beta,j,\delta}$, is obtained from a two-dimensional matrix, $|E| \times |\Delta|$, which stores time-dependent travel times. For example, if $\beta = 1, j = 2$, and the time at which α is ready to travel from β is $\delta = 15$, the travel time between nodes 1 and 2, $c^{\alpha}_{1,2,15}$, is a component in the matrix identified by arc $(1, 2)$ and $\delta = 15$. Set the value to infinity for all other arcs $(i,j), (i,j) \in E$ and $i \neq \beta$;

(2) Mark all (i,j) unvisited;

(3) Identify the unvisited (i,j) with the smallest value. Mark (i,j) visited. Set the destination node, i.e. the second node j, in (i,j) as the current node. If j is the desired demand point $\gamma, \gamma \in \Gamma$, stop. The value set for (i, γ) is the earliest arrival time of α traveling from β at γ;

(4) For each unvisited (i,j) whose i is the current node, compare the arrival time at j and the value set for (i,j). The arrival time at j is the summation of time at which α arrives at i and travel time from i to j. The arrival time at i is the value set for the arc marked as visited in Step 3. The travel

time from i to j, $c^{\alpha}_{i,j,\delta}$, is obtained from the matrix $|E| \times |\Delta|$. δ is the value set for the arc marked as visited in Step 3;

(5) For each unvisited (i,j) whose i is the current node, set its value as the smaller one between the arrival time at j calculated in Step 4 and the value set for (i,j). Go to Step 3.

3.4. Assignment algorithm

After calculating all the shortest paths between depots and demand points, an assignment algorithm needs to be implemented to determine which vehicles from a depot will travel to a demand point to meet the demand. There are several variants of APs (Pentico, 2007), e.g. generalized AP, bottleneck AP, quadratic AP, and semi-AP. A semi-AP is a modified classic AP which differs in the disproportionality between demand points and depots. The TDVRP is a semi-AP except that a vehicle is not required to be assigned to a demand point. The Hungarian method (Kuhn, 1955) is used as the assignment algorithm to solve the TDVRP.

4. Experiments and validation

4.1. Case study of a small road network

A road network in Midwest of the United States of America (Figure 2) is analyzed to validate the TDSP and assignment algorithms. In this case study, a no-notice evacuation prompts the need for transportation agencies to deploy vehicles and personnel to specific locations for traffic control. The network has 346 nodes (diamonds in Figure 2), out of which six are depots (red diamonds) and 15 are demand points (blue diamonds). The depots are locations where a state or federal transportation agency, e.g. the state department of transportation, dispatches service vehicles and equipment. The demand points are locations where the agency identifies a need for traffic control during a no-notice evacuation event. Overall, there are 654 roads (arcs in Figure 2) connecting all the nodes; thus the

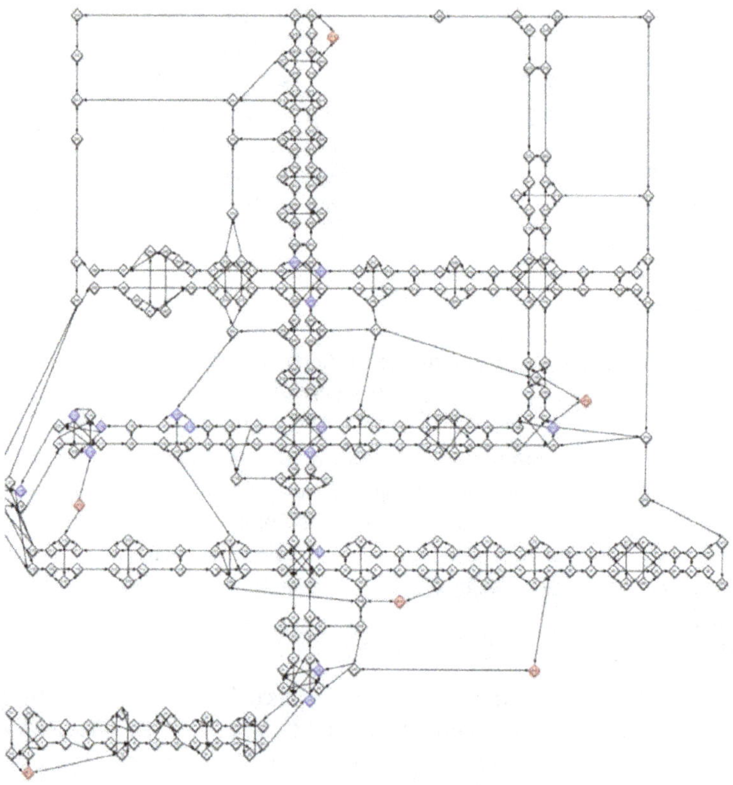

Figure 2. A road network in Midwest of the United States of America.

Table 1. Algorithms run times (in seconds) for the road network in Figure 2			
Planning period (min)	**GAMS**	**VBA® node labeling**	**VBA® arc labeling**
120	191	9	13
240	449	11	13
360	681	13	13
480	872	15	15
1,440	1,954	33	24

mean degree is $\frac{654\times2}{346}=3.78$. There are a total of 92 vehicles available at six depots of a public transportation agency. Each time interval is one minute. The TDSP and assignment algorithms are implemented in VBA® and the GAMS (GAMS Development Corporation, 2013). All computation results are obtained using a Windows 7 ×64 PC, as described previously. Table 1 summarizes run times.

Algorithms implemented in VBA® and the GAMS provide the same optimal solutions and values. Table 1 shows, however, that there is a substantial advantage in using VBA® to identify the optimal solution and value for real-time TDVRPs. The GAMS employs a suite of solvers, e.g. CPLEX, and algorithms, e.g. simplex and branch-and-cut algorithms, to solve linear programming and mixed integer programming problems, but lack efficiency in computer memory management. To compute the results in Table 1, both the TDSP algorithm with arc labeling and assignment algorithm are coded in the GAMS to identify the optimal solution and value. It takes even more computation time when models in Equations (1) and (2) are directly used in the GAMS to find the optimal solution and value. The TDSP and assignment algorithms implemented in VBA® directly manipulate data and are customized to solve the TDVRPs; they are more efficient in terms of run time and computer storage space. Table 1 does not show a substantial difference in run times between node labeling and arc labeling for the TDSP algorithm. The difference between these two methods needs to be further investigated using larger road networks.

4.2. Node labeling and arc labeling in VBA® for a large road network

Figure 3 shows a road network in the City of San Francisco (Brinkhoff, 2002), which includes 174,956 nodes and 223,001 arcs with a mean degree of 2.55 $\left(=\frac{223{,}001\times2}{174{,}956}\right)$. A portion of the network (area inside the red box at the lower left corner of Figure 3), including 492 nodes and 559 arcs that connect the nodes, is selected to validate the TDSP and assignment algorithms. The selected portion has a mean degree of 2.27 $\left(=\frac{559\times2}{492}\right)$. Six depots and 15 demand points are randomly and uniformly selected among the 492 nodes. Total 92 vehicles are randomly and uniformly stationed at six depots. Each time interval is one minute. The GAMS requires more than 10 h computing the optimal solution to the TDVRP for this network and sometimes stops unexpectedly due to insufficient computer memory. Run times for VBA® are summarized in Table 2, which does not include run times for the GAMS since they are too large and the authors choose not to complete the runs in the GAMS. The authors choose five different planning periods, each of which specifies a time period during which vehicles must be dispatched to demand points to meet the demand. Run time increases as the planning period increases, indicating more computation time is needed for a larger planning period.

Arc labeling takes more time to compute the optimal solution than node labeling when the planning period is less than 1,440 min. The difference in data processing time between arc labeling and node labeling is relatively insignificant when the planning period is short. Once the TDSP is identified, the optimal solution, i.e. a sequence of nodes on the shortest path, must be determined by backtracking from the destination to origin. Arc labeling is expected to take more time in backtracking since it must find the two end nodes of a visited arc. The advantage of arc labeling becomes evident when the number of nodes or planning period increases. In Table 2, when the planning period is 1, 440 min (24 h), node labeling could not compute the optimal solution due to insufficient memory;

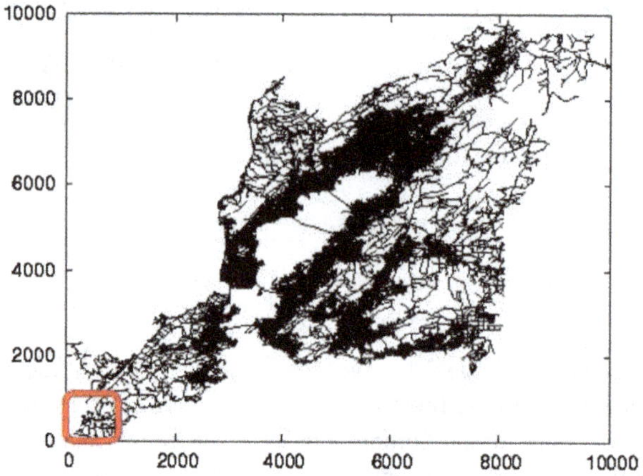

Figure 3. A road network in the City of San Francisco (Brinkhoff, 2002).

Table 2. Algorithms run times (in seconds) for the road network in Figure 3

Planning period (min)	VBA® node labeling	VBA® arc labeling
120	14	26
240	19	28
360	22	30
480	27	32
1,440	N/A	48

run time is not applicable (N/A). It takes 48 s for arc labeling to compute the optimal solution, which is acceptable for real-time TDVRPs.

4.3. Run times of TDSP and assignment algorithms for simulated road networks

The road network in Section 4.1 has a mean degree of 3.78 whereas the two road networks in Section 4.2 have a mean degree of 2.55 and 2.27, respectively. To further validate the algorithms implemented in VBA®, simulated road networks are generated for a given degree distribution. A random network, one of the most studied networks, is described using a degree distribution $\binom{n-1}{d}p^d(1-p)^{n-1-d}$, where n is the total number of nodes, d is node degree, and p is the probability that an arc between two nodes exists. The mean degree $\bar{d} = (n-1)p$. Random networks were used to describe the structure of the US highway system (Barabasi, 2002; Jeong, 2003). Five random road networks with total 500, 1,000, 1,200, 1,500, and 2,000 nodes are generated using a mean degree of 2.5.

The TDSP and assignment algorithms in VBA® are applied to the five simulated networks and their run times are summarized in Table 3, which reveals several important observations:

(a) Run times increase as the total number of nodes increases;

(b) Run times increase in most cases as the planning period increases. This is consistent with results in Tables 1 and 2;

(c) If node labeling can compute an optimal solution to the TDVRP, it takes less time than arc labeling. This is due to the additional time required for arc labeling to backtrack from the destination to origin and identify the optimal solution;

(d) Node labeling cannot compute an optimal solution when size of the matrix that stores travel times between nodes becomes too large. For instance, when total number of nodes is 1,500 and 2,000, node labeling cannot compute an optimal solution for any planning period; and

Table 3. Algorithms run times (in seconds) in VBA® for five simulated road networks

Number of nodes	Labeling	Planning period (min)				
		120	240	360	480	1,440
500	Node	18	19	24	28	N/A
	Arc	34	36	39	41	62
1,000	Node	50	N/A	N/A	N/A	N/A
	Arc	92	91	96	100	129
1,200	Node	74	N/A	N/A	N/A	N/A
	Arc	117	123	128	133	170
1,500	Node	N/A	N/A	N/A	N/A	N/A
	Arc	234	237	246	252	301
2,000	Node	N/A	N/A	N/A	N/A	N/A
	Arc	388	449	414	410	479

(e) Arc labeling can efficiently compute optimal solutions to the TDVRP for large road networks over a long planning period. The largest run time in Table 3 is 479 s, which is about 8 min and acceptable for real-time TDVRPs.

5. Conclusions and future research

This paper develops a new TDSP algorithm with arc labeling and integrates it with the assignment algorithm to identify the optimal solution and value for the TDVRP in real time. The experiments show that the algorithms implemented in VBA® may be used for real-time vehicle routing in large road networks. The case study presented in Section 4.1 implements the algorithms in a no-notice evacuation for a metropolitan area with acceptable performance and does not require above average computing power. Both the TDSP and assignment algorithms are validated using a large road network and five simulated networks. The TDSP with arc labeling may be used for real-time TDVRP over a long planning period. Future research will investigate the performance of the TDSP and assignment algorithms in a variety of real-world road networks with various degree distributions, and explore certain heuristic algorithms, e.g. genetic algorithm, to solve the TDVRP. In addition, other programming languages, e.g. C++, Java, and Julie, will be used to further improve the efficiency of the algorithms.

Funding
The authors received no direct funding for this research.

Author details
Xin Chen[1]
E-mail: xchen@siue.edu
Zhu Zhang[1]
E-mail: zhuzhang1989@gmail.com
Ryan Fries[2]
E-mail: rfries@siue.edu
[1] Department of Industrial Engineering, Southern Illinois University Edwardsville, Edwardsville, IL 62026-1805, USA.
[2] Department of Civil Engineering, Southern Illinois University Edwardsville, Edwardsville, IL 62026-1800, USA.

References

Almoustafa, S., Hanafi, S., & Mladenović, N. (2013). New exact method for large asymmetric distance-constrained vehicle routing problem. *European Journal of Operational Research, 226*, 386–394. http://dx.doi.org/10.1016/j.ejor.2012.11.040

Ando, N., & Taniguchi, E. (2006). Travel time reliability in vehicle routing and scheduling with time windows. *Networks and Spatial Economics, 6*, 293–311. http://dx.doi.org/10.1007/s11067-006-9285-8

Andres Figliozzi, M. (2012). The time dependent vehicle routing problem with time windows: Benchmark problems, an efficient solution algorithm, and solution characteristics. *Transportation Research Part E: Logistics and Transportation Review, 48*, 616–636. http://dx.doi.org/10.1016/j.tre.2011.11.006

Balas, E., & Toth, P. (1985). Branch and bound methods. In E. L. Lawler, J. K. Lenstra, A. H. G. Rinnooy Kan, & D. G. Shmoys (Eds.), *The traveling salesman problem* (pp. 361–401). Chichester: Wiley.

Barabasi, A. L. (2002). *Linked: The new science of networks*. Cambridge, MA: Perseus.

Bräysy, O., & Gendreau, M. (2005). Vehicle routing problem with time windows, Part I: Route construction and local search algorithms. *Transportation Science, 39*, 104–118. http://dx.doi.org/10.1287/trsc.1030.0056

Brinkhoff, T. (2002). A framework for generating network-based moving objects. *GeoInformatica, 6*, 153–180. http://dx.doi.org/10.1023/A:1015231126594

Chen, H. K., Hsueh, C. F., & Chang, M. S. (2006). The real-time time-dependent vehicle routing problem. *Transportation Research Part E: Logistics and Transportation Review, 42*, 383–408. http://dx.doi.org/10.1016/j.tre.2005.01.003

Chen, B. Y., Lam, W. H. K., Sumalee, A., Li, Q., Shao, H., & Fang, Z. (2013). Finding reliable shortest paths in road networks under uncertainty. *Networks and Spatial Economics, 13*, 123–148. http://dx.doi.org/10.1007/s11067-012-9175-1

Clarke, G., & Wright, J. V. (1964). Scheduling of vehicles from a central depot to a number of delivery points. *Operations Research, 12*, 568–581. http://dx.doi.org/10.1287/opre.12.4.568

Dantzig, G. B., Fulkerson, D. R., & Johnson, S. M. (1954). Solution of a large-scale traveling salesman problem. *Operations Research, 2*, 393–410.

Dreyfus, S. E. (1969). An appraisal of some shortest-path algorithms. *Operations Research, 17*, 395–412. http://dx.doi.org/10.1287/opre.17.3.395

GAMS Development Corporation. (2013). *GAMS Distribution 24.0.2*. Retrieved from https://www.gams.com/

Haghani, A., & Jung, S. (2005). A dynamic vehicle routing problem with time-dependent travel times. *Computers & Operations Research, 32*, 2959–2986.

Haimovich, M., Rinnooy Kan, A. H. G., & Stougie, L. (1988). Analysis of heuristic routing problems. In B. L. Golden & A. Assad (Eds.), *Vehicle routing: Methods and studies* (pp. 47–61). Amsterdam: North Holland.

Ichoua, S., Gendreau, M., & Potvin, J. Y. (2003). Vehicle dispatching with time-dependent travel times. *European Journal of Operational Research, 144*, 379–396. http://dx.doi.org/10.1016/S0377-2217(02)00147-9

Jeong, H. (2003). Complex scale-free networks. *Physica A: Statistical Mechanics and its Applications, 321*, 226–237. http://dx.doi.org/10.1016/S0378-4371(02)01774-0

Kaufman, D. E., & Smith, R. L. (1993). Fastest paths in time-dependent networks for intelligent vehicle-highway systems application. *Journal of Intelligent Transportation Systems, 1*(1), 1–11.

Kritzinger, S., Doerner, K. F., Hartl, R. F., Kiechle, G. Ÿ., Stadler, H., & Manohar, S. S. (2012). Using traffic information for time-dependent vehicle routing. *Procedia—Social and Behavioral Sciences, 39*, 217–229. http://dx.doi.org/10.1016/j.sbspro.2012.03.103

Kuhn, H. W. (1955). The Hungarian method for the assignment problem. *Naval Research Logistics Quarterly, 2*, 83–97. http://dx.doi.org/10.1002/(ISSN)1931-9193

Laporte, G., Nobert, Y., & Taillefer, S. (1987). A branch-and-bound algorithm for the asymmetrical distance-constrained vehicle routing problem. *Mathematical Modelling, 9*, 857–868. http://dx.doi.org/10.1016/0270-0255(87)90004-2

Laporte, G., Nobert, Y., & Taillefer, S. (1988). Solving a family of multi-depot vehicle routing and location-routing problems. *Transportation Science, 22*, 161–172. http://dx.doi.org/10.1287/trsc.22.3.161

Lecluyse, C., Van Woensel, T., & Peremans, H. (2009). Vehicle routing with stochastic time-dependent travel times. *4OR-Q Journal of Operations Research, 7*, 363–377.

Maden, W., Eglese, R., & Black, D. (2010). Vehicle routing and scheduling with time-varying data: A case study. *Journal of the Operational Research Society, 61*, 515–522. http://dx.doi.org/10.1057/jors.2009.116

Montoya-Torres, J. R., Franco, J. L., Isaza, S. N., Jiménez, H. F., & Herazo-Padilla, N. (2015). A literature review on the vehicle routing problem with multiple depots. *Computers & Industrial Engineering, 79*, 115–129.

Pentico, D. W. (2007). Assignment problems: A golden anniversary survey. *European Journal of Operational Research, 176*, 774–793. http://dx.doi.org/10.1016/j.ejor.2005.09.014

Rakha, H., El-Shawarby, I., Arafeh, M., & Dion, F. (2006). *Estimating path travel time reliability*. In *Proceedings of 2006 IEEE Intelligent Transportation Systems Conference* (pp. 236–241), Toronto.

Spliet, R., & Gabor, A. F. (2012). The time window assignment vehicle routing problem. *Erasmus School of Economics (ESE), EI2012_07*, 1–19.

Van Woensel, T., Kerbache, L., Peremans, H., & Vandaele, N. (2008). Vehicle routing with dynamic travel times: A queueing approach. *European Journal of Operational Research, 186*, 990–1007. http://dx.doi.org/10.1016/j.ejor.2007.03.012

Vidal, T., Crainic, T. G., Gendreau, M., & Prins, C. (2014). A unified solution framework for multi-attribute vehicle routing problems. *European Journal of Operational Research, 234*, 658–673. http://dx.doi.org/10.1016/j.ejor.2013.09.045

Zare-Reisabadi, E., & Mirmohammadi, S. H. (2015). Site dependent vehicle routing problem with soft time window: Modeling and solution approach. *Computers & Industrial Engineering, 90*, 177–185.

11

A nonparametric approach to the estimation of jump-diffusion models with asymmetric kernels

Muhammad Hanif[1]*

*Corresponding author: Muhammad Hanif, Department of Mathematics and Statistics, PMAS-Arid Agriculture University, Rawalpindi, Pakistan

E-mail: mhpuno@hotmail.com

Reviewing editor: Zudi Lu, University of Southampton, UK

Abstract: This paper presents the nonparametric estimation of first and second infinitesimal moments of the underlying jump-diffusion model with asymmetric kernel functions. In particular, we use asymmetric kernel estimators characterized by the gamma distribution. This approach allows to conciliate the idea of using the asymmetric kernel with jump-diffusion models. We show that the proposed estimators are consistent and asymptotically follow normal distribution under the conditions of recurrence and stationarity.

Subjects: Economics, Finance, Business & Industry; Mathematics & Statistics; Science; Social Sciences

Keywords: gamma Nadaraya–Watson estimators; Harris recurrence; jump-diffusion model; local time; nonparametric estimation; stochastic differential equation

AMS subject classifications: 62G99; 60J75

1. Introduction

Over the past few decades, nonparametric methods have become increasingly popular research fields e.g. statistics, economics, and probability communities, and such a trend undoubtedly will continue in future. The main advantage of the nonparametric methods is that they are helpful to estimate the coefficients of underlying model in a flexible way. In this paper, we study the nonparametric estimation of infinitesimal moments of jump-diffusion models with asymmetric kernel functions. Our approach allows us to conciliate the idea of using the asymmetric kernel functions and jump-diffusion models. We show that the proposed estimators are consistent and asymptotically follow normal distribution under the conditions of recurrence as well as for stationarity.

In nonparametric estimation, the traditional kernel smoothing has been playing a wide role in estimating continuous-time diffusion processes since many years. Most of the kernels which are used in the literature are symmetric kernels and fixed. Several researchers have used nonparametric

ABOUT THE AUTHOR

Muhammad Hanif is an associate professor in the Department of Mathematics and Statistics, PMAS-Arid Agriculture University, Rawalpindi, Pakistan. His major research interests are Statistical Inference for Stochastic Processes, Mathematical Statistics, Nonparametric Estimation.

PUBLIC INTEREST STATEMENT

Nonparametric methods have become increasingly popular research fields e.g. statistics, economics, and probability communities. The main advantage of the nonparametric methods is that they are helpful to estimate the coefficients of underlying model in a flexible way. This study concerned with the nonparametric estimation of infinitesimal moments of jump-diffusion models with asymmetric kernel functions. The author showed that the proposed estimators are consistent and asymptotically follow normal distribution under the conditions of recurrence as well as for stationarity.

methods based on traditional kernel smoothing and studied the diffusion model and jump-diffusion models (cf. Arapis & Gao, 2006; Bandi & Phillips, 2003; Nicolau, 2003; Stanton, 1997). Furthermore, the bias of the standard kernel estimator is larger near the boundary than in the interior. Previously, many authors have proposed the methods to remove the boundary bias (see e.g. Fan & Gijbels, 1992; Rice, 1984; Zhang, Karunamuni, & Jones, 1999 and many others).

Asymmetric kernels are useful for the density estimation with a number of advantages such as they are free from boundary bias and are always nonnegative. They also achieve the optimal rate of convergence within the class of nonnegative kernel density estimators. Chen (2000) studied that replacing fixed kernels with asymmetric kernels substantially had increased the precision of density estimation close to boundary. Hanif (2015) studied the nonparametric estimation of drift coefficient and diffusion coefficient in the second-order diffusion equation by using the asymmetric kernel functions. He proved that the proposed estimators of second-order diffusion equation are consistent and asymptotically follow normal distribution. Other successful applications of asymmetric kernels include Bertin and Klutchnikoff (2011), Bouezmarni and Scaillet (2005), Gustafsonn, Hagmann, Nielsen, and Scaillet (2009), Hagmann and Scaillet (2007), Scaillet (2004), and Shunpu (2010); where such kernels are shown to lead desirable properties.

A continuous-time model governed by the stochastic differential equation has the following form:

$$dX_t = \mu(X_t)dt + \sigma(X_t)dW_t, \tag{1.1}$$

where $\{W_t,\ 0 \leq t \leq T\}$ is a standard Brownian motion. The functions $\mu(x)$ and $\sigma^2(x)$ are drift and diffusion functions of the process. The diffusion model defined in Equation (1.1) is widely used to describe the properties of underlying economic variables. Gospodinovy and Hirukawa (2012) studied the gamma kernel for diffusion model and obtained the asymptotic normality for the drift and the diffusion functions. While, we have studied the asymmetric kernel functions for the jump-diffusion models. The purpose of the present paper is to study gamma and beta Nadaraya–Watson estimators of first and second infinitesimal moments of jump-diffusion models. We obtain the consistency and asymptotic normality of estimators under recurrence and stationarity. Furthermore, we study the infinite jump activity case in finite interval. In other words, the jumps parts in this paper can be compensated.

The rest of the paper is organized as follows: Section 2 introduces the jump-diffusion models and asymmetric kernel functions. This Section defines the conditional moments of the jump-diffusion model and derive the gamma and beta kernel estimators for the conditional moments and Section 3 presents assumptions, some useful preliminary results about the local time and the main results of our paper. Section 4 contains the proofs of the presented results.

2. Jump-diffusion model and asymmetric kernel functions

During the last decade, jump-diffusion models have been considered as a valuable tool in the modeling of stochastic differential equations. The jump-diffusion models have been widely used in a variety of financial applications. Usually, jump-diffusion models consist of two parts, the Brownian motion and the Poisson process. The first attempt to incorporate jumps into diffusion model was made by Merton (1976). Later on, jump-diffusion models have been extended in various directions by specifying different structures for the drift, diffusion, and jump components. Kou (2002) and Ramezani and Zeng (2007) used an asymmetric double exponential distribution for log of jump part and showed that the resulting model could capture asymmetric leptokurtic features and volatility smile features frequently observed in financial data. For other contributions on jump-diffusion processes and their applications, see, Aït-Sahalia and Jacod (2009), Bandi and Nguyen (2003), Beckers (1981), and the references therein.

Our motivation is to study the nonparametric estimation for jump-diffusion model stated below:

$$dX_t = [\mu(X_{t-}) - \lambda(X_{t-})\int_Y c(X_{t-}, y)\Pi(dy)]dt + \sigma(X_{t-})dW_t + dJ_t, \tag{2.1}$$

where $\mu(x)$ and $\sigma(x)$ are smooth functions, $\{W_t, t \geq 0\}$ is a standard Brownian motion, $\{J_t, t \geq 0\}$ is a jump process independent of $\{W_t, t \geq 0\}$, and $\lambda(x)$ is the conditional intensity of the jumps. Furthermore, we use $c(\cdot, \xi)$ to characterize the impact of jumps, where ξ is a random variable with range Y and $\Pi(y)$ is the probability distribution function of ξ. There are several studies on specification of J_t. For example, $J_t = \xi N$, where N is a Poisson process with an intensity $\lambda(X_t)$ and ξ is the jump size having the time-invariant distribution $\Pi(y)$. We denote $\Delta X_t = X_t - X_{t-}$, and have

$$dJ_t = \Delta X_t = \int_Y c(X_{t-}, y)N(dt, dy). \tag{2.2}$$

The integral form is as follows:

$$X_{t+\Delta} = X_t + \int_t^{t+\Delta} \mu(X_{s-})ds + \int_t^{t+\Delta} \sigma(X_{s-})dW_s + \int_t^{t+\Delta}\int_Y c(X_{t-}, y)\bar{\nu}(ds, dy), \tag{2.3}$$

where

$$\bar{\nu}(ds, dy) = N(dt, dy) - E(N(dt, dy)) = N(dt, dy) - \lambda(X_{t-})\Pi(dy)dt \tag{2.4}$$

is a compensated Poisson random measure. From Equations (2.2) and (2.4), we have

$$\int_t^{t+\Delta}\int_Y c(X_{s-}, y)\bar{\nu}(ds, dy) = \int_t^{t+\Delta} dJ_s - \int_t^{t+\Delta} \lambda(X_{s-})E_Y[c(X_{s-}, \xi)]ds, \tag{2.5}$$

and it represents the variation between t and $t + \Delta$ due to discontinuous jumps. The conditional moments of the jump-diffusion models are defined by the following relations (cf. Bandi & Nguyen, 2003):

$$M^1(x) := \lim_{\Delta\to 0}\frac{1}{\Delta}E[X_{t+\Delta} - X_t | X_t = x] = \mu(x), \tag{2.6}$$

$$M^2(x) := \lim_{\Delta\to 0}\frac{1}{\Delta}E[(X_{t+\Delta} - X_t)^2 | X_t = x] = \sigma^2(x) + \lambda(x)E_Y[c^2(x, \xi)] \tag{2.7}$$

Many researchers have used nonparametric techniques to study the jump-diffusion models. Schmisser (2014) discussed the nonparametric adaptive estimation of the drift for a jump-diffusion process and provided the bounds for the risks of the two estimators for ergodic and strictly stationary. Lin and Wang (2010) considered the empirical likelihood inference for jump-diffusion model. Hanif, Lin, and Wang (2012) discussed nonparametric estimation of the second infinitesimal moment by using the reweighted Nadaraya–Watson approach of the underlying jump-diffusion model and established the strong consistency and asymptotic normality of the second infinitesimal moment. Hanif (2012) studied the nonparametric estimations of first and second infinitesimal moments based on the local linear estimators for jump-diffusion models.

In fixed symmetric kernel, the allocation of weights is outside the density support when smoothing is made and it causes the boundary bias. To avoid the boundary bias, Chen (2000) proposed a gamma kernel function as the gamma density function that has flexible shapes and lies within [0, ∞). The gamma kernel function is defined as

$$K_{G(x/b+1,\, b)}(a) = \frac{a^{x/b}\exp(-a/b)}{b^{x/b+1}\Gamma(x/b+1)},$$

where $\Gamma(m) = \int_0^\infty y^{m-1}exp(-y)dy,\ m > 0$ is the gamma function, x is the design point, and b the smoothing parameter. The density function of gamma distribution having support $[0, \infty)$ implies that gamma kernel function does not generate the boundary bias. Furthermore, the asymptotic properties of the gamma estimators depend on the position of the design point x.

Beta kernel function can be written as

$$K_{B(x,\ b)}(a) = \frac{a^{x/b}(1-a)^{(1-x)/b}}{B\{x/b+1,(1-x)/b+1\}},$$

where $B(l,m) = \int_0^1 y^{l-1}(1-y)^{m-1},\ l,m > 0$ is the beta function and b is the smoothing parameter. The density function of beta distribution has support [0, 1]. Beta kernel function has unique features as the shape and the amount of smoothing change according to the position of the design points x. Beta kernel assigns no weight outside the data support and is free from the boundary bias.

Remark 1 In fixed kernel function, one works by using $K(\frac{X_{i\Delta_{n,T}}-a}{h})$, whereas in asymmetric kernel uses $K_{G(x/b+1,\ b)}(X_{i\Delta_{n,T}})$. Furthermore, $K_{G(x/b+1,\ b)}(X_{i\Delta_{n,T}})$ is not fixed and varies according to the design point x. The Gamma Nadaraya-Watson estimators of $M^1(x)$ and $M^2(x)$ are denoted by $\hat{M}^1_{NW}(x,b)$ and $\hat{M}^2_{NW}(x,b)$ defined by

$$\hat{M}^1_{NW}(x,b) = \frac{\sum_{i=1}^{n-1} K_{G(x/b+1,\ b)}(X_{i\Delta_{n,T}})\left(X_{(i+1)\Delta_{n,T}} - X_{i\Delta_{n,T}}\right)}{\Delta_{n,T}\sum_{i=1}^{n} K_{G(x/b+1,\ b)}(X_{i\Delta_{n,T}})} \tag{2.8}$$

and

$$\hat{M}^2_{NW}(x,b) = \frac{\sum_{i=1}^{n-1} K_{G(x/b+1,\ b)}(X_{i\Delta_{n,T}})\left(X_{(i+1)\Delta_{n,T}} - X_{i\Delta_{n,T}}\right)^2}{\Delta_{n,T}\sum_{i=1}^{n} K_{G(x/b+1,\ b)}(X_{i\Delta_{n,T}})}, \tag{2.9}$$

respectively. Whereas, $T \to \infty$, $\Delta_{n,T} := T/n \to 0$ and $b = b_{n,T} \to 0$ as $n \to \infty$. Similarly beta Nadaraya-Watson estimators of $M^1(x)$ and $M^2(x)$ are obtained by replacing $K_{G(x/b+1,\ b)}$ with $K_{B(x,\ b)}$ in Equations (2.8) and (2.9). We obtain the consistency and asymptotic normality of the proposed estimators under recurrence and stationarity.

3. Assumptions and main results

The assumptions we need for the proofs of the main results are listed below for convenient reference. Assume that $\mathfrak{I} = [0, \infty)$ is the range of the process X_t. Let

$$s(\alpha) = \int_{z_0}^{\alpha} \exp\left\{\int_{z_0}^{z} -\frac{2\mu(x)}{\sigma^2(x)}dx\right\}dz$$

be the scale density function and $m(x) = 2(\sigma^2(x)s'(x))^{-1}$ denotes the speed function of the process X_t (see Bandi & Phillips, 2003). Moreover, z_0 is any fixed point belonging to the interval specified above.

A1.

(i) The coefficients $\mu(\cdot)$, $\sigma(\cdot)$, $c(\cdot,y)$, and $\lambda(\cdot)$ have continuous derivatives of order two and satisfy

$$|\mu(x)-\mu(z)| + |\sigma(x)-\sigma(z)| + \lambda(x)\int_Y |c(x,y)-c(z,y)|\Pi(dy) \le C_1|x-z| \tag{3.1}$$

and

$$|\mu(x)| + |\sigma(x)| + \lambda(x)\int_Y |c(x,y)|\Pi(dy) \le C_2\{1+|x|\}. \tag{3.2}$$

(ii) There exist constants C_3 and $\alpha > 2$ such that

$$\lambda(x)\int_Y |c(x,y)|^\alpha \Pi(dy) \leq C_3\{1+|x|^\alpha\}. \tag{3.3}$$

(iii) $\lambda(\cdot) \geq 0$ and $\sigma^2(\cdot) > 0$.

Remark 2 Assumption A1 assures existence and uniqueness of a strong solution to the stochastic differential Equation (2.1).

A2.

The solution to Equation (2.1) is positive Harris recurrent.

Remark 3 Harris recurrence guarantees the existence and uniqueness of invariant measure $s(dx)$. This assumption is much weaker than the usual assumption.

A3.

$T \to \infty, \Delta_{n,T}: = T/n \to 0$ and $b = b_{n,T} \to 0$ as $n \to \infty$ such that

$$\left(\frac{\bar{L}_X(T,x)}{b}\right)\left(\Delta_{n,T}\log(\frac{1}{\Delta_{n,T}})\right)^{\frac{1}{2}} = o_{a.s.}(1), \tag{3.4}$$

where $\bar{L}_X(T,x)$ is defined in Lemma 3.

A4.

$T \to \infty, \Delta_{n,T}: = T/n \to 0$ and $b = b_{n,T} \to 0$ as $n \to \infty$ such that

$$\left(\frac{T}{b}\right)\left(\Delta_{n,T}\log(\frac{1}{\Delta_{n,T}})\right)^{\frac{1}{2}} = o_{a.s.}(1).$$

Now we describe results regarding the local time that will be useful in obtaining the main results of this paper (see, Revuz & Yor, 2003).

LEMMA 1 *Let the process X_t be a semimartingale with quadratic variation $[X]_s^c$ and $L_X(t,a)$ be local time at a. Then for every positive bounded Borel measurable function g*

$$\int_0^\infty L_X(t,a)g(a)da = \int_0^t g(X_{s-})d[X]_s^c \text{ a.s.}$$

LEMMA 2 *Let X be a semimartingale satisfying $\sum_{0<s\leq t}|\Delta X_s| < \infty$ a.s. $\forall t > 0$. Then, for any a and t, we have*

$$\bar{L}_X(t,a): = \lim_{\varepsilon\to 0}\frac{1}{2\varepsilon}\int_0^t 1_{(|X_t-a|\leq\varepsilon)}d[X]_s^c = \frac{1}{\sigma^2(a)}L_X(t,a) \text{ a.s.}$$

LEMMA 3 *Let X_t be the process defined by Equation (2.1) and $b = b_{n,T} \to 0$ as $n \to \infty$ see, e.g. Bandi and Nguyen (2003). If*

$$\frac{1}{b}\left(\Delta_{n,T}\log(\frac{1}{\Delta_{n,T}})\right)^{\frac{1}{2}} = o(1), \tag{3.5}$$

then

$$\hat{L}_X(T,x) \xrightarrow{a.s.} \bar{L}_X(T,x). \tag{3.6}$$

It is well known that for recurrent diffusion process, the local time for a fixed T can be consistently estimated by

$$\hat{L}_X(T,x) = \Delta_{n,T} \sum_{i=1}^{n} K_{G(x/b+1,\, b)}(X_{i\Delta_{n,T}}) \tag{3.7}$$

based on asymmetric kernel (gamma kernel). Similarly, we can obtain results for beta kernel function by replacing $K_{G(x/b+1,\, b)}$ with $K_{B(x,\, b)}$.

We now give the main results of our paper.

THEOREM 1 *Under assumptions A1–A3,*

$$\hat{M}^1_{NW}(x,b) \xrightarrow{a.s.} M^1(x).$$

Furthermore, if $b^{5/2}\bar{L}_X(T,x) = O_{a.s.}(1)$, then for "interior x"

$$\sqrt{b^{1/2}\bar{L}_X(T,x)}\left(\hat{M}^1_{NW}(x,b) - M^1(x) - bB_{\hat{M}^1_{NW}(x,b)}\right) \Rightarrow N\left(0, \frac{M^2(x)}{2\sqrt{\pi}}x^{1/2}\right), \tag{3.8}$$

where $B_{\hat{M}^1_{NW}(x,b)}$ denotes bias of the estimator $\hat{M}^1_{NW}(x,b)$, given by

$$B_{\hat{M}^1_{NW}(x,b)} = (M^1(x))'\left\{1 + \frac{xs'(x)}{s(x)}\right\} + \frac{x}{2}(M^1(x))''.$$

Similarly for "boundary x", if $b^3\bar{L}_X(T,x) = O_{a.s.}(1)$, then

$$\sqrt{b\bar{L}_X(T,x)}\left(\hat{M}^1_{NW}(x,b) - M^1(x) - bB_{\hat{M}^1_{NW}(x,b)}\right) \Rightarrow N\left(0, \frac{M^2(x)\Gamma(2d+1)}{2^{2d+1}\Gamma^2(d+1)}\right). \tag{3.9}$$

THEOREM 2 *Under assumptions A1–A3,*

$$\hat{M}^2_{NW}(x,b) \xrightarrow{a.s.} M^2(x).$$

Furthermore, if $\frac{b^{5/2}\bar{L}_X(T,x)}{\Delta_{n,T}} = O_{a.s.}(1)$, then for "interior x"

$$\sqrt{\frac{b^{1/2}\bar{L}_X(T,x)}{\Delta_{n,T}}}\left(\hat{M}^2_{NW}(x,b) - M^2(x) - bB_{\hat{M}^2_{NW}(x,b)}\right) \Rightarrow N\left(0, \frac{2(M^2(x))^2}{\sqrt{\pi}x^{1/2}}\right), \tag{3.10}$$

where

$$B_{\hat{M}^2_{NW}(x,b)} = (M^2(x))'\left\{1 + \frac{xs'(x)}{s(x)}\right\} + \frac{x}{2}(M^2(x))''.$$

Similarly for "boundary x", if $\frac{b^3\bar{L}_X(T,x)}{\Delta_{n,T}} = O_{a.s.}(1)$, then

$$\sqrt{\frac{b\bar{L}_X(T,x)}{\Delta_{n,T}}}\left(\hat{M}^1_{NW}(x,b) - M^2(x) - bB_{\hat{M}^2_{NW}(x,b)}\right) \Rightarrow N\left(0, \frac{4(M^2(x))^2\Gamma(2d+1)}{2^{2d+1}\Gamma^2(d+1)}\right). \tag{3.11}$$

In available literature, the asymptotic law of gamma and beta Nadaraya–Watson estimators can also be obtained by assuming the stationarity of the process X_t. We assume that the speed function $m(x)$ satisfies $\int_0^\infty m(x)dx < \infty$, where $f(x)$ is the stationary distribution. Furthermore, it requires that $\bar{L}_X(T,x)/T \xrightarrow{a.s.} f(x)$ and $s'(x)/s(x) \xrightarrow{a.s.} f'(x)/f(x)$. The results are given for such an idea in the following corollaries for "interior x" and "boundary x", respectively.

COROLLARY 1 Under assumptions A1 and A4,

$\hat{M}^1_{NW}(x,b) \xrightarrow{a.s.} M^1(x).$

Then for "interior x", if $b = O(T^{-2/5})$, then

$$\sqrt{Tb^{1/2}}\left(\hat{M}^1_{NW}(x,b) - M^1(x) - bB_{\hat{M}^1_{NW}(x,b)}\right) \Rightarrow N\left(0, \frac{M^2(x)}{2\sqrt{\pi}x^{1/2}f(x)}\right), \tag{3.12}$$

where $B_{\hat{M}^1_{NW}(x,b)}$ denotes bias of the estimator $\hat{M}^1_{NW}(x,b)$, given by

$$B_{\hat{M}^1_{NW}(x,b)} = (M^1(x))'\left\{1 + \frac{xf'(x)}{f(x)}\right\} + \frac{x}{2}(M^1(x))''.$$

Similarly for "boundary x", if $b = O(T^{-1/3})$, then

$$\sqrt{Tb}\left(\hat{M}^1_{NW}(x,b) - M^1(x) - bB_{\hat{M}^1_{NW}(x,b)}\right) \Rightarrow N\left(0, \frac{M^2(x)\Gamma(2d+1)}{2^{2d+1}\Gamma^2(d+1)f(x)}\right). \tag{3.13}$$

COROLLARY 2 Under assumptions A1 and A4,

$\hat{M}^2_{NW}(x,b) \xrightarrow{a.s.} M^2(x).$

Then for "interior x", if $b = O(T^{-2/5})$,

$$\sqrt{Tb^{1/2}}\left(\hat{M}^2_{NW}(x,b) - M^2(x) - bB_{\hat{M}^2_{NW}(x,b)}\right) \Rightarrow N\left(0, \frac{2(M^2(x))^2}{\sqrt{\pi}x^{1/2}f(x)}\right), \tag{3.14}$$

where

$$B_{\hat{M}^2_{NW}(x,b)} = (M^2(x))'\left\{1 + \frac{xf'(x)}{f(x)}\right\} + \frac{x}{2}(M^2(x))''.$$

Furthermore, for "boundary x", if $b = O(T^{-1/3})$, then

$$\sqrt{Tb}\left(\hat{M}^2_{NW}(x,b) - M^2(x) - bB_{\hat{M}^2_{NW}(x,b)}\right) \Rightarrow N\left(0, \frac{4(M^2(x))^2\Gamma(2d+1)}{2^{2d+1}\Gamma^2(d+1)f(x)}\right). \tag{3.15}$$

An important feature of the gamma kernel estimators is that the variance coefficient decreases as x increases. This property does not hold with any of the fixed symmetric kernels. The large value of x reduces the variance while it increases the bias of gamma kernel function. The shape of gamma kernel varies along with the position of x and the smoothing bandwidth can be changed from one place to another.

The results for the beta kernel are obtained by replacing x with $x(1-x)$ in variance and bias of gamma kernel. For "interior x" design point x satisfies $x/b \to \infty, (1-x)/b \to \infty$ and for "boundary x" it satisfies $x/b \to d$, $(1-x)/b \to d$ for some $d > 0$ as $n \to \infty$. Furthermore, the standard theory on beta distribution has mean $(x+b)/(1+2b)$ and variance $b(x+b)(1-x+b)/\{(1+2b)^2(1+3b)\}$, respectively.

4. Proofs

Proof of Lemma 3. By Lemma 1 and 2, we obtain

$$\begin{aligned}\int_0^T K_{G(x/b+1,\,b)}(X_{s-})\frac{d[X,X]_s}{\sigma^2(X_{s-})} &= \int_0^\infty K_{G(x/b+1,\,b)}(a)\frac{L_X(T,a)}{\sigma^2(a)}da \\ &= \bar{L}_X(T,x)\end{aligned} \tag{4.1}$$

with probability one, around $a = x$.

In order to obtain the main results, it is necessary to show

$$\Delta_{n,T}\sum_{i=1}^{n} K_{G(x/b+1,\ b)}(X_{i\Delta_{n,T}}) - \int_0^T K_{G(x/b+1,\ b)}(X_{s-})ds \to 0 \tag{4.2}$$

which is equivalent to

$$\sum_{i=0}^{n-1}\int_{iT/n}^{(i+1)T/n}\left[K_{G(x/b+1,\ b)}(X_{i\Delta_{n,T}}) - K_{G(x/b+1,\ b)}(X_{s-})\right]ds$$
$$-\Delta_{n,T}K_{G(x/b+1,\ b)}(X_0) + \Delta_{n,T}K_{G(x/b+1,b)}(X_{n\Delta_{n,T}}) \to 0. \tag{4.3}$$

The left of Equation (4.3) is bounded by

$$\left|\sum_{i=0}^{n-1}\int_{iT/n}^{(i+1)T/n}\left[K_{G(x/b+1,b)}(X_{i\Delta_{n,T}}) - K_{G(x/b+1,\ b)}(X_{s-})\right]ds\right|$$
$$+\Delta_{n,T}\left|K_{G(x/b+1,\ b)}(X_0)\right| + \Delta_{n,T}\left|K_{G(x/b+1,\ b)}(X_{n\Delta_{n,T}})\right|$$
$$\le \sum_{i=0}^{n-1}\int_{iT/n}^{(i+1)T/n}\left|K'_{G(x/b+1,\ b)}(\tilde{X}_{is-})(X_{s-} - X_{i\Delta_{n,T}})\right|ds$$
$$+\Delta_{n,T}\left|K_{G(x/b+1,\ b)}(X_0)\right| + \Delta_{n,T}\left|K_{G(x/b+1,\ b)}(X_{n\Delta_{n,T}})\right|$$
$$\le \sum_{i=0}^{n-1}\int_{iT/n}^{(i+1)T/n}\left|K'_{G(x/b+1,\ b)}(\tilde{X}_{is-})\right|\,|X_{s-} - X_{i\Delta_{n,T}}|ds$$
$$+\Delta_{n,T}\left|K_{G(x/b+1,\ b)}(X_0)\right| + \Delta_{n,T}\left|K_{G(x/b+1,\ b)}(X_{n\Delta_{n,T}})\right| \tag{4.4}$$

where $\tilde{X}_{is-}$ is the some value between $X_{i\Delta_{n,T}}$ and X_{s-}. Let

$$k_{n,T} = \max_{i\le n}\sup_{i\Delta_{n,T}\le s\le (i+1)\Delta_{n,T}}|X_{i\Delta_{n,T}} - X_{s-}|.$$

It follows from the Lévy continuity of modulus for diffusion process that

$$k_{n,T} = O_{a.s.}\left(\sqrt{\Delta_{n,T\log\frac{1}{\Delta_{n,T}}}}\right). \tag{4.5}$$

Using Equation (4.5) and A3, we acquire

$$\left|K'_{G(x/b+1,b)}(\tilde{X}_{is-})\right| = \left|K'_{G(x/b+1,\ b)}(X_{s-} + O_{a.s.}(\sqrt{\Delta_{n,T\log\frac{1}{\Delta_{n,T}}}}))\right|$$
$$= \left|K'_{G(x/b+1,\ b)}(X_{s-} + o_{a.s.}(1)\right|.$$

The first term on the right-hand side of Equation (4.4)

$$\le k_{n,T}\sum_{i=0}^{n-1}\int_{iT/n}^{(i+1)T/n}\left|K'_{G(x/b+1,\ b)}(X_{s-} + o_{a.s.}(1)\right|ds$$
$$= k_{n,T}\int_0^T\left|K'_{G(x/b+1,\ b)}(X_{s-} + o_{a.s.}(1)\right|ds$$
$$= \frac{k_{n,T}}{b}\int_0^\infty b\left|K'_{G(x/b+1,\ b)}(a + o_{a.s.}(1)\right|\bar{L}_X(T,a)da$$
$$\le \frac{k_{n,T}}{b}O_{a.s.}(\bar{L}_X(T,x))$$

around $a = x$. Similarly we can also show the remaining quantities in Equation (4.4). It completes the proof of Lemma 3.

4.1. Proof of Theorem 1

To prove the consistency of $\hat{M}^1_{NW}(x, b)$, we begin by writing

$$\begin{aligned}
\hat{M}^1_{NW}(x,b) &= \frac{\sum_{i=1}^{n-1} K_{G(x/b+1,\, b)}(X_{i\Delta_{n,T}})\left(X_{(i+1)\Delta_{n,T}} - X_{i\Delta_{n,T}}\right)}{\Delta_{n,T}\sum_{i=1}^{n} K_{G(x/b+1,\, b)}(X_{i\Delta_{n,T}})} \\
&= \frac{\sum_{i=1}^{n-1} K_{G(x/b+1,\, b)}(X_{i\Delta_{n,T}}) \int_{i\Delta_{n,T}}^{(i+1)\Delta_{n,T}} \mu(X_{s-})ds}{\Delta_{n,T}\sum_{i=1}^{n} K_{G(x/b+1,\, b)}(X_{i\Delta_{n,T}})} \\
&\quad + \frac{\sum_{i=1}^{n-1} K_{G(x/b+1,\, b)}(X_{i\Delta_{n,T}}) \int_{i\Delta_{n,T}}^{(i+1)\Delta_{n,T}} \sigma(X_{s-})dW_s}{\Delta_{n,T}\sum_{i=1}^{n} K_{G(x/b+1,\, b)}(X_{i\Delta_{n,T}})} \qquad (4.6)\\
&\quad + \frac{\sum_{i=1}^{n-1} K_{G(x/b+1,\, b)}(X_{i\Delta_{n,T}}) \int_{i\Delta_{n,T}}^{(i+1)\Delta_{n,T}} \int_Y c(X_{s-}, y)\bar{v}(ds, dy)}{\Delta_{n,T}\sum_{i=1}^{n} K_{G(x/b+1,\, b)}(X_{i\Delta_{n,T}})} \\
&=: A_{11} + B_{11} + C_{11}.
\end{aligned}$$

We start with

$$\begin{aligned}
A_{11} &= \frac{\sum_{i=1}^{n-1} K_{G(x/b+1,\, b)}(X_{i\Delta_{n,T}}) \int_{i\Delta_{n,T}}^{(i+1)\Delta_{n,T}} \mu(X_{s-})ds}{\Delta_{n,T}\sum_{i=1}^{n} K_{G(x/b+1,\, b)}(X_{i\Delta_{n,T}})} \\
&= \frac{\Delta_{n,T}\sum_{i=1}^{n-1} K_{G(x/b+1,\, b)}(X_{i\Delta_{n,T}})(\mu(X_{i\Delta_{n,T}} + o_{a.s.}(1))}{\Delta_{n,T}\sum_{i=1}^{n} K_{G(x/b+1,\, b)}(X_{i\Delta_{n,T}})} \\
&= \frac{\bar{L}_X(T,a)\mu(a)}{\bar{L}_X(T,a)} \\
&\xrightarrow{a.s.} \mu(x)
\end{aligned}$$

which implies that $A_{11} \xrightarrow{a.s.} M^1(x)$, around $a = x$. Now consider B_{11}, the strong law of large number for martingale difference with zero mean and finite variance Hall and Heyde (1986), yields $B_{11} \xrightarrow{a.s.} 0$ which can be proved by using Knight's embedding theorem Revuz and Yor (2003). Since C_{11} is martingale, it implies that $C_{11} \xrightarrow{a.s.} 0$. It completes the proof of the consistency of $\hat{M}^1_{NW}(x, b)$.

To derive the asymptotic normality of $\hat{M}^1_{NW}(x, b)$, write

$$\begin{aligned}
\hat{M}^1_{NW}(x,b) - M^1(x) &= \frac{\sum_{i=1}^{n-1} K_{G(x/b+1,\, b)}(X_{i\Delta_{n,T}})\left(X_{(i+1)\Delta_{n,T}} - X_{i\Delta_{n,T}}\right)}{\Delta_{n,T}\sum_{i=1}^{n} K_{G(x/b+1,\, b)}(X_{i\Delta_{n,T}})} - M^1(x) \\
&= \frac{\Delta_{n,T}\sum_{i=1}^{n-1} K_{G(x/b+1,\, b)}(X_{i\Delta_{n,T}})\frac{1}{\Delta_{n,T}}\left(\int_{i\Delta_{n,T}}^{(i+1)\Delta_{n,T}} \mu(X_{s-})ds - M^1(x)\Delta_{n,T}\right)}{\Delta_{n,T}\sum_{i=1}^{n} K_{G(x/b+1,\, b)}(X_{i\Delta_{n,T}})} \\
&\quad + \frac{\sum_{i=1}^{n-1} K_{G(x/b+1,\, b)}(X_{i\Delta_{n,T}}) \int_{i\Delta_{n,T}}^{(i+1)\Delta_{n,T}} \sigma(X_{s-})dW_s}{\Delta_{n,T}\sum_{i=1}^{n} K_{G(x/b+1,\, b)}(X_{i\Delta_{n,T}})} \\
&\quad + \frac{\sum_{i=1}^{n-1} K_{G(x/b+1,\, b)}(X_{i\Delta_{n,T}}) \int_{i\Delta_{n,T}}^{(i+1)\Delta_{n,T}} \int_Y c(X_{s-}, y)\bar{v}(ds, dy)}{\Delta_{n,T}\sum_{i=1}^{n} K_{G(x/b+1,\, b)}(X_{i\Delta_{n,T}})} \\
&=: A_{22} + B_{11} + C_{11}.
\end{aligned}$$

We evaluate these terms separately.

Using A2 and Quotient Limit Theorem for Harris recurrent Markov processes Aźema, Kaplan-Duflo, and Revuz (1967), A_{22} can be written as

$$\frac{\Delta_{n,T}\sum_{i=1}^{n-1}K_{G(x/b+1,\ b)}(X_{i\Delta_{n,T}})\left(\mu(X_{i\Delta_{n,T}})-M^1(x)\right)+o_{a.s.}(1)}{\Delta_{n,T}\sum_{i=1}^{n}K_{G(x/b+1,\ b)}(X_{i\Delta_{n,T}})}$$
$$=\frac{\int_0^T K_{G(x/b+1,\ b)}(X_{s-})(\mu(X_{s-})-M^1(x))ds+o_{a.s.}(1)}{\int_0^T K_{G(x/b+1,\ b)}(X_{s-})ds+o_{a.s.}(1)}$$
$$=\frac{\int_0^\infty K_{G(x/b+1,\ b)}(a)(\mu(a)-M^1(x))s(a)da+o_{a.s.}(1)}{\int_0^\infty K_{G(x/b+1,\ b)}(a)s(a)da+o_{a.s.}(1)},$$

where $s(dx)$ is the $\sigma-$finite invariant measure. The standard theory on gamma distribution has mean $x+b$ and variance $xb+b^2$. Using the Taylor expansion for $\mu(a)-M^1(x)$ around $a=x$, we get

$$\frac{\int_0^\infty K_{G(x/b+1,\ b)}(a)(\mu(a)-M^1(x))s(a)da}{\int_0^\infty K_{G(x/b+1,\ b)}(a)s(a)da}$$
$$=\left[(M^1(x))'\left\{1+\frac{xs'(x)}{s(x)}\right\}+\frac{x}{2}(M^1(x))''\right]b+o(b). \tag{4.7}$$

The involvement of s' in the bias is less desirable, which is due to the fact that x is not the mean of gamma kernel rather it is the mode.

Now we consider B_{11} and C_{11}. To obtain the results, we use the quadratic variation process. We take numerator of $[B_{11}, B_{11}]$

$$\sum_{i=1}^{n-1}K^2_{G(x/b+1,\ b)}(X_{i\Delta_{n,T}})\int_{i\Delta_{n,T}}^{(i+1)\Delta_{n,T}}\sigma^2(X_{s-})ds$$
$$=\Delta_{n,T}\sum_{i=1}^{n-1}K^2_{G(x/b+1,\ b)}(X_{i\Delta_{n,T}})\frac{1}{\Delta_{n,T}}\int_{i\Delta_{n,T}}^{(i+1)\Delta_{n,T}}\sigma^2(X_{s-})ds \tag{4.8}$$
$$=\int_0^T K^2_{G(x/b+1,\ b)}(X_{s-}+o_{a.s.}(1))(\sigma^2(X_{s-}+o_{a.s.}(1))ds.$$

Define

$$A_b(x)=b^{-1}\Gamma(2x/b+1)/2^{2x/b+1}\Gamma^2(x/b+1).$$

Then (4.8) can be written as

$$\int_0^T K^2_{G(x/b+1,\ b)}(X_{s-})\sigma^2(X_{s-})ds=A_b(x)\int_0^T K_{G(2x/b+1,\ b)}(X_{s-})\sigma^2(X_{s-})ds$$
$$=A_b(x)\int_0^\infty K_{G(2x/b+1,\ b)}(a)\sigma^2(a)\bar{L}_X(T,a)da$$

by using Lemma 1. Similarly the numerator of $[C_{11}, C_{11}]$ yields

$$\sum_{i=1}^{n-1}K^2_{G(x/b+1,\ b)}(X_{i\Delta_{n,T}})\int_{i\Delta_{n,T}}^{(i+1)\Delta_{n,T}}\int_Y c^2(X_{s-},y)\bar{v}(ds,dy)$$
$$=A_b(x)\int_0^\infty K_{G(2x/b+1,\ b)}(a)\left(\lambda(a)\int_Y c^2(a,y)\Pi(y)dy\right)\bar{L}_X(T,a)da. \tag{4.9}$$

From Chen (2000), we have

$$A_b(x) = \begin{cases} \frac{b^{-1/2}}{2\sqrt{\pi}x^{1/2}} + o(b^{-1/2}) & \text{for interior x} \\ \frac{b^{-1}\Gamma(2d+1)}{2^{2d+1}\Gamma^2(d+1)} + o(b^{-1}) & \text{for boundary x.} \end{cases} \quad (4.10)$$

Hence, for "interior x"

$$b^{1/2}[B_{11}, B_{11}] \xrightarrow{a.s.} \frac{\sigma^2(x)}{2\sqrt{\pi}x^{1/2}\bar{L}_X(T,x)}$$

and

$$b^{1/2}[C_{11}, C_{11}] \xrightarrow{a.s.} \frac{\left(\lambda(x)\int_Y c^2(x,y)\Pi(y)dy\right)}{2\sqrt{\pi}x^{1/2}\bar{L}_X(T,x)}.$$

Therefore

$$\sqrt{b^{1/2}\bar{L}_X(T,x)}B_{11} \Rightarrow N\left(0, \frac{\sigma^2(x)}{2\sqrt{\pi}x^{1/2}}\right) \quad (4.11)$$

and

$$\sqrt{b^{1/2}\bar{L}_X(T,x)}C_{11} \Rightarrow N\left(0, \frac{\left(\lambda(x)\int_Y c^2(x,y)\Pi(y)dy\right)}{2\sqrt{\pi}x^{1/2}}\right). \quad (4.12)$$

If $b^{5/2}\bar{L}_X(T,x) = O_{a.s.}(1)$, we obtain

$$\sqrt{b^{1/2}\bar{L}_X(T,x)}\left(\hat{M}^1_{NW}(x,b) - M^1(x) - bB_{\hat{M}^1_{NW}(x,b)}\right) \Rightarrow N\left(0, \frac{M^2(x)}{2\sqrt{\pi}x^{1/2}}\right),$$

where $B_{\hat{M}^1_{NW}(x,b)}$ denotes bias of the estimator $\hat{M}^1_{NW}(x,b)$, given by

$$B_{\hat{M}^1_{NW}(x,b)} = (M^1(x))'\left\{1 + \frac{xs'(x)}{s(x)}\right\} + \frac{x}{2}(M^1(x))''.$$

The results for "boundary x" can be established similarly. It completes the Proof of Theorem 1.

4.2. Proof of Theorem 2

To prove the consistency of $\hat{M}^2_{NW}(x,b)$, note that

$$\begin{aligned}(X_{(i+1)\Delta_{n,T}} - X_{i\Delta_{n,T}})^2 &= X^2_{(i+1)\Delta_{n,T}} - X^2_{i\Delta_{n,T}} - 2X_{i\Delta_{n,T}}[X_{(i+1)\Delta_{n,T}} - X_{i\Delta_{n,T}}] \\ &= \int_{i\Delta_{n,T}}^{(i+1)\Delta_{n,T}} (\sigma^2(X_{s-}) + (\int_Y c^2(X_{s-},y)\Pi(dy))\lambda(X_{s-}))ds \\ &+ 2\int_{i\Delta_{n,T}}^{(i+1)\Delta_{n,T}} (X_{s-} - X_{i\Delta_{n,T}})\mu(X_{s-})ds \\ &+ 2\int_{i\Delta_{n,T}}^{(i+1)\Delta_{n,T}} (X_{s-} - X_{i\Delta_{n,T}})\sigma(X_{s-})dW_s \\ &+ \int_{i\Delta_{n,T}}^{(i+1)\Delta_{n,T}} \int_Y ((X_{s-} + c)^2 - X^2_{s-} - 2X_{i\Delta_{n,T}}c(X_{s-},y))\bar{\nu}(ds,dy).\end{aligned}$$

Inserting the four terms into $\hat{M}^2_{NW}(x,b)$, we seprate $\hat{M}^2_{NW}(x,b)$ into four parts $A_{33}+B_{33}+C_{33}+D_{33}$.

The first term

$$A_{33}=\frac{\sum_{i=1}^{n-1}K_{G(x/b+1,\,b)}(X_{i\Delta_{n,T}})\int_{i\Delta_{n,T}}^{(i+1)\Delta_{n,T}}(\sigma^2(X_{s-})+(\int_Y c^2(X_{s-},y)\Pi(dy))\lambda(X_{s-}))ds}{\Delta_{n,T}\sum_{i=1}^{n}K_{G(x/b+1,b)}(X_{i\Delta_{n,T}})} \tag{4.13}$$

$$\xrightarrow{\text{a.s.}}\frac{\bar{L}_X(T,a)(\sigma^2(a)+\int_Y c^2(a,y)\Pi(dy)\lambda(x))}{\bar{L}_X(T,a)}$$

$$\xrightarrow{\text{a.s.}}\sigma^2(x)+\int_Y c^2(x,y)\Pi(dy)\lambda(x)=\sigma^2(x)+E_Y[c^2(x,\xi)]\lambda(x)=M^2(x) \tag{4.14}$$

around $a=x$.

We now consider B_{33}. Previous arguments imply that

$$\int_{i\Delta_{n,T}}^{(i+1)\Delta_{n,T}}(X_{s-}-X_{i\Delta_{n,T}})\mu(X_{s-})ds=O_{a.s.}(\Delta_{n,T}\log(\frac{1}{\Delta_{n,T}}))^{\frac{1}{2}}\int_{i\Delta_{n,T}}^{(i+1)\Delta_{n,T}}\mu(X_{s-})ds,$$

then

$$B_{33}=\frac{2\sum_{i=1}^{n-1}K_{G(x/b+1,\,b)}(X_{i\Delta_{n,T}})\int_{i\Delta_{n,T}}^{(i+1)\Delta_{n,T}}(X_{s-}-X_{i\Delta_{n,T}})\mu(X_{s-})ds}{\Delta_{n,T}\sum_{i=1}^{n}K_{G(x/b+1,\,b)}(X_{i\Delta_{n,T}})} \tag{4.15}$$

$$\xrightarrow{\text{a.s.}}0.$$

Similarly, we can also obtain

$$C_{33}\xrightarrow{\text{a.s.}}0 \tag{4.16}$$

and

$$D_{33}\xrightarrow{\text{a.s.}}0. \tag{4.17}$$

Combining Equations (4.14), (4.15), (4.16), and (4.17), we get

$$\hat{M}^2_{NW}(x,b)\xrightarrow{\text{a.s.}}M^2(x).$$

It shows the consistency of estimator $\hat{M}^2_{NW}(x,b)$.

For asymptotic normality of $\hat{M}^2_{NW}(x,b)$, we proceed as

$$
\begin{aligned}
&\hat{M}^2_{NW}(x,b)-M^2(x)\\
&=\frac{\Delta_{n,T}\sum_{i=1}^{n-1}K_{G(x/b+1,\,b)}(X_{i\Delta_{n,T}})\frac{1}{\Delta_{n,T}}\left[\int_{i\Delta_{n,T}}^{(i+1)\Delta_{n,T}}(\sigma^2(X_{s-})+(\int_Y c^2(X_{s-},y)\Pi(dy))\lambda(X_{s-}))ds-M^2(x)\Delta_{n,T}\right]}{\Delta_{n,T}\sum_{i=1}^{n}K_{G(x/b+1,\,b)}(X_{i\Delta_{n,T}})}\\
&+\frac{2\sum_{i=1}^{n-1}K_{G(x/b+1,\,b)}(X_{i\Delta_{n,T}})\int_{i\Delta_{n,T}}^{(i+1)\Delta_{n,T}}(X_{s-}-X_{i\Delta_{n,T}})\mu(X_{s-})ds}{\Delta_{n,T}\sum_{i=1}^{n}K_{G(x/b+1,\,b)}(X_{i\Delta_{n,T}})}\\
&+\frac{2\sum_{i=1}^{n-1}K_{G(x/b+1,\,b)}(X_{i\Delta_{n,T}})\int_{i\Delta_{n,T}}^{(i+1)\Delta_{n,T}}(X_{s-}-X_{i\Delta_{n,T}})\sigma(X_{s-})dW_s}{\Delta_{n,T}\sum_{i=1}^{n}K_{G(x/b+1,\,b)}(X_{i\Delta_{n,T}})}\\
&+\frac{\sum_{i=1}^{n-1}K_{G(x/b+1,\,b)}(X_{i\Delta_{n,T}})+\int_{i\Delta_{n,T}}^{(i+1)\Delta_{n,T}}\int_Y((X_{s-}+c)^2-X_{s-}^2-2X_{i\Delta_{n,T}}c(X_{s-},y))\bar{v}(ds,dy).}{\Delta_{n,T}\sum_{i=1}^{n}K_{G(x/b+1,\,b)}(X_{i\Delta_{n,T}})}\\
&=:A_{44}+B_{44}+C_{44}+D_{44}.
\end{aligned}
$$

First, we consider

$$
\begin{aligned}
A_{44}&=\frac{\Delta_{n,T}\sum_{i=1}^{n-1}K_{G(x/b+1,\,b)}(X_{i\Delta_{n,T}})\frac{1}{\Delta_{n,T}}[\int_{i\Delta_{n,T}}^{(i+1)\Delta_{n,T}}(\sigma^2(X_{s-})+(\int_Y c^2(X_{s-},y)\Pi(dy))\lambda(X_{s-}))ds-M^2(x)\Delta_{n,T}]}{\Delta_{n,T}\sum_{i=1}^{n}K_{G(x/b+1,\,b)}(X_{i\Delta_{n,T}})}\\
&=\frac{\Delta_{n,T}\sum_{i=1}^{n-1}K_{G(x/b+1,\,b)}(X_{i\Delta_{n,T}})[(\sigma^2(X_{i\Delta_{n,T}})+(\int_Y c^2(x,y)\Pi(dy))\lambda(x))-M^2(x)]+o_{a.s.}(1)}{\Delta_{n,T}\sum_{i=1}^{n}K_{G(x/b+1,\,b)}(X_{i\Delta_{n,T}})}\\
&=\frac{\int_0^\infty K_{G(x/b+1,\,b)}(a)[(\sigma^2(a)+(\int_Y c^2(x,y)\Pi(dy))\lambda(x))-M^2(x)]s(a)da+o_{a.s.}(1)}{\int_0^\infty K_{G(x/b+1,\,b)}(a)s(a)da+o_{a.s.}(1)},
\end{aligned}
$$

where $s(dx)$ is the $\sigma-$ finite invariant measure. Taylor expansion for $(\sigma^2(a)+(\int_Y c^2(x,y)\Pi(dy))\lambda(x))-M^2(x)$ around $a=x$ gives

$$
\left[(M^2(x))'\left\{1+\frac{xs'(x)}{s(x)}\right\}+\frac{x}{2}(M^2(x))''\right]b+o(b). \tag{4.18}
$$

The quadratic variation process of C_{44} provides that

$$
\begin{aligned}
[C_{44},C_{44}]&=\frac{4\sum_{i=1}^{n-1}K^2_{G(x/b+1,\,b)}(X_{i\Delta_{n,T}})(\int_{i\Delta_{n,T}}^{(i+1)\Delta_{n,T}}(X_{s-}-X_{i\Delta_{n,T}}))^2\sigma^2(X_{s-})ds}{[\Delta_{n,T}\sum_{i=1}^{n}K_{G(x/b+1,\,b)}(X_{i\Delta_{n,T}})]^2}\\
&=\frac{4\Delta^2_{n,T}\sum_{i=1}^{n-1}K^2_{G(x/b+1,\,b)}(X_{i\Delta_{n,T}})(\sigma^4(X_{i\Delta_{n,T}})+o_{a.s.}(1))}{[\Delta_{n,T}\sum_{i=1}^{n}K_{G(x/b+1,\,b)}(X_{i\Delta_{n,T}})]^2}.
\end{aligned}
$$

Following Equation (4.8), for "interior x"

$$
\frac{b^{1/2}}{\Delta_{n,T}}[C_{44},C_{44}]\xrightarrow{a.s.}\frac{2\sigma^4(x)}{\sqrt{\pi}x^{1/2}\bar{L}_X(T,x)}
$$

and

$$
\sqrt{\frac{b^{1/2}\bar{L}_X(T,x)}{\Delta_{n,T}}}C_{44}\Rightarrow N\left(0,\frac{2\sigma^4(x)}{\sqrt{\pi}x^{1/2}}\right). \tag{4.19}
$$

For D_{44}, it is similar to (4.9). We omit it here.

If $\frac{b^{5/2}\bar{L}_X(T,x)}{\Delta_{n,T}} = O_{a.s.}(1)$, then we get

$$\sqrt{\frac{b^{1/2}\bar{L}_X(T,x)}{\Delta_{n,T}}}\left(\hat{M}^2_{NW}(x,b) - M^2(x) - bB_{\hat{M}^2_{NW}(x,b)}\right) \Rightarrow N\left(0, \frac{2(M^2(x))^2}{\sqrt{\pi}x^{1/2}}\right),$$

where

$$B_{\hat{M}^2_{NW}(x,b)} = (M^2(x))'\left\{1 + \frac{xs'(x)}{s(x)}\right\} + \frac{x}{2}(M^2(x))''.$$

Similarly, we can obtain the results for "boundary x". This leads to the Proof of Theorem 2.

For beta kernel function, Chen (2000) provided that

$$A_b(x) = \begin{cases} \frac{b^{-1/2}}{2\sqrt{\pi}\sqrt{x(1-x)}} + o(b^{-1/2}) & \text{if} x/b \text{ and } (1-x)/b \to \infty \\ \frac{b^{-1}\Gamma(2d+1)}{2^{2d+1}\Gamma^2(d+1)} + o(b^{-1}) & \text{if} x/b \text{ and} (1-x)/b \to d, \end{cases}$$

where d is a nonnegative constant. Furthermore, to get the results, it needs to define

$$A_b(x) = B\{(2x/b+1), 2(1-x)/b+1\}/B^2\{(x/b+1), (1-x)/b+1\}.$$

The rest of the procedure is similar to gamma kernel as prescribed above.

Our findings indicate that the corollaries can be proved on the same steps as followed in the case of theorems.

Funding
The author received no direct funding for this research.

Author details
Muhammad Hanif[1]
E-mail: mhpuno@hotmail.com
[1] Department of Mathematics and Statistics, PMAS-Arid Agriculture University, Rawalpindi, Pakistan.

References

Aït-Sahalia, Y., & Jacod, J. (2009). Testing for jumps in a discretely observed process. *Annals of Statistics, 37*, 84-222.

Arapis, M., & Gao, J. (2006). Empirical comparisons in short-term interest rate models using nonparametric methods. *Journal of Financial Econometrics, 4*, 310-345.

Aźema, J., Kaplan-Duflo, M., & Revuz, D. (1967). Measure invariante sur les classes recurrents des processus de Markov. *Zeit Wahrsh Verwand Gebie, 8*, 157-181.

Bandi, F. M., & Nguyen, T. H. (2003). On the functional estimation of jump-diffusion models. *Journal of Econometrics, 116*, 293-328.

Bandi, F. M., & Phillips, P. (2003). Fully nonparametric estimation of scalar diffusion models. *Econometrica, 71*, 241-283.

Beckers, S. (1981). A note on estimating the parameters of the diffusion jump model of stock returns. *Journal of Financial and Quantitative Analysis, 16*, 127-140.

Bertin, K., & Klutchnikoff, N. (2011). Minimax properties of beta kernel estimators. *Journal of Statistical Planning and Inferences, 124*, 2287-2297.

Bouezmarni, T., & Scaillet, O. (2005). Consistency of asymmetric kernel density estimators and smoothed histograms with application to income data. *Econometric Theory, 21*, 390-412.

Chen, S. X. (2000). Probability density function estimation using gamma kernels. *Annals of the Institute of Statistical Mathematics, 52*, 471-480.

Fan, J., & Gijbels, I. (1992). Variable bandwidth and local linear regression smoothers. *Annals of Statistics, 20*, 2008-2036.

Gospodinovy, N., & Hirukawa, M. (2012). Nonparametric estimation of scalar diffusion processes of interest rates using asymmetric kernels. *Journal of Empirical Finance, 19*, 595-609.

Gustafsonn, J., Hagmann, M., Nielsen, J. P., & Scaillet, O. (2009). Local transformation kernel density estimation of loss distributions. *Journal of Business and Economic Statistics, 27*, 161-175.

Hall, P., & Heyde, C. C. (1986). *Martingale limit theory and its applications*. New York, NY: Academic Press.

Hagmann, M., & Scaillet, O. (2007). Local multiplicative bias correction for asymmetric kernel density estimators. *Journal of Econometrics, 141*, 213-249.

Hanif, M. (2012). Local linear estimation of recurrent jump-diffusion models. *Communications in Statistics: Theory and Methods, 41*, 4142-4163.

Hanif, M., Lin, Z. Y., & Wang, H. C. (2012). Reweighted Nadaraya--Watson estimator of jump-diffusion models. *Science China Mathematics, 55*, 1005–1016.
Hanif, M. (2015). Nonparametric estimation of second order diffusion models by using asymmetric kernels. *Communications in Statistics: Theory and Methods, 44*, 1896–1910.
Kou, S. G. (2002). A jump-diffusion model for option pricing. *Management Science, 48*, 1086–1101.
Lin, Z. Y., & Wang, H. C. (2010). Empirical likelihood inference for diffusion processes with jumps. *Science China Mathematics, 53*, 1805–1816.
Merton, R. C. (1976). Option pricing when underlying stock returns are discontinuous. *Journal of Financial Economics, 3*, 125–144.
Nicolau, J. (2003). Bias reduction in nonparametric diffusion coefficient estimation. *Econometric Theory, 19*, 754–777.
Ramezani, C. A., & Zeng, Y. (2007). Maximum likelihood estimation of the double exponential jump-diffusion process. *Annals of Finance, 3*, 487–507.
Revuz, D., & Yor, M. (2003). *Continuous Martingales and Brownian motion*. New York, NY: Springer.
Rice, J. (1984). Boundary modification for kernel regression. *Communications in Statistics: Theory and Methods, 13*, 893–900.
Scaillet, O. (2004). Density estimation using inverse and reciprocal inverse Gaussian kernels. *Journal of Nonparametric Statistics, 16*, 217–226.
Schmisser, É. (2014). Non-parametric adaptive estimation of the drift for a jump diffusion process. *Stochastic Processes and Their Applications, 124*, 883–914.
Shunpu, Z. (2010). A note on the performance of the gamma kernel estimators at the boundary. *Statistics and Probability Letters, 80*, 548–557.
Stanton, R. (1997). A nonparametric model of term structure dynamics and the market price of interest rate risk. *Journal of Finance, 52*, 1973–2002.
Zhang, S., Karunamuni, R. J., & Jones, M. C. (1999). An improved estimator of the density function at the boundary. *Journal of the American Statistical Association, 94*, 1231–1241.

12

Stress pulsation mechanism during filling and discharging granular materials form silos

Waseem Ghazi Alshanti[1]*

*Corresponding author: Waseem Ghazi Alshanti, Department of Mathematics, University of Hail, Hail, Kingdom of Saudi Arabia

E-mails: w.alshanti@uoh.edu.sa, waseemalshanti@yahoo.com

Reviewing editor: Benchawan Wiwatanapataphee, Curtin University, Australia

Abstract: In this paper, we investigate numerically the dynamic pulsation nature of stresses acting on hopper walls during filling and discharging granular materials from silos. From the results, at the end of filling process, it can be concluded that the larger the frictional wall coefficient is, the less the ultimate stress values acting on the hopper wall will be. Furthermore, for smaller values of internal friction, a consistent stress fluctuation appears for a slightly long period of time before the stress trace turns to a straight line. For the discharging process, the required time for the discharging of the silo is almost the same for different values of wall frictions. Moreover, for large values of internal friction coefficients, the discharging process takes more time.

Subjects: Engineering & Technology; Mathematics & Statistics; Science; Technology

Keywords: granular material; discrete element method; stress pulsation

1. Introduction

The silo system has been considered as one of the emerging issues in the granular materials research over many years. Past and current research efforts focus on investigating many aspects relevant to the silo system. Problems in silo, such as flow blockage, jamming, segregation, dead zones, and silo collapse attract the interest of researchers and raise the importance to study the storage vessels. Great efforts have been made to closely study aspects such as flow pattern of granular material through silos, stress field propagation along silo walls, force fluctuation at the boundary of granular flow in silo, and the effect of particle shape on silo discharge.

ABOUT THE AUTHOR

Waseem Ghazi Alshanti received the PhD degree in numerical analysis from the Department of Mathematics and Statistics, Curtin University, Australia, in 2010 where he worked as a lecturer from 2007 to 2010. Based on his excellent results he achieved in his PhD studies, he was awarded Curtin University completion scholarship as well as the mathematics honours and postgrad prize. Currently, he is working as Assistant Professor of Mathematics at the University of Hail. His research interests and areas of expertise include: Applied mathematical modeling (Fluid dynamics, granular flow, geomechanics theory, continmuum mechanics theory), Image processing, Difference equations, and advanced real analysis.

PUBLIC INTEREST STATEMENT

The silo system has been considered as one of the emerging issues in the granular materials research over many years. It is known that granular materials cannot be characterized as gas and fluid or solid. Discrete element methods (DEM's) are numerical techniques capable of simulating the entire behavior of systems of discrete interacting elements. In the present paper, by using this numerical simulation technique, we study the effects of the friction between the particles themselves and between the particles and the boundary walls in a silo system during filling and discharging processes. It is found that, for large values of friction between the particles' system, the discharging process takes more time. Moreover, the required time for the discharging of the silo is almost the same for different values of particle/wall frictions.

Recently, many experimental methods have been conducted in the analysis of wall stress measurements. For instance, Chou, Tzeng, Smid, Kuo, and Hsiau (2000) investigated the pressure of moving granular bed acting on the side walls. Force fluctuation at the boundary of a two-dimensional granular flow was investigated by Longhi, Easwar, and Menon (2002). Simultaneous measurements of both the flow pattern and wall pressure in a full scale silo storing granular solids were conducted by Chen, Rotter, Ooi, and Zhong (2006).

Extensive discrete approaches in two and three dimensions have been conducted to investigate problems relevant to silo operation. Many aspects such as filling and discharging under gravity of non-cohesive discs in two and three dimensions from silos have been carried out via discrete element Newtonian dynamics simulations (Langston, Tüzün, et al., 1995). Pressure and shear forces that acting on the wall of an out-flowing hopper using discrete element simulations has been measured (Ristow & Herrmann, 1995). Depending on various material and geometrical parameters, the discharge process has been investigated in two-dimensional hopper on various material and geometrical parameters (Cleary & Sawley, 2002). It was shown that the outflow rate does not depend on the restitution coefficient and the transition from funnel flow to mass flow behavior is clearly observed when the hopper angle increases. Discrete element method supported by an averaging method has been employed to quantitatively investigate the velocity and stress distribution of a hopper flow (Zhu & Yu, 2003). Their results indicate that the discrete element technique is a powerful tool for macroscopic analysis of the dynamic behavior of hopper flow. By using large-scale discrete element computer simulations, the change of the vertical stress profiles with dimensionality has been investigated by Landry, Grest, et al. (2004). A general method was proposed by Ristow (1997) for performing discrete element simulations of huge system of particles through partitioning them into sub-domains that are analyzed separately. The effect of wide selection of particle shapes on hopper discharging has been investigated in many attempts. It was shown that the circular particles have little resistance and shear forces such results are due to Masson and Martinez (2000). Further attempts that concern with the effect of particle mechanical properties on filling and discharging granular materials from silos have been conducted (Goda & Ebert, 2005; Yang & Hsiau, 2001). More and more attempts have been directed toward studying the behavior of granular flow during filling and discharging granular materials from silos (Hirshfeld & Rapaport, 2001; Nguyen, Cogné, Guessasma, Bellenger, & Fortin, 2009; Parisi, Masson, & Martinez, 2004; Sykut, Molenda, & Horabik, 2008). In the present work, we establish a discrete element model and numerical algorithm to investigate the pulsation profiles of stresses acting on hopper walls during filling and discharging granular materials from silos. In order to determine the points at which the stress pulsation exhibits more vacillations, different values of coefficients for both wall and internal frictions are considered to study the influences of these parameters on the stress pulsation nature during the processes of filling and discharging.

2. Implementation of DEM model

The present numerical simulations are based on the soft-particle discrete method which originally was proposed by Cundall and Strack (1979). In our simulation model, the granular particles are subjected, during simulation, to two types of forces including contact forces and gravitational force. Unlike the gravitational body force, the particle contact forces act only when the particle is in contact with other particles and/or with a boundary wall. The normal contact force is determined by a damped linear spring model, while the tangential contact force is determined by a linear spring in series with a frictional sliding element model. The interaction forces developed between two particles or between particle and boundary wall are calculated based on the physical properties and the relative velocities. The dynamic equations for the motion for each of the particles under these forces are

$$m_i \frac{d^2\vec{r}_i(t)}{dt^2} = \vec{F}_{gi} + \sum_{j=1, j\neq i}^{N_p} \vec{F}_{c,ij} + \sum_{k=1}^{N_w} \vec{F}_{c,iw_k}, \tag{1}$$

$$I_i \frac{d^2\vec{\theta}_i(t)}{dt^2} = \sum_{j=1, j\neq i}^{N_p} \left(\vec{r}_i \times \vec{F}_{c,ij}\right) + \sum_{k=1}^{N_w} \left(\vec{r}_i \times \vec{F}_{c,iw_k}\right), \tag{2}$$

where m_i, I_i, $\vec{r}_i$, and $\vec{\theta}_i(t)$ are, respectively, the mass, rotational moment of inertia, position, and rotational vectors of the centre of particle i, $\vec{F}_{gi}$, $\vec{F}_{c,ij}$ and $\vec{F}_{c,iw_k}$ are, respectively, gravitational body force and contact forces acting on particle i due to particle j and boundary wall w_k, N_p and N_w are, respectively, the number of particles and walls in the simulation. For simplicity, we denote

$$\sum_{j=1, j\neq i}^{N_p} \vec{F}_{c,ij} + \sum_{k=1}^{N_w} \vec{F}_{c,iw_k} = \vec{F}_{c,i}, \tag{3}$$

$$\sum_{j=1, j\neq i}^{N_p} \left(\vec{r}_i \times \vec{F}_{c,ij}\right) + \sum_{k=1}^{N_w} \left(\vec{r}_i \times \vec{F}_{c,iw_k}\right) = \vec{M}_{c,i}. \tag{4}$$

Therefore, Equations (1 and 2) become

$$m_i \frac{d^2\vec{r}_i}{dt^2}(t) = \vec{F}_{gi} + \vec{F}_{c,i}, \tag{5}$$

$$I_i \frac{d^2\vec{\theta}_i}{dt^2}(t) = \vec{M}_{c,i}. \tag{6}$$

Unlike the gravitational body force, the particle contact forces act only when the particle is in contact with other particles and/or with a boundary wall. In the analysis of the system, any contact force is decomposed into normal and tangential components. For any typical pair of particles i and j, if they are in contact, then the contact force is determined by the following formula

$$\vec{F}_{ij}(t) = \vec{F}_{ij,\hat{n}}(t)\hat{n} + \vec{F}_{ij,\hat{s}}(t)\hat{s}, \tag{7}$$

where $\hat{n}$ and $\hat{s}$ are unit vectors in the normal and shear directions of the contact plane, $\vec{F}_{\hat{n},ij}(t)$ and $\vec{F}_{\hat{s},ij}(t)$ are, respectively, the magnitudes of the normal contact force and shear contact force, namely,

$$\vec{F}_{ij,\hat{n}}(t) = -k_{\hat{n}}\, \vec{\delta}_{ij,\hat{n}}(t), \tag{8}$$

$$\vec{F}_{ij,\hat{s}}(t) = -\text{sign}\left[\vec{\delta}_{ij,\hat{s}}(t)\right] \cdot \min\left\{ k\hat{s}\left|\vec{\delta}_{ij,\hat{s}}(t)\right|, \mu\vec{F}^{e}_{ij,\hat{n}}(t) \right\} \tag{9}$$

where $k_{\hat{n}}$ and $k_{\hat{s}}$ are, respectively, the particle-particle normal and tangential spring coefficients, $\vec{F}^{e}_{ij,\hat{n}}(t)$ and μ are, respectively, the elastic contribution of the contact force between the particles i and j in the normal direction ($\hat{n}$ direction) and the friction coefficient of the granular particles, $\vec{\delta}_{ij,\hat{n}}(t) = \left(R_i + R_j\right) - \left|\vec{r}_i(t) - \vec{r}_j(t)\right|$ and $\vec{\delta}_{ij,\hat{s}}(t) = \int_{t_0}^{t} \left(\dot{\vec{r}}_i(\eta) - \dot{\vec{r}}_j(\eta)\right) \cdot \hat{s}\, d\eta$ are, respectively, the normal compression and the tangential displacement between the particles and over the time step $\Delta t = t - t_0$, R_i and R_j are the radii of the particles i and j. In order to reduce the number of contact checks for each particle, a neighboring cell technique is implemented in the present simulations. Once all forces and moments acting on particle due to contacts are identified, new velocities and positions of all particles are computed by numerical integration of Newton's equation of motion. Setting

$$\vec{v}_i = \frac{d\vec{r}_i}{dt}, \quad \vec{\omega}_i = \frac{d\vec{\theta}_i}{dt}. \tag{10}$$

Therefore, $\forall i = 1, 2, 3, \ldots, N_p$, we have a system of first-order ordinary differential equations as follows

$$\frac{d\vec{r}_i(t)}{dt} = \vec{v}_i(t), \tag{11}$$

$$\frac{d\vec{\theta}_i(t)}{dt} = \vec{\omega}_i(t), \tag{12}$$

$$\frac{d\vec{v}_i(t)}{dt} = \frac{\vec{F}_{gi}(t) + \vec{F}_{c,i}(t)}{m_i} \tag{13}$$

$$\frac{d\vec{\omega}_i(t)}{dt} = \frac{\vec{M}_{c,i}(t)}{I_i}. \tag{14}$$

Thus, setting

$$\Psi(t) = \begin{bmatrix} \vec{r}_1(t) \\ \vec{\theta}_1(t) \\ \vdots \\ \vec{r}_{N_p}(t) \\ \vec{\theta}_{N_p}(t) \end{bmatrix}, \quad \Phi(t) = \begin{bmatrix} \vec{v}_1(t) \\ \vec{\omega}_1(t) \\ \vdots \\ \vec{v}_{N_p}(t) \\ \vec{\omega}_{N_p}(t) \end{bmatrix}, \quad F(t) = \begin{bmatrix} \left(\vec{F}_{gi}(t) + \vec{F}_{c,1}(t)\right)/m_1 \\ \vec{M}_{c,1}(t)/I_1 \\ \vdots \\ \left(\vec{F}_{gN_p}(t) + \vec{F}_{c,N_p}(t)\right)/m_{N_p} \\ \vec{M}_{c,N_p}(t)/I_{N_p} \end{bmatrix}, \quad \forall i = 1, 2, \ldots, N_p,$$

yields a system of $4N_p$ first-order ordinary differential equations that represents the dynamic equations for the motion of the N_p particles and can be expressed by

$$\frac{d\Psi(t)}{dt} = \Phi(t), \tag{15}$$

$$\frac{d\Phi(t)}{dt} = F(t). \tag{16}$$

By using the central difference scheme and considering that $\vec{F}_{c,i}$, $\vec{M}_{c,i}$, $\vec{v}_i$, $\vec{\omega}_i$, $\dot{\vec{v}}_i$, and $\dot{\vec{\omega}}_i$ are all constants over the time interval $\left[t_{N-\frac{1}{2}}, t_{N+\frac{1}{2}}\right]$, (16) becomes

$$\Phi^{N+\frac{1}{2}} = \Phi^{N-\frac{1}{2}} + F^N \cdot \Delta t. \tag{17}$$

Over $\left[t_N, t_{N+1}\right]$, further application of the central difference scheme to (15) at time $t_{N+\frac{1}{2}}$ yields,

$$\Psi^{N+1} = \Psi^N + \Phi^{N+\frac{1}{2}} \cdot \Delta t. \tag{18}$$

Substituting (17) into (18), one can obtain new translational and rotational displacements at time $t = N + 1$ as follows

Table 1. Baseline parameters used in the simulation	
Parameter	**Value (dimensionless)**
Number of particles, N_p	7000
Particle diameter, d/d_0 (where d_0 = 1.0 mm)	0.81–0.99 (uniform distribution)
Particle density, ρ/ρ_0 (where $\rho_0 = 1.0 \times 10^3$ kg / m^3)	2.5
Particle-particle normal spring stiffness, $k_{n,pp}/(\rho_0 \vec{g} d_0^2)$	1.266×10^5
Particle-particle normal dashpot coefficient, $C_{n,pp}/(\rho_0 \vec{g}^{\,0.5} d_0^{2.5})$	2.524×10^2
Particle-particle tangential spring stiffness, $k_{s,pp}/(\rho_0 \vec{g} d_0^2)$	1.266×10^5
Particle-particle friction coefficient, μ_{pp}	1.0
Particle-wall normal spring stiffness, $k_{n,pw}$	1.566×10^5
Particle-wall normal dashpot coefficient, $c_{n,pw}/(\rho_0 \vec{g}^{\,0.5} d_0^{2.5})$	2.01×10^2
Particle-wall tangential spring stiffness, $k_{s,pw}/(\rho_0 \vec{g} d_0^2)$	3.006×10^4
Particle-wall friction coefficient, μ_{pw}	1.0
Gravity in x-direction, $\|\vec{g}_x\|/\|\vec{g}\|$	0.0
Gravity in y-direction, $\|\vec{g}_y\|/\|\vec{g}\|$	-9.8
Time step, $\Delta t \sqrt{\|\vec{g}\|/d_0}$	5.37×10^{-6}

$$\Psi^{N+1} = \Psi^N + \Phi^{N-\frac{1}{2}} \cdot \Delta t + F^N \cdot (\Delta t)^2. \quad (19)$$

The new translational and rotational velocities $\vec{v}_i$ and $\vec{\omega}_i$, at time $t = N + 1$, can be determined by,

$$\frac{d\Psi^{N+1}}{dt} = \Phi^{N+1}. \quad (20)$$

In order to reduce the number of contact checks for each particle, neighboring cell technique is implemented in the present simulations. Once the particle contact forces are calculated, the new velocity and position can be obtained by integrating the differential equations for particle motion over small simulation time step. The input parameters in our simulations include: physical properties of the granular particles under consideration, initial conditions, and boundary conditions. The initial conditions include: the initial positions and velocities of all particles and the geometry of the boundary conditions. All parameters are non-dimensionalized using the density of particle, gravitational acceleration, and particle diameter. A list of used mechanical and environmental dimensionless simulation parameters are given in Table 1.

3. Model setup and simulation schemes

The silo system under consideration consists of two connected silos one above the other as shown in Figure 1. To simulate the real silo filling process, the particles are modeled as a conglomeration of discs with a uniform distribution of diameters, $0.81 \le d \le 0.99$ mm. The discs dispersed inside the upper silo, with given random initial positions and velocities, and are allowed to fall under gravity down into the upper silo. During the filling process, the outlet of the upper silo is kept closed. Hence, due to damping forces, the particles ultimately come to rest at the bottom of the silo. Once the upper silo is almost filled with particles and the system has achieved a nearly steady-state condition, the outlet of the upper silo is opened and the particles start to flow out of the silo under gravity toward the lower silo. To study the nature of normal stresses exerted on the hopper wall of the lower silo, we divide the hopper wall into five equidistance segments as shown in Figure 1. The segment length is approximately five times the particle diameter.

During typical simulation, each segment element of the hopper's wall bears stresses that are exerted by the granular material. To analyze these stresses, we suppose that segment j (where $1 \le j \le 5$) has m contact points $c_1, c_2, \ldots, c_m$ as shown in Figure 2.

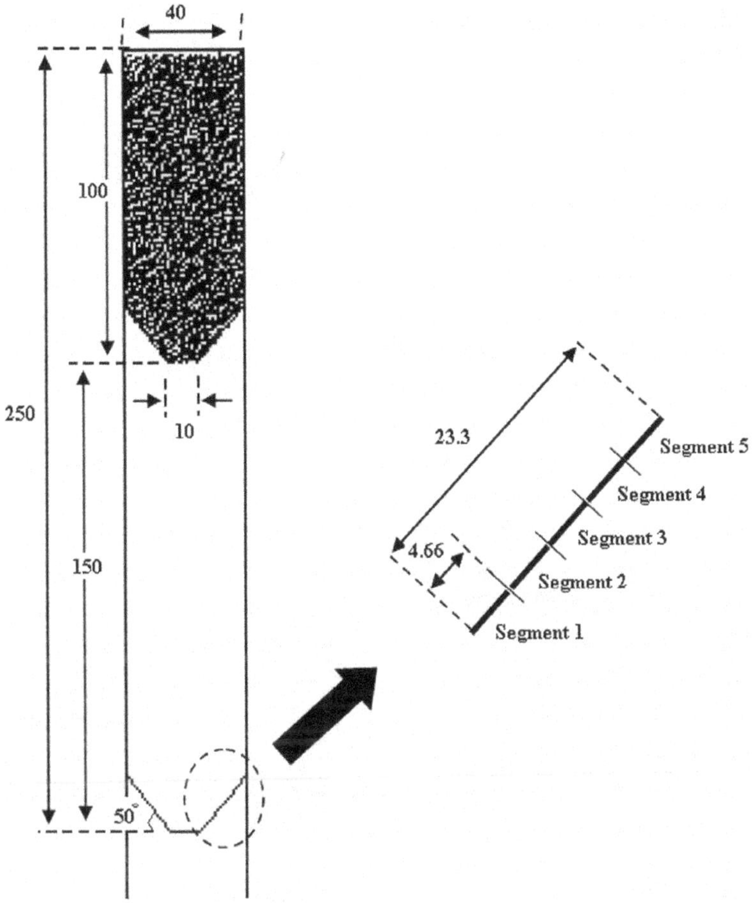

Figure 1. Dimensionless silo geometry and hopper wall segments.

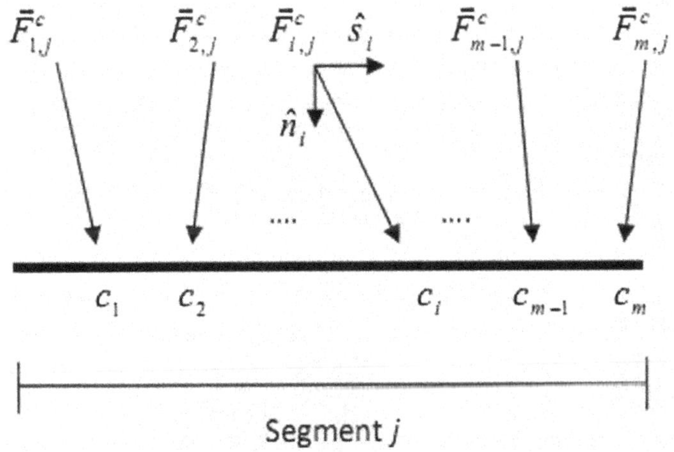

Then the corresponding contact forces acting on segment *j* are

$$\vec{F}^{c}_{i,j}(t) = \left(\vec{F}^{c_i}_{x,j}(t), \vec{F}^{c_i}_{y,j}(t)\right), \quad i = 1, 2, \ldots, m. \tag{21}$$

Figure 2. Contact forces acting on segment *j*.

We define the average contact force, $\vec{F}_j^{ave}(t)$, acting on segment j(with length $l_{seg\,j} = 4.66$ mm) due to the m contact points by

$$\vec{F}_j^{ave}(t) = \frac{1}{l_{seg\ j}} \sum_{i=1}^{m} \vec{F}_{i,j}^{c}(t), \tag{22}$$

such that

$$\vec{F}_{j,\hat{n}}^{ave}(t) = \frac{1}{l_{seg j}} \sum_{i=1}^{m} \left(\vec{F}_{i,j}^{c} \cdot \hat{n}_i \right)$$

and

$$\vec{F}_{j,\hat{s}}^{ave}(t) = \frac{1}{l_{seg j}} \sum_{i=1}^{m} \left(\vec{F}_{i,j}^{c} \cdot \hat{s}_i \right) \qquad i = 1, 2, 3, \ldots, m \tag{23}$$

are, respectively, the normal and tangential components of $\vec{F}_j^{ave}(t)$ and $(\hat{n}_i, \hat{s}_i)$ represents normal and tangential unit vectors corresponding to the segment j at the contact points. Therefore, the total stresses acting on segment j, namely, $\vec{S}_j^T(t)$ can be obtained by

$$\vec{S}_j^T = \frac{1}{l_{seg j}} \left(\sqrt{(\vec{F}_{j,\hat{n}}^{ave})^2 + (\vec{F}_{j,\hat{s}}^{ave})^2} \right) + \left(\sqrt{(\vec{F}_{j,g_x})^2 + (\vec{F}_{j,g_y})^2} \right) \tag{24}$$

where $\vec{F}_{j,g_x}(t)$ and $\vec{F}_{j,g_y}(t)$ are, respectively, the gravity force acting on segment j in the x and y directions.

4. Flow pattern and velocity field

Once the outlet of the upper hopper is opened, the particles start to flow out of the silo under gravity toward the lower silo. The flow pattern and the velocity field of the granular material at different simulation time steps are shown in Figure 3. Mass flow can be observed to occur in the entire upper silo. At the point of transition where the silo wall meet the hopper wall, the particle moving direction changes sharply from straight down to a direction parallel to the hopper inclined wall. Consequently, during the discharging process, the stress on the hopper wall reaches the maximum at these two points and decreases toward the outlet. Hence, we expect that the value of the acting stress on segment five of the hopper wall to be the maximum. However, it has been observed that the more particles reach the point close to the outlet the more they gain room to move. Thus, the stress profile of segment one may more likely show an oscillating nature than other segments of the hopper's wall.

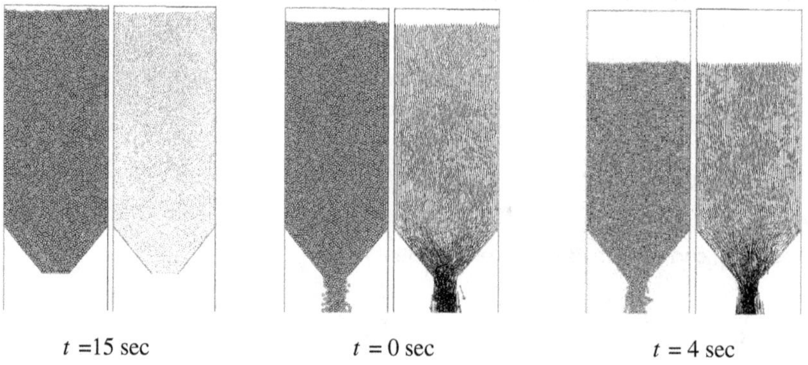

Figure 3. Snapshots for various simulation time instants showing the flow pattern and the velocity field of granular material.

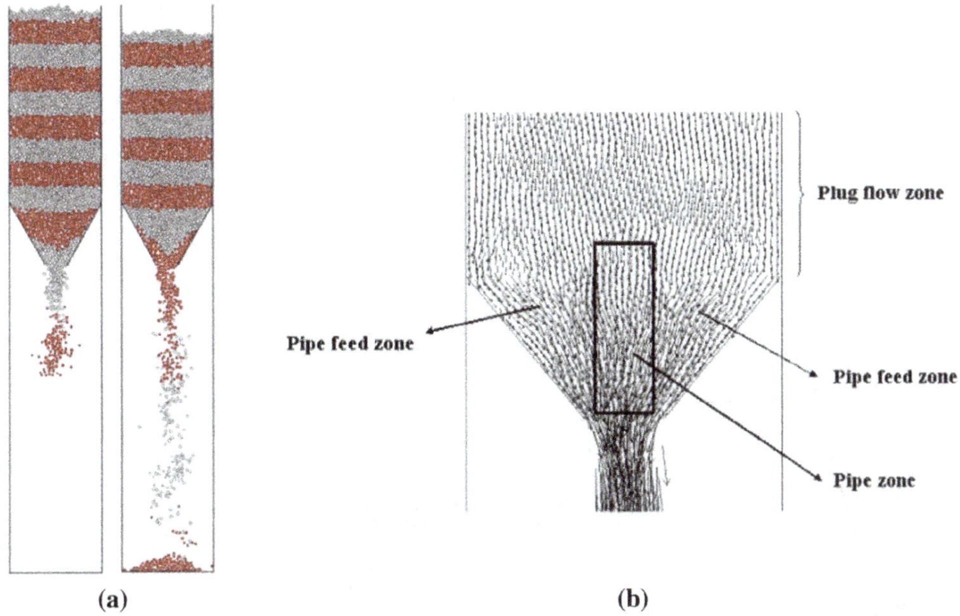

Figure 4. (a) Mass flow of bulk material (b) Flow zones.

Even though mass flow can be seen to occur in the entire silo as can be seen in Figure 4(a), it is possible to recognize different zones with different flow patterns. These flow zones can be divided into: pipe, pipe feed, and plug flow zones. Figure 4(b) shows the three different flow zones.

5. Simulation results and discussion

In this section, we investigate the stress fluctuating nature developed along the hopper wall during filling and discharging of a silo. A simulated filling and discharging processes were carried out to study the variation in the stress that acts on the hopper wall, with time. Moreover, we investigate the dynamic nature of these stresses at various values of the wall and the internal friction coefficients.

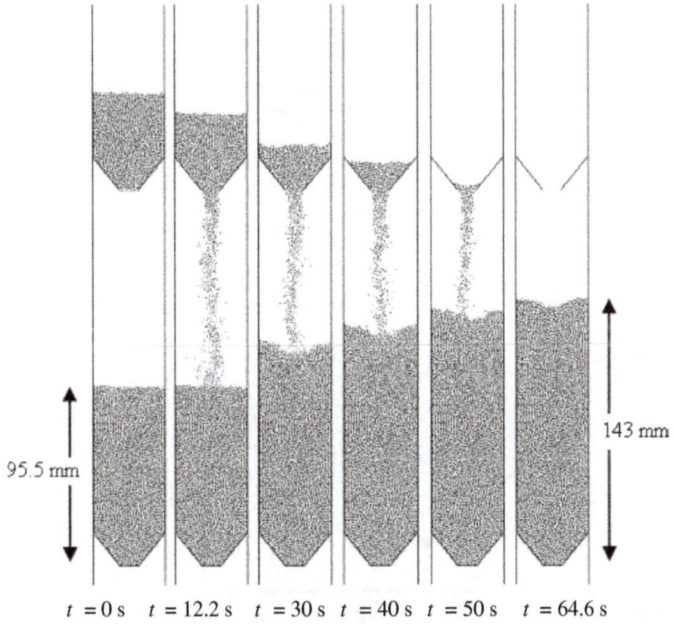

Figure 5. Simulation snapshots showing the discharging of granules at various simulation time instants.

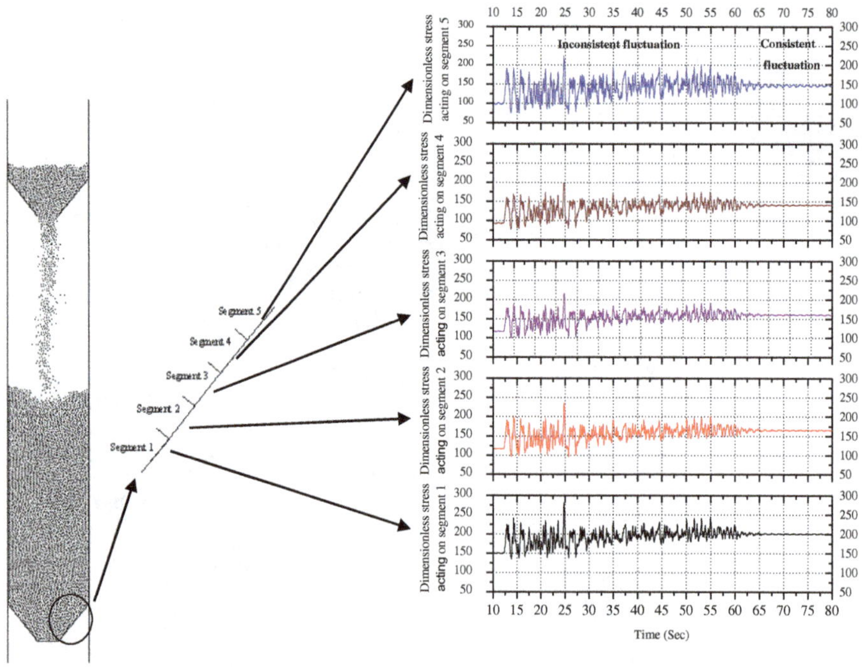

Figure 6. The variations of stress with time along the hopper wall segments.

5.1. Stress pulsation mechanism

We, partially, filled both the upper and the lower silos by allowing particles, having random initial positions and velocities, to fall down under gravity. The upper silo was filled with 2,500 particles while 4,500 particles were placed in the lower silo. Then the outlet of the upper silo was opened and the particles start to flow down from the upper silo under gravity toward the lower silo. Snapshots of the process at different time steps are shown in Figure 5.

At $t = 0$ s, each of the five segments of the lower silo wall, bears an initial stress value due to the initial existing particles in the lower silo. The particle packing plays an important role on the initial stress profile on these segments. Once the particles that flow out from the upper silo strike the static

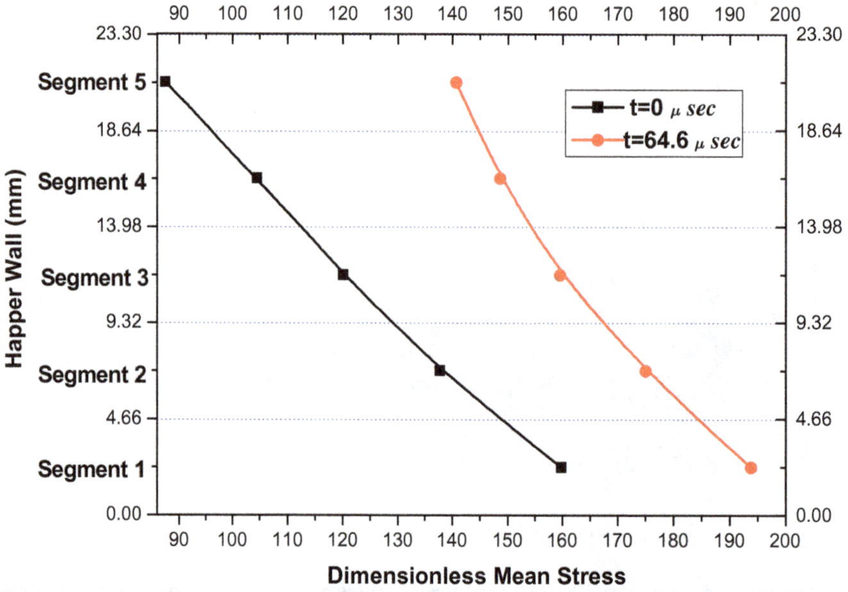

Figure 7. Mean stresses acting on hopper's wall of the lower silo at $t = 0$ s and $t = 64.6$ s (Figure 4).

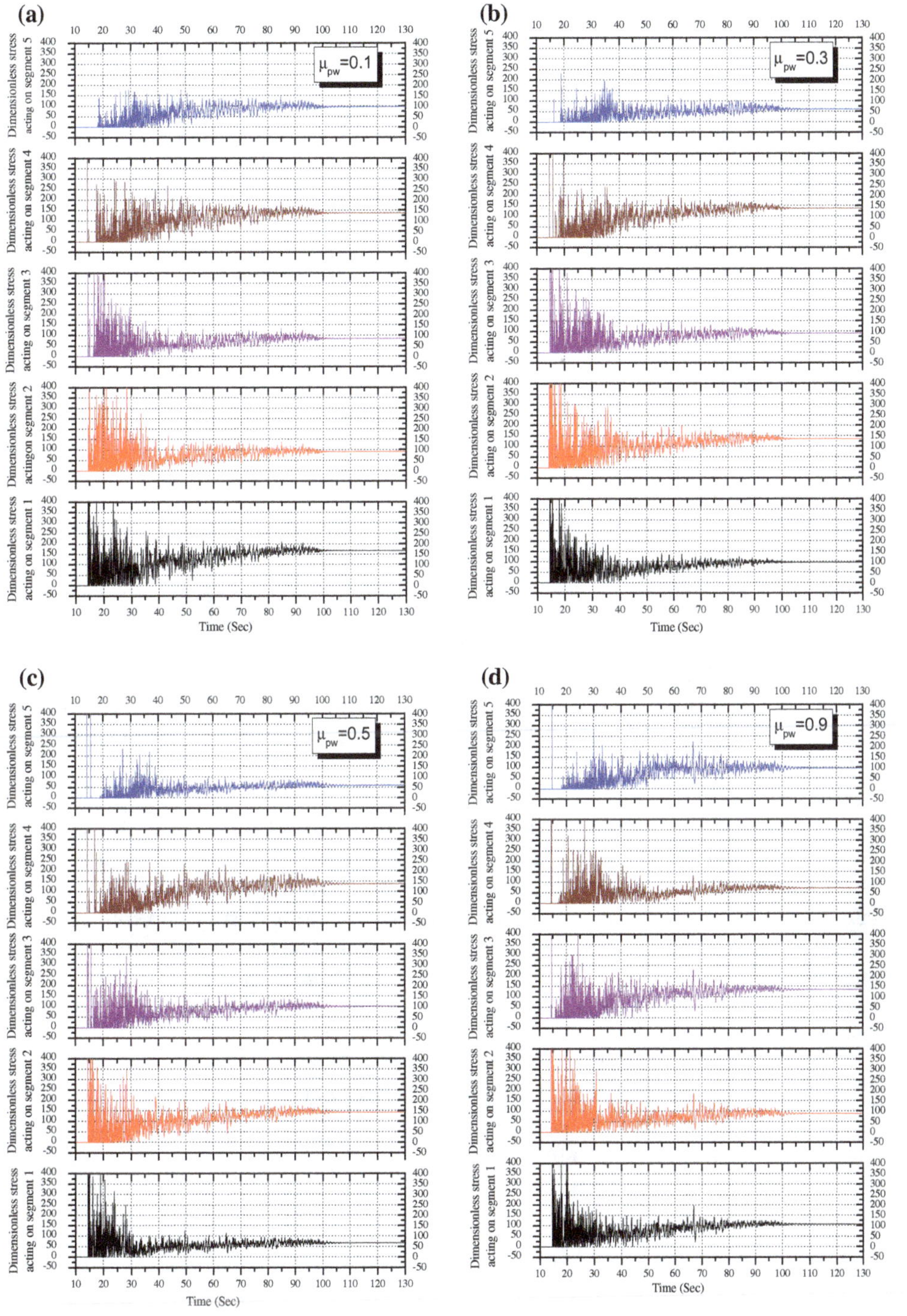

Figure 8. Variations of stresses acting on the hopper's wall during filling (a) μ_{pw} = 0.1, (b) μ_{pw} = 0.3, (c) μ_{pw} = 0.5, (d) μ_{pw} = 0.9.

surface of the granular material in the lower silo, the stress value acting on each segment starts to increase. Figure 6 shows the stress profile acting on each segment of the hopper's wall of the lower silo throughout the simulation time.

At $t \cong 12.2$ s, particles from the upper silo started to hit the static surface of granular material in the lower silo. Consequently, stress trace of each segment starts to fluctuate as the height of bulk material in the lower silo starts to increase. For each stress's profile, two fluctuation traces can be identified, namely, inconsistent and consistent stress fluctuation traces. The inconsistent stress trace dominates as long as there are particles falling down from the upper silo and hitting the granular surface in the lower silo. As shown in Figure 6, it extends from $t \cong 12.2 - 64.6$ s. Once all the particles of the upper silo reach the lower silo and settle down, the stress trace on each segment does not turn to a straight line directly. A temporary consistent fluctuation which has a harmonic motion-like trace starts to dominate before the stress profile turns to a straight line. Therefore, the system possesses a temporary dynamic pulsation stress-like nature on the hopper wall after it appears to reach a near static state. The stress profile on segment five which is the closest to the surface of bulk material appears to have larger trace fluctuation amplitude than that on other segments of the hopper wall during the process. Figure 7 shows the dimensionless mean stress acting on the hopper wall of the lower silo at two different simulation stages, namely, at $t = 0$ s and $t = 64.6$ s. It is obvious that the stress value on each segment increases and almost on all segments the stress increases monotonically throughout the simulation time. In both cases, the stress acting on segment one appears to be the largest and the wall stress decreases gradually along the hopper wall toward segment five. This result is, generally, in agreement with existing experiment results (Cleary & Sawley, 2002; Goda & Ebert, 2005; Masson & Martinez, 2000).

5.2. Influence of wall friction

In this section, we investigate the effect of the wall friction on the dynamic nature of the wall stress developed during filling and discharging of the lower silo. A sequence of test simulations was carried out. In each simulation, we vary the value of the wall frictional coefficient μ_{pw} and maintain other simulation parameters the same. The experiments were conducted using the parameter values shown in Table 1, except for the value of μ_{pw} which varies from 0.1 to 0.9. The methodology was to, completely, fill the upper silo with 7,000 particles and let them to settle down till they come to rest. Then the outlet was opened and the particles flow down under the gravitational force into the lower silo. The stresses acting on each segment of the hopper's wall, during filling the lower silo for various values of μ_{pw}, are shown in Figure 8.

Our results reveal that, after the filling process, granular systems with small wall fictional coefficients, in general, achieve static state faster than those with large values. For the cases of $\mu_{pw} = 0.1$ and 0.3, there are no stress fluctuation after $t = 100$ s which means that the two systems are stable and there is no stress variation after this time. However, the stress fluctuation continues after $t = 100$ s for the cases of and $\mu_{pw} = 0.5$ and $\mu_{pw} = 0.9$, which means that there is still a stress variation occurring beyond this time. Nevertheless, the amplitude of the stress fluctuation tends to be smaller as the wall frictional coefficient increases. This is because of the rapid dissipation of the kinetic

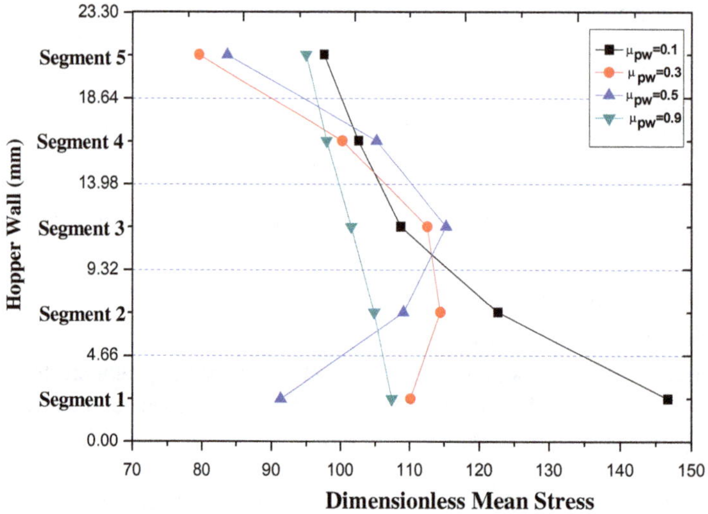

Figure 9. Mean stress acting on hopper's wall at the end of filling process for $\mu_{pw} = 0.1$, $\mu_{pw} = 0.3$, $\mu_{pw} = 0.5$, and $\mu_{pw} = 0.9$.

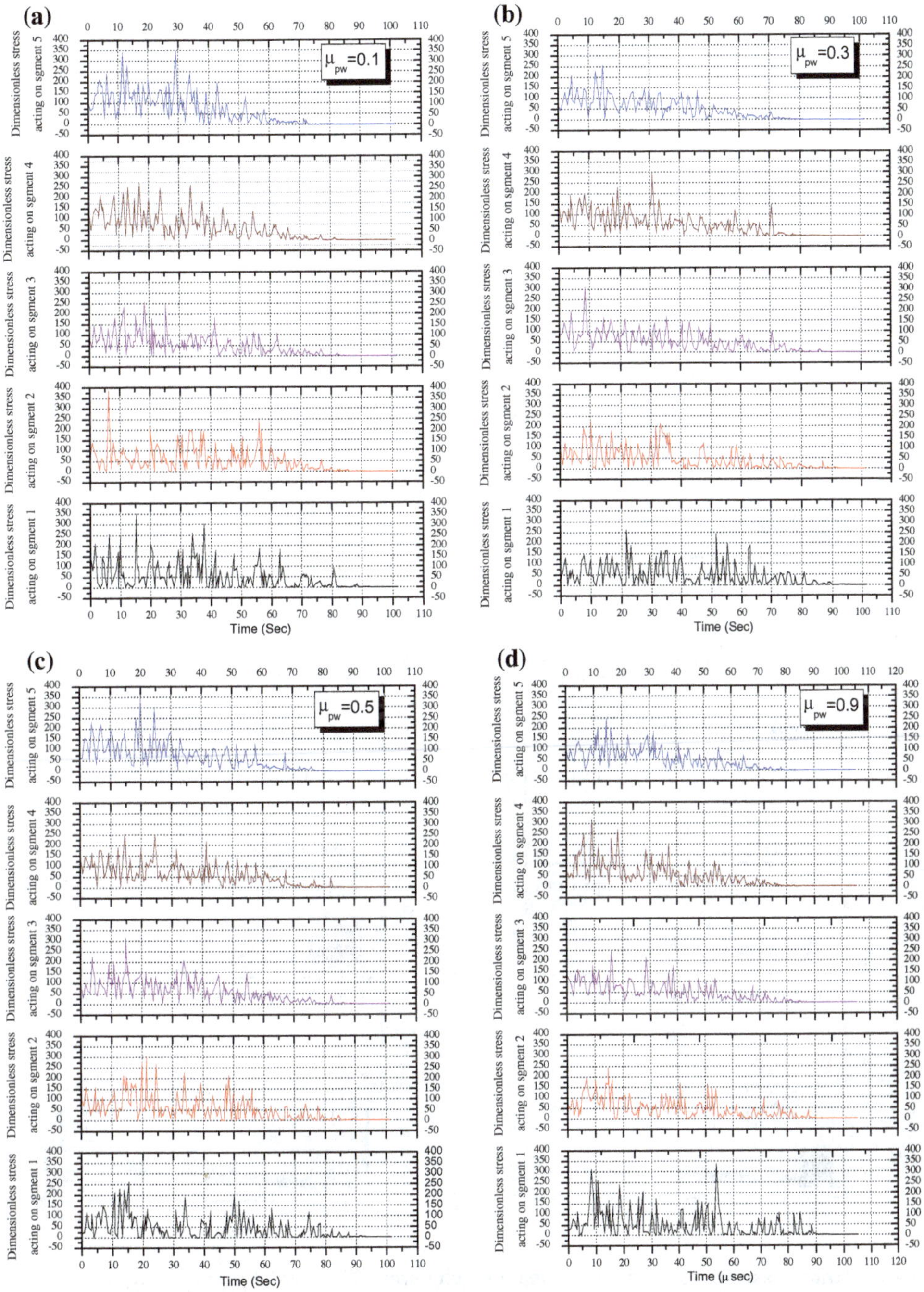

Figure 10. Variations of stresses acting on the hopper's wall during discharging for (a) μ_{pw} = 0.1, (b) μ_{pw} = 0.3, (c) μ_{pw} = 0.5, (d) μ_{pw} = 0.9.

energy of particles in direct contact with the hopper wall with high friction. At the end of the filling process, in general, we observe that the ultimate stress values acting on each segment decrease as the value of μ_{pw} increases. This can be explained by the mechanism that the larger is μ_{pw}, the more percentage of material weight the silo walls will carry. Figure 9 shows the mean stress acting on each segment of the hopper wall in the lower silo at the end of the filling process. Except the case when μ_{pw} = 0.5, stresses exerted on segment one were the maximum while those exerted on segment five were the minimum. However, no considerable changes were observed on the stresses that acting on other segments.

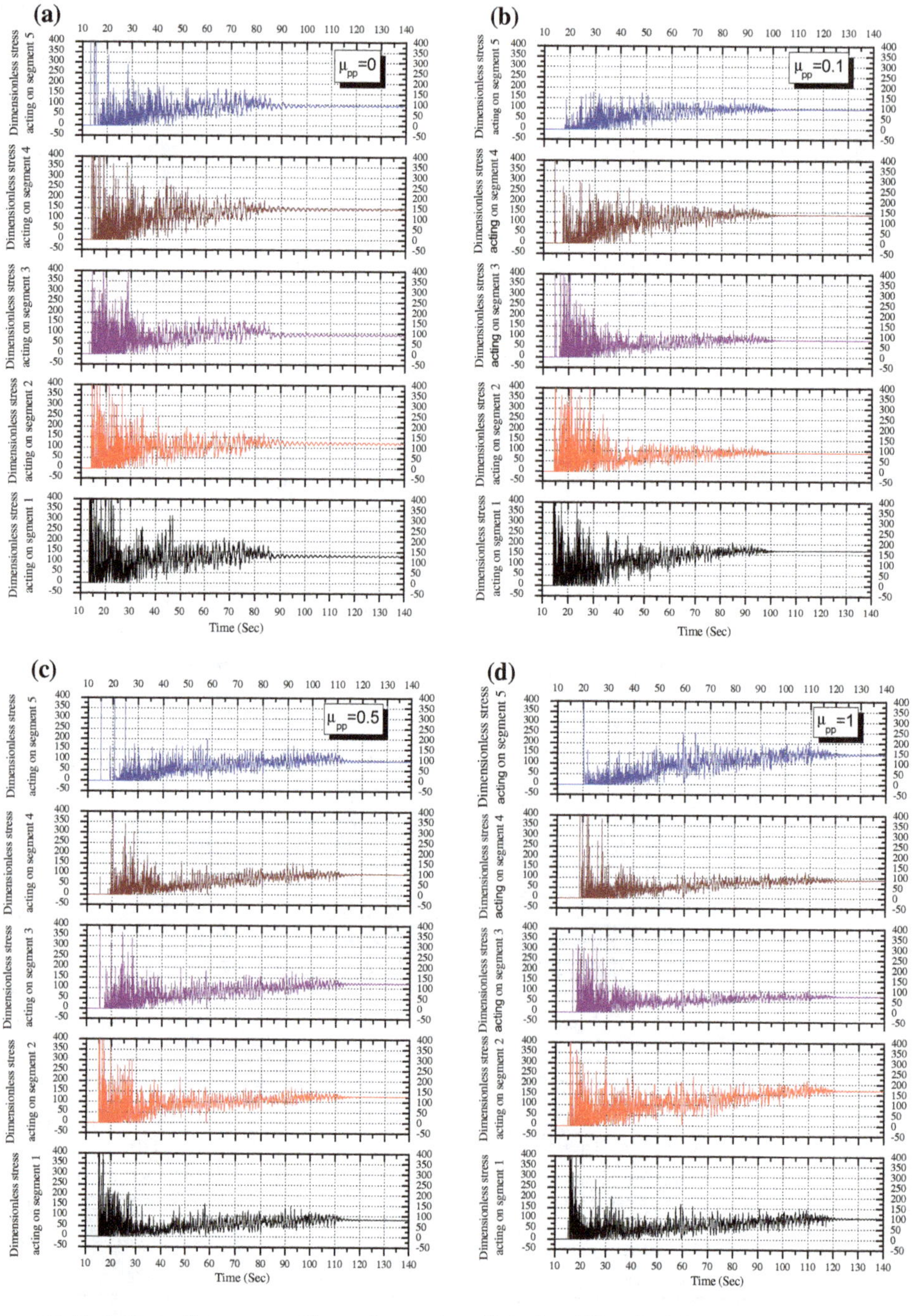

Figure 11. Variations of stresses acting on hopper's wall during filling for (a) μ_{pp} = 0, (b) μ_{pp} = 0.1, (c) μ_{pp} = 0.5, (d) μ_{pp} = 1.

To investigate the influence of μ_{pw} on the discharging process, another series of simulations was performed for discharging the particles from the lower silo this time. In these simulations, we also use the mechanical and environmental parameter values listed in Table 1. Different values of μ_{pw} were considered in each simulation, namely, 0.1, 0.3, 0.5, and 0.9. During the process, the stresses acting on the hopper's segments were measured. At t = 0 s, the lower silo outlet opened and the particles started to flow out of the silo under gravity. As the particles flow out form the silo, the values of the stresses acting on each segment start to fluctuate and decrease gradually till the end of the discharging process. Figure 10 shows the stress traces for each segment along the hopper's wall of the lower silo during the discharging process. Each graph corresponds to a different value of μ_{pw}

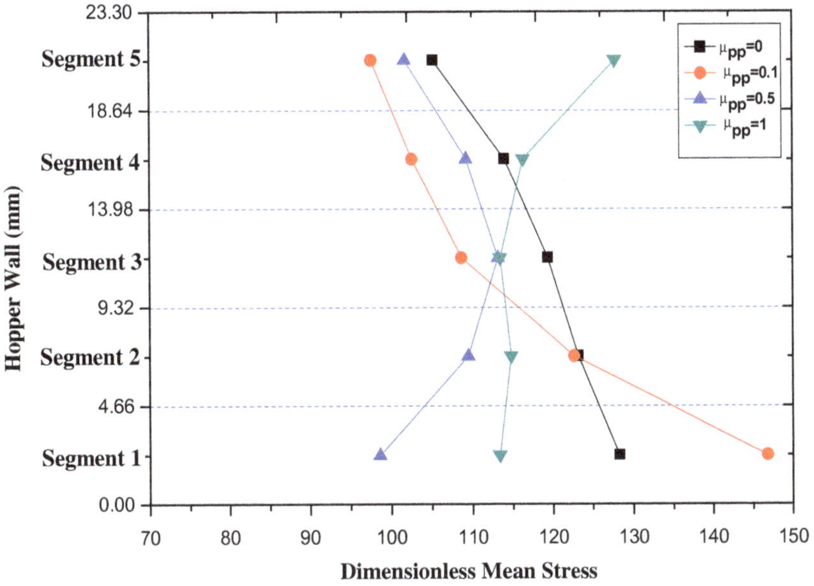

Figure 12. Mean stress acting on hopper wall at the end of filling process for μ_{pp} = 0, μ_{pp} = 0.1, μ_{pp} = 0.5, and μ_{pp} = 1.

while other simulation parameters are kept the same. The results show that, for different values of μ_{pw}, the required time for the discharging of the silo is almost the same. Moreover, the fluctuation amplitude of the stress that acting on segment five is larger than that acting on the other segments. In general, the graph shows that the stress fluctuation amplitude decreases as the value of μ_{pw} increases.

According to our simulation outputs, the stress fluctuation profile, during the discharging process, is not considerably affected by the value of μ_{pw}. However, in general, the fluctuation amplitude for larger values of μ_{pw} tends to be smaller than that for small values of μ_{pw}. On segment five the stress fluctuation trace has the largest amplitude compared to those on all the other segments. Therefore, the points of transition from vertical silo wall to the hopper wall bear the maximum value of the acting stresses during the discharging process.

5.3. Influence of internal friction

To investigate the effect of the internal friction coefficient μ_{pp} on the stresses profile, we follow the same procedure as that in 5.2. We run a series of simulations in which all simulation parameters are fixed while μ_{pp} is given different values each time. We consider four different values of μ_{pp}, namely, 0, 0.1, 0.5, and 1. For the filling process, the outlet of the filled upper silo was opened and the particles started to fall down into the lower silo. Figure 11 shows the resulting stresses profiles on the segments of the hopper's wall for the four different values of μ_{pp}. Results show that granular systems with small values of μ_{pp} comes to rest faster than those with large values of μ_{pp}. Moreover, at the end of the filling process and when the assembly eventually comes to rest at the base of the lower silo, stress profiles appear to have different traces. For the case of μ_{pp}, a consistent stress fluctuation that has a simple harmonic motion-like trace appears for a slightly long period of time before the stress trace turns to a straight line. On the other hand, for other cases i.e. μ_{pp} = 0.1, 0.5, and 1, no long consistent stress fluctuation has been observed and the stress trace turns to a straight line rapidly. This phenomenon can be explained as flows, for μ_{pp} = 0, there is no internal friction between particles and the system has a fluid-like properties. In this case particles take more time to settle down and achieve a real steady state. Another observation, that can be obtained from Figure 11, is that the stress fluctuation amplitude for the systems with smaller value of μ_{pp} is greater than that for the systems with larger values of μ_{pp}. This is reasonable for the same reason that has been mentioned earlier.

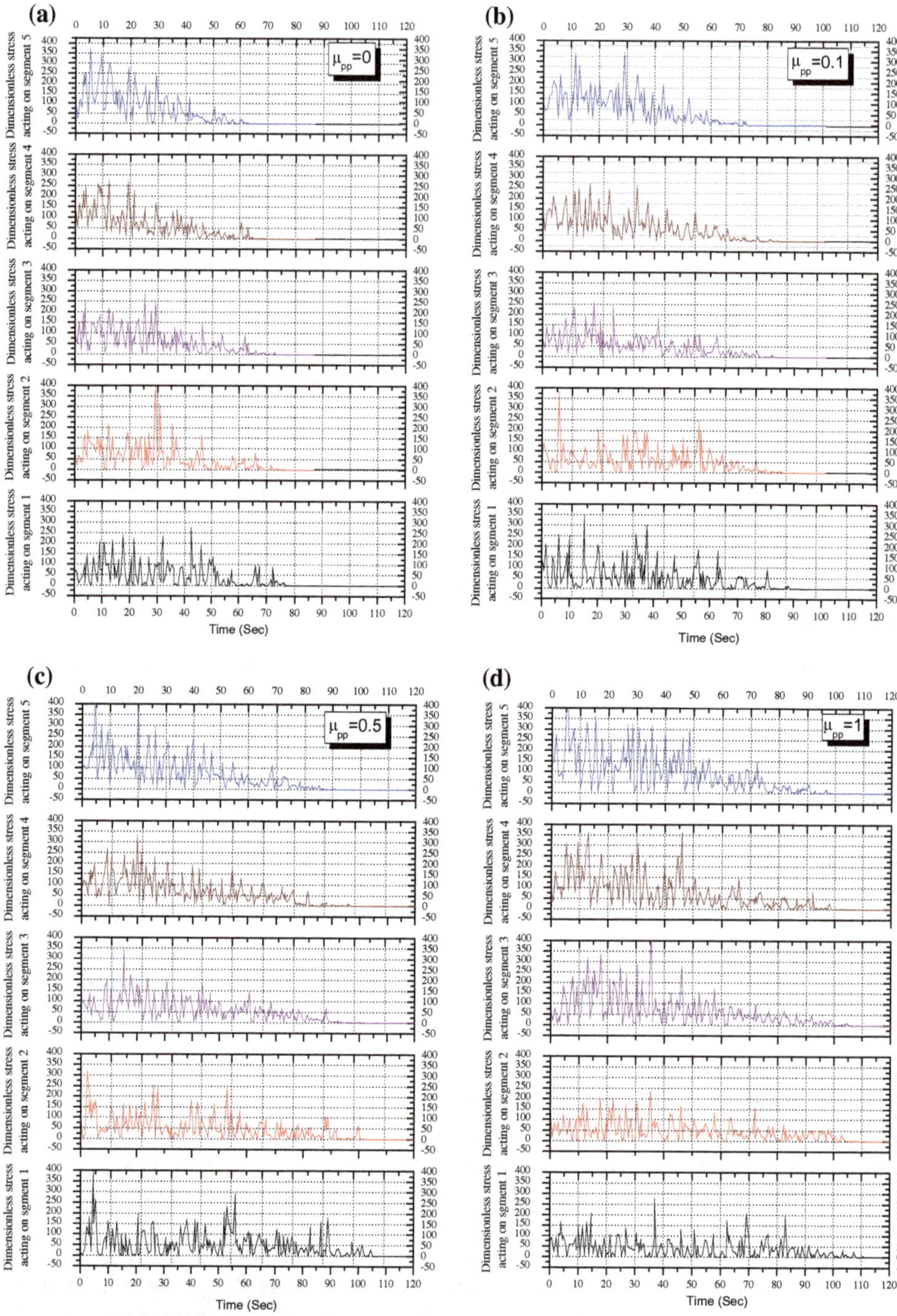

Figure 13. Variations of stresses acting on hopper's wall during discharging (a) μ_{pp} = 0, (b) μ_{pp} = 0.1, (c) μ_{pp} = 0.5, and (d) μ_{pp} = 1.

At the end of the filling process, the mean stress acting on each segment is plotted in Figure 12. For μ_{pp} = 0, the ultimate stress profile is reasonably close to a straight line as the bulk material has a fluid-like property. It is also found that at the end of filling process, for $\mu_{pp} = 0.5$ and 1, stresses exerted on segment five were the maximum while those exerted on segment one were the minimum. In these cases, the transition's points of the silo system carry most of the load of the bulk material due to the large internal friction coefficients. Conversely, segment 1 bears the maximum stress for $\mu_{pp} = 0$ and 0.1 due to the low internal friction between particles.

For the effect of the internal friction coefficient μ_{pp} on the stresses during the discharging process, we performed a series of simulations using the same four different values of μ_{pp}. Other parameters were preserved the same. Figure 13 shows the stresses profile developed on each segment of the hopper's wall during discharging of the lower silo for the four different values of μ_{pp}.

According to our results, the discharging process takes more time for large values of μ_{pp} as would be expected. Moreover, due to dilation of the particles at the outlet zone of the silo system during flow, segment one almost maintains the same fluctuation amplitude throughout the discharging process. For other segments, the fluctuation stress trace gradually decreases till it turns to a straight line at the end of the process. On the transition between the hopper wall and the vertical bin wall (segment 5), the stress appears to have the maximum fluctuation.

6. Concluding remarks

Two-dimensional soft particle discrete element simulations have been conducted to simulate the filling and discharging of granular particles from a silo system. The simulated filling and discharging processes have been carried out to study the variation in stress, which acts on the hopper wall, with time. The effects of wall and internal friction coefficients on the dynamic nature of stresses acting on hopper wall have also been investigated. From the results, it can be concluded that the stresses that act on the hopper wall have a temporary dynamic pulsation-like nature at the end of the filling process when the granular particles appear to reach a near static state. Accordingly, under the condition that the particle-wall contact force, in the normal direction, is modeled as visco-elastic (damped harmonic oscillator), the proposed soft particle discrete element model is capable of simulating the silo system. The influences of wall and internal friction coefficients on the oscillating nature have been investigated during filling and discharging processes. At the end of filling process, it can be concluded that the larger the μ_{pw} is, the less the ultimate stress values acting on the hopper wall will be. Moreover, for smaller values of μ_{pp}, the system has fluid-like properties and hence, the particles take more time to settle down and achieve a real steady state. Therefore, a consistent stress fluctuation appears for a slightly long period of time before the stress trace turns to a straight line. For the discharging process, the results show that for different values of μ_{pw}, the required time for the discharging of the silo is almost the same. Furthermore, the fluctuation amplitude of stress acting on the upper part of the hopper wall is larger than that on the other parts. Also, it is found that the discharging process takes more time for large values of μ_{pp}. Due to the particles' dilation at the zone of the silo's outlet, the lower part of the hopper wall maintains almost the same fluctuation amplitude throughout the discharging process. On the transition between vertical bin and hopper's wall, the stress appears to have the maximum fluctuation during the discharging process for all different values of μ_{pp}.

Funding
The author received no direct funding for this research.

Author details
Waseem Ghazi Alshanti[1]
E-mails: w.alshanti@uoh.edu.sa, waseemalshanti@yahoo.com
[1] Department of Mathematics, University of Hail, Hail, Kingdom of Saudi Arabia.

References

Chen, J. F., Rotter, J. M., Ooi, J. Y., & Zhong, Z. (2006). Correlation between flow pattern and wall pressures in a full scale silo. *Engineering Structures, V29*, 2308–2320.

Chou, C. S., Tzeng, C.Y., Smid, J., Kuo, J. T., & Hsiau, S. S. (2000, May 21–24). *Experimental study of moving granular bed: Wall stress measurement*. 14th ASCE Engineering Mechanics Division Conference, Austin, TX.

Cleary, P. W., & Sawley, M. L. (2002). DEM modelling of industrial granular flows: 3D case studies and the effect of particle shape on hopper discharge. *Applied Mathematical Modelling, 26*, 89–111. http://dx.doi.org/10.1016/S0307-904X(01)00050-6

Cundall, P. A., & Strack, O. D. L. (1979). A discrete numerical model for granular assemblies. *Géotechnique, 29*, 47–65. http://dx.doi.org/10.1680/geot.1979.29.1.47

Goda, T. J., & Ebert, F. (2005). Three-dimensional discrete element simulations in hoppers and silos. *Powder Technology, 158*, 58–68. http://dx.doi.org/10.1016/j.powtec.2005.04.019

Hirshfeld, D., & Rapaport, D. C. (2001). Granular flow from a silo: Discrete-particle simulations in three dimensions. *The European Physical Journal E, 4*, 193–199. http://dx.doi.org/10.1007/s101890170128

Landry, J. W., Grest, G. S., et al. (2004). Discrete element simulations of stress distributions in silos: Crossover from two to three dimensions. *Powder Technology, 139*, 233–239.http://dx.doi.org/10.1016/j.powtec.2003.10.016

Langston, P. A., Tüzün, U., et al. (1995). Discrete element simulation of granular flow in 2D and 3D hoppers: Dependence of discharge rate and wall stress on particle interactions. *Chemical Engineering Science, 50*, 967–987. http://dx.doi.org/10.1016/0009-2509(94)00467-6

Longhi, E., Easwar, N., & Menon, N. (2002). Large force fluctuations in a flowing granular medium. *Physical Review Letters, 89*, 045501. http://dx.doi.org/10.1103/PhysRevLett.89.045501

Masson, S., & Martinez, J. (2000). Effect of particle mechanical properties on silo flow and stresses from distinct element simulations. *Powder Technology, 109*, 164–178. http://dx.doi.org/10.1016/S0032-5910(99)00234-X

Nguyen, V. D., Cogné, C., Guessasma, M., Bellenger, E., & Fortin, J. (2009). Discrete modeling of granular flow with thermal transfer: Application to the discharge of silos. *Applied Thermal Engineering, 29*, 1846–1853. http://dx.doi.org/10.1016/j.applthermaleng.2008.09.009

Parisi, D. R., Masson, S., & Martinez, J. (2004). Partitioned distinct element method simulation of granular flow within industrial silos. *Journal of Engineering Mechanics, 130*, 71–779.

Ristow, G. H. (1997). Outflow rate and wall stress for two-dimensional hoppers. *Physica A: Statistical Mechanics and its Applications, 235*, 319–326. http://dx.doi.org/10.1016/S0378-4371(96)00365-2

Ristow, G. H., & Herrmann, H. J. (1995). Forces on the walls and stagnation zones in a hopper filled with granular material. *Physica A: Statistical Mechanics and its Applications, 213*, 474–481. http://dx.doi.org/10.1016/0378-4371(94)00249-S

Sykut, J., Molenda, M., & Horabik, J. (2008). DEM simulation of the packing structure and wall load in a 2-dimensional silo. *Granular Matter, 10*, 273–278. http://dx.doi.org/10.1007/s10035-008-0089-z

Yang, S. C., & Hsiau, S. S. (2001). The simulation and experimental study of granular materials discharged from a silo with the placement of inserts. *Powder Technology, 120*, 244–255. http://dx.doi.org/10.1016/S0032-5910(01)00277-7

Zhu, H., & Yu, A. (2003). A numerical study of the stress distribution in hopper flow. *China Particuology, 1*, 57–63. http://dx.doi.org/10.1016/S1672-2515(07)60109-2

13

An ecogenetic disease-affected predator–prey model

Ilaria Usville[1], Cristina Paola Viola[1] and Ezio Venturino[1*]

*Corresponding author: Ezio Venturino, Dipartimento di Matematica "Giuseppe Peano", Università di Torino, Torino, Italy

E-mail: ezio.venturino@unito.it

Reviewing editor: Benchawan Wiwatanapataphee, Curtin University, Australia

Abstract: A nonlinear ecoepidemic model of new type is introduced here, in that it contains genetically distinguishable subpopulations. Further, in the system a predator is present, that hunts these two disease-affected genotypes. Under the assumptions of the model, the disease cannot endemically survive in the predator-free environment. The healthy prey can thrive in the absence of the predators, but this is in line with previous results and does not appear to be due to the effects of the epidemics. On the other hand, the disease affects the stability of the purely demographic equilibria.

Subject: Applied Mathematics; Dynamical Systems; Mathematical Biology; Mathematics & Statistics; Science

Keywords: predator–prey; epidemics; transmissible diseases; genetic variability

AMS subject classifications: Primary: 92D25; Secondary: 92D10; 92D40

1. Introduction

Mathematical ecogenetic models have recently been introduced (Venturino, 2012; Viberti & Venturino, 2014). They seem to be a particular example of what are nowadays commonly called "structured populations models," that have been studied for quite some time, starting from the

ABOUT THE AUTHOR

Ezio Venturino, professor of Mathematics, holds a PhD in Applied Mathematics from SUNY at Stony Brook. His research interests concern numerical analysis and mainly mathematical modeling (biological population theory, epidemiology, and socioeconomic applications). He has authored about 40 research papers in numerical analysis and 160 in mathematical modeling. He has coauthored two books for CRC and Birkhaeuser, and serves in the Editorial Board of several international Journals. This paper is part of his scientific interests in mathematical biology, in which he has contributed to found a specific research field, now called ecoepidemiolgy, dealing with interacting populations that are in addition subject to a disease. The latter can deeply influence the ecosystem behavior. Another fairly recent idea accounts for models of herd behavior and pack hunting. Sometimes a richer dynamics is discovered. More recently, he introduced models for genetically distinguishable species interactions, of which this paper is an instance.

PUBLIC INTEREST STATEMENT

Evolutionary adaptation in population genetics is observed in the peppered moth, Biston betularia, possessing melanic and non-melanic morphs. As a mimetic feature against predators, 200 years ago, it was light colored like trees trunks. But in the following century industrial pollution darkened the trees; the melanic moth, carbonaria, thrived instead. Parasites respond differently to host genotypes. Experiments with Plodia interpuncella and the granulosis virus, report an increased viral resistance in the virus-affected subpopulation. A predator–prey mathematical model is introduced, where the prey are genetically distinguishable and one genetic subpopulation suffers an epidemic. Specifically, the prey is Drosophila suzukii, one of its genotype being affected by nematodes; its predators are frogs, but also the commercially available Orius majusculus and Orius laevigatus. The ecosystem thrives, the predators are instrumental in disease eradication, both genotypes of the healthy prey can survive in a predator-free environment, predators elimination is unrelated to the disease presence.

seminal work of Kostitzin in the 30s of the past century; on this, see the nice review paper (Scudo, 1976). In fact, for structured populations, now, one understands mainly populations that are not just dependent on time, but also on age, distinguishing therefore several age cohorts (Cushing, 1998). Human survival tables based on age cohorts have been in use among accountants for centuries, in order to calculate life insurance premiums. In ecology, in the simplest case of insects, for instance, these could be the larval, pupa, and adult life stages; for fishes instead, we would have eggs, larvae, immature juveniles, and adults. Structure therefore adds to the population description a new dimension, be it age, as depicted and analyzed in the classical papers (Gurtin & McCamy, 1974, 1979), and for which analytical approaches are available (Webb, 1985), or instead size, for which more sophisticated numerical methods need to be introduced (Angulo & López-Marcos, 1999). Indeed, the basic age description corresponds to a linear, or more generally nonlinear (Gurtin & Levine, 1979), wave equation, with an initial and a boundary condition at age zero, for which standard numerical schemes are available (Smith, 1985). These considerations could be extended to interacting populations (Venturino, 1984, 1987) A structure can also be superimposed on diseases: the classical Kermack-McKendrick model (Kermack & McKendrick, 1927) considers the population partitioned among susceptibles, infected, and recovered individuals. In addition, one could also include a latent period of exposed individuals in which the disease is incubating, but they are not yet able to spread it. A more general description in which all these stages are envisioned in a continuum is contained in Venturino (1985).

In the of models (Venturino, 2012; Viberti & Venturino, 2014), the main idea is to consider a population which has two identifiable genotypes, and in some way intermingles with another population. In Venturino (2012) the case of predator–prey interactions in considered. Here too, we examine a highly nonlinear system, containing a basic demographic situation of the same type as Venturino (2012), which differs however in a major feature.

In the past 20 years or so, a lot of effort in mathematical biology has been devoted to the study of ecosystems in which also epidemics spread. Thus, in addition to their demographic interactions, populations also experience transmissible diseases. The presence of the latter profoundly influences the system's outcomes, so that a stable population configuration from the purely demographic point of view becomes instead unattainable in presence of the disease (Beltrami & Carroll, 1994; Hadeler & Freedman, 1989; Venturino, 1994). The basic demographic models can be of the predator–prey type with disease affecting the prey (Chattopadhyay & Arino, 1999; Venturino, 1995), or also the predators (Auger et al., 2009; Venturino, 2002), but can include also e.g. competition (Saenz & Hethcote, 2006). Some of the earlier developments of this discipline are contained in Chapter 7 of Malchow, Petrovskii, and Venturino (2008). More recent contributions consider issues such as the study of equilibria (Delgado, Molina-Becerra, & Suarez, 2005) and their global stability (Zhen & Haque, 2006). Viruses in planktonic and other ecosystems have been considered in Beretta and Kuang (1998) and Singh, Chattopadhyay, and Sinha (2004), but the effect of viruses could also be favorably exploited to control the spread of pests (Bhattacharyya & Bhattacharya, 2006). More recent investigations include findings of complex dynamics (Bate & Hilker, 2013), and of chaotic behavior (Kooi, van Voorn, & Das, 2011), predation switching (Hotopp, Malchow, & Venturino, 2010), an idea taken from the purely demographic models (Khan, Balakrishnan, & Wake, 2004), applications of harvesting and control theory as measures to contain epidemics (Bairagi, Chaudhuri, & Chattopadhyay, 2009; Jana & Kar, 2013; Kar & Jana, 2013), more sophisticated models including delays (Bairagi, Sarkar, & Chattopadhyay, 2008), various functional response functions (Bairagi, Roy, & Chattopadhyay, 2007). These ecoepidemic models have also been reformulated as intraguild predations (Sieber & Hilker, 2011). A recent, rather comprehensive, review of the field is provided in Venturino (2016).

It is well known that genetic variability may influence the response to pathogens and diseases. For instance, an example of this kind of situation is provided in humans by sickle cell anemia, which affects mainly individuals belonging to a recessive blood type.

In the animal realm, the classical example of evolutionary adaptation in population genetics is provided for instance by the peppered moth *Biston betularia*. It is well-known that this insect has several melanic and non-melanic morphs that are genetically controlled. About 200 years ago the *B. betularia* was light colored and this helped against predators, mainly birds, since the trunk of the trees where they rested were mainly of light colors (Grant, 1999). In the following century, however, smoke and pollution due to the newly implanted industries in the UK rendered the trunks of dark colors, causing higher moth mortality due to predation, as they were more easily spotted. Conversely, the melanic moth, carbonaria, thrived because with its dark color it could easily hide among the trees. Rather recently, there has been a hot debate on whether this phenomenon is attributable or not to evolutionary mechanisms (Clarke, 2003; Hooper, 2002).

In Lively and Apanius (1995) it is remarked that parasites respond in different ways to various host genotypes, this being supported by some field evidence, for which parasites attack the most abundant genotype. Evidence, Lively, Craddock, and Vrijenhoek (1990) and Vrijenhoek (1993) shows that the parthenogenetic fish *Poeciliopsis* spp., which has a sexual and a clonal form coexisting in the population, infected by the trematode larvae *Uvulifer* spp., harbored more parasites than coexisting outcrossed fish. The result was the opposite in case of a highly inbred population, but later, when sexual fish were reintroduced into the initial inbred population, the clonal fish turned out to be more infected than the sexual ones (Vrijenhoek, 1993). This demonstrates that parasites can rapidly spot changes in the genotype frequency in a population.

Also, one strain of nematode that originally was able to infect four species of *Drosophila*, after being exposed to only one of them for a couple of years, was unable to infect one of the original species (Jaenike, 1993). Similarly, experiments with *Plodia interpuncella* and the granulosis virus (Read, 1991) report an increased viral resistance that was much higher in the subpopulation maintained in the presence of the virus than in the virus-free control subpopulation. This evidence indicates that host specificity may change in time due also to genetic changes.

Motivated by the fact that different genotypes may experience different responses to external interferences therefore also in animals, as mentioned above, and thus not only to predators, but also to pathogens, we consider here a hypothetical ecogenetic model, similar to some other systems already studied, that in comparison with the current literature (Venturino, 2012; Viberti & Venturino, 2014), makes a step forward, namely it introduces a disease among the genetically distinguishable population, that affects only one genotype.

Specifically, we consider two genotypes of the *Drosophila suzukii*, one of which is affected by nematodes, that are however not explicitly built into the model. They only partition the insects among healthy and infected classes. This fruit fly has several predators, the most common of which are the frogs. In addition, there are some that are even commercially available (Cuthbertson, Blackburn, & Audsley, 2014), such as *Orius majusculus*, *Orius laevigatus*, *Atheta coriaria*, *Hypoaspis miles*, and *Anthocoris nemoralis*.

The ecosystem under investigation then contains the prey, with two distinct genotypes, of which only one is subject to a disease. In addition, the predators hunt these subpopulations. Possible questions to which such a model could provide an answer is related to whether the presence of the predators is able to extinguish the disease, or whether the infection in the prey can wipe out the predators. The answers are tied to the interpretation of the actual situation at hand. For instance, we would like to get rid of the predators if they are seen as a pest; on the other hand, if the prey are a valuable resource, to eliminate the disease among them would certainly enhance their survival.

The presentation is organized as follows. In the next section, we describe mathematically the system in consideration. Section 3 studies the possible long-term behavior of the model, assessing its equilibria, and in Section 4 their stability is investigated. A final discussion of the findings concludes the paper.

2. The Model

Let X and Y be the genotypes of the prey population. We assume that one of them is prone to a disease, for which we partition its individuals in the two subpopulations of susceptibles S and infected I, so that $X = S + I$. Let Z denote the predators.

We make the standard assumptions of mathematical ecogenetic models (Venturino, 2012), without using the more sophisticated HTII response terms considered in Viberti and Venturino (2014). The two prey genotypes reproduce exponentially, a fact that is modeled via logistic terms, and produce offsprings of both genotypes with probabilities p and q, $p + q = 1$.

The reproduction at rate is r, all the offsprings are born healthy, i.e. they belong to class S, in other words the disease is not vertically transmitted. The susceptibles feel the population pressure of the similar individuals and those of the other genotype, but not of the infected. Here also the infected recovering from the disease that re-enter into this class are accounted for. Its losses are due to predation and possible contagion of the disease.

The infected are recruited only via these successful contacts, are hunted and can recover. We assume the disease to be mild and also predation to occur at a fast rate, for which the disease-induced mortality can be disregarded. We also disregard the intraspecific population pressure on the infected.

The second prey genotype population Y reproduces also logistically, feeling the intraspecific population pressure, as well as the one of the other genotype. Furthermore, contacts of Y with infected constitute an additional mortality μIY, which we take at the same rate as the intraspecific population pressure, thus setting $\mu = b$. This asymmetry in the infected behavior is ascribed to the fact that they are poisonous to the predators Z, as we will see below, and in that sense they are also harmful to the genotype Y that is not their own.

The predators hunt the first genotype, independently of whether it is infected or not, and the second one Y possibly at a different rate. Their natural mortality is m, to which a mortality j due to interaction with the infected is added. Note that in describing the hunting of the predators on the infected, we separate the influences that the latter have on the predator: there is a positive effect due to the feeding, but also the contact with them might lead to the predators' death.

The model is thus:

$$\begin{aligned} S' &= [rp - aS](S+Y) - hSZ - \lambda SI + \gamma I \\ I' &= I[\lambda S - hZ - \gamma] \\ Y' &= [rq - bY](S+I+Y) - gYZ \\ Z' &= Z[e(h(S+I)+gY) - m - jI] \end{aligned} \tag{1}$$

where r is the prey reproduction rate, a and b are the intraspecific competition rates of genotypes X and Y, h and g the predators' hunting rates on genotypes X and Y, respectively. Note that we have also implicitly assumed that it is equally likely for the predators to capture a healthy or an infected individual of genotype X. Further, $e < 1$ denotes the conversion factor of captured prey into new predators, λ is the disease contact rate and γ represents its recovery rate.

3. Equilibria

We find the following system's equilibria $E_k = (S_k, I_k, Y_k, Z_k)$. The origin E_0, the predator- and disease-free equilibria E_1 and the disease-free point E_2:

$$E_1 = \left(\frac{rp}{a}, 0, \frac{rq}{b}, 0\right), \quad E_2 = \left(\frac{m - egY_2}{eh}, 0, Y_2, \frac{H_2Y_2 + K_2}{g^2eh(ag - hb)Y_2}\right),$$

where $H_2 = egr(h-g)(qag+phb) - hbm(ag-hb)$, $K_2 = rqm(bh^2 - ag^2)$ and Y_2 solves

$$A_2Y^2 + A_1Y + A_0 = 0, \tag{2}$$

with $A_2 = eg(ag-hb)$, $A_1 = eghr + m(hb-ag)$, $A_0 = -hrmq < 0$. There are two cases, whether this quadratic Equation 2 has one or two nonnegative roots. In the former, by Descartes' rule, to have a nonnegative root it is enough to impose positivity of the first coefficient, from which we have $ag > hb$. In the second case, we have instead two positive roots if $A_2 < 0$, $A_1 > 0$ and $\Delta = A_1^2 - 4A_0A_2 > 0$. Positivity of A_1 is ensured by $ag < hb$, and this entails the negativity of A_2 and in turn the fact that the roots are real, recalling that $q < 1$:

$$\begin{aligned}\Delta &= [eghr - m(ag-hb)]^2 + 4eghmrq(ag-hb)\\ &> (eghr)^2 + m^2(ag-hb)^2 + 2eghmr(ag-bh) = [eghr + m(ag-hb)]^2 > 0.\end{aligned}$$

We then need the nonnegativity of Z_2, which leads to

$$Y_2H_2 + K_2 \equiv Y_2[egr(h-g)(qag+phb) - hbm(ag-hb)] + rqm(bh^2 - ag^2) > 0$$

in the former case, and to the opposite inequality in the latter, in view of the sign of A_1. Now looking once again for sufficient conditions, $K_2 > 0$ implies $h > g$ in the former case and then asking $H_2 > 0$ yields $egr(h-g)(qag+phb) > hbm(ag-hb)$. For $ag < bh$ instead, we find that requesting $K_2 < 0$ gives $h < g$ and then $H_2 < 0$ is ensured by $egr(h-g)(qag+phb) < hbm(ag-hb)$.

In summary, one equilibrium is feasible if

$$Y_2 \le \frac{m}{eg}, \quad ag > hb, \quad h > g, \quad egr(h-g)(qag+phb) > hbm(ag-hb); \tag{3}$$

two such points instead arise and are feasible if

$$Y_2 \le \frac{m}{eg}, \quad ag < hb, \quad h < g, \quad egr(h-g)(qag+phb) < hbm(ag-hb). \tag{4}$$

For the coexistence equilibrium $E_3 = (\bar{S}, \bar{I}, \bar{Y}, \bar{Z})$, we solve the second and fourth equations of (1) for S and I, to get

$$\bar{S} = \frac{hZ+\gamma}{\lambda}, \quad \bar{I} = \frac{\lambda(m-egY) - eh(hZ+\gamma)}{\lambda(eh-j)}.$$

Substitution into the remaining equations gives the following two conic sections

$$\begin{aligned}f(Y,Z) = &\left[\frac{eh^3}{\lambda(eh-j)} - \frac{ah^2}{\lambda^2} - \frac{h^2}{\lambda}\right]Z^2 + \left[\frac{heg}{eh-j} - \frac{ah}{\lambda}\right]YZ + \left[rp - \frac{a\gamma}{\lambda}\right]Y\\ &+ \left[\frac{rph}{\lambda} - \frac{2a\gamma h}{\lambda^2} - \frac{h\gamma}{\lambda} - \frac{hm}{eh-j} + \frac{eh^2\gamma}{\lambda(eh-j)}\right]Z + \frac{rp\gamma}{\lambda} - \frac{a\gamma^2}{\lambda^2} = 0\end{aligned} \tag{5}$$

$$\begin{aligned}g(Y,Z) = &\left[\frac{beg}{eh-j} - b\right]Y^2 + \left[rq - \frac{egrq}{eh-j} - \frac{b\gamma}{\lambda} - \frac{bm}{eh-j} + \frac{ehb\gamma}{\lambda(eh-j)}\right]Y+\\ &+ \left[\frac{ebh^2}{\lambda(eh-j)} - \frac{bh}{\lambda} - g\right]YZ + \left[\frac{rqh}{\lambda} - \frac{rqeh^2}{\lambda(eh-j)}\right]Z\\ &+ \frac{rq\gamma}{\lambda} + \frac{rqm}{eh-j} - \frac{ehrq\gamma}{\lambda(eh-j)} = 0\end{aligned} \tag{6}$$

Both are conic sections of the type described by the generic equation $AZ^2 + 2HYZ + BY^2 + 2GZ + 2FY + C = 0$, with, using obvious notations, $A_f \neq 0$ and $B_f = 0$ for the first one and for $A_g = 0$ and $B_g \neq 0$ the second one.

Now the invariants of Equation 5 are

$$\Gamma_f = -A_f F_f^2 - C_f H_f^2 + 2F_f G_f H_f = \frac{h^2(a\gamma - rp\lambda)}{4\lambda^2(eh-j)^2}\Big[e^2g^2\gamma - e^2ghpr$$
$$+ egjpr + ehjpr - aehm - eg\gamma j + eg\lambda m - j^2pr + ajm\Big],$$
$$\Pi_f = -H_f^2 = -\left(\frac{heg}{2(eh-j)} - \frac{ah}{2\lambda}\right)^2,$$

so that it is a hyperbola since $\Pi_f < 0$, assuming nondegeneracy, $\Gamma_f \neq 0$. Also Equation 6 is a hyperbola, since

$$\Gamma_g = -B_g G_g^2 - C_g H_g^2 + 2F_g G_g H_g = \frac{rq(heg\lambda - bhj - gj\lambda)}{4\lambda^2(eh-j)^3}\Big[eh^2jqr$$
$$- eghjqr + eg\gamma hj - egh\lambda m - hj^2qr - g\gamma j^2 + gj\lambda m\Big],$$
$$\Pi_g = -H_g^2 = -\left(\frac{ebh^2}{2\lambda(eh-j)} - \frac{bh}{2\lambda} - g\right)^2,$$

with once again $\Pi_g < 0$ and assuming nondegeneracy, $\Gamma_g \neq 0$.

These curves are better investigated by solving for each variable, to get:

$$Y_f = -\frac{1}{2}\frac{\rho(Z)}{\theta(Z)} \equiv -\frac{1}{2}\frac{A_fZ^2 + 2G_fZ + C_f}{H_fZ + F_f}, \quad Z_g = -\frac{1}{2}\frac{\Psi(Y)}{\phi(Y)} \equiv -\frac{1}{2}\frac{B_gY^2 + 2F_gY + C_g}{H_gY + G_g}.$$

We seek now sufficient conditions ensuring the existence of the equilibrium where all population thrive.

For Z_g, note that if $B_g < 0$ it follows $G_g > 0$, $H_g < 0$. If we take also $C_g > 0$, then $Z_g(0) < 0$ and $\Psi(Y)$ is a quadratic with one positive root in view of Descartes' rule of signs. These roots $Y_g^- < 0 < Y_g^+$ always exist since $\Delta_g > 0$. This statement follows observing that $2F_g = -b(rq)^{-1}C_g - rqb^{-1}B_g$ so that easily $\Delta_g = 4F^2 - 4B_gC_g = [b(rq)^{-1}C_g - rqb^{-1}B_g]^2 > 0$. Since $\phi(Y) \geq 0$ whenever $Y < Y_g^\infty \equiv -2G_gH_g^{-1}$, with $Y_g^\infty > 0$, the function Z_g is positive only in between Y_g^+ and Y_g^∞, independently of their order.

A similar analysis on $Y_f(Z)$ shows the same result. Only in the absence of more specific information on the respective hunting rates on the two genotypes, g and h, we assume both $H_f < 0$ and $A_f < 0$, given their asymmetry; the latter would however be a direct consequence of the former if $g \geq h$. Note also that taking $C_f > 0$ we obtain $F_f > 0$. The discriminant then is always positive in view of the opposite signs of the coefficients A_f and C_f, $\Delta_f = G^2 - A_fC_f > 0$. The function $Y_f(Z)$ then is positive only in the interval between the positive root Z_f^+ and the vertical asymptote, Z_f^∞, independently of which one is the smallest.

Evidently since these curves are surjective on their respective half ranges, the half lines $Z \geq 0$ and $Y \geq 0$, respectively, they always meet at a feasible point. Therefore, in summary, sufficient conditions for the feasibility of the equilibrium E_3 are provided by the following inequalities:

$$\frac{eg}{eh-j} < 1, \quad \gamma + \frac{m\lambda}{eh-j} > \frac{eh\gamma}{eh-j}; \quad \frac{eg}{eh-j} < \frac{a}{\lambda}, \quad \frac{eh}{eh-j} < \frac{a}{\lambda} + 1, \quad \frac{rp}{\gamma} > \frac{a}{\lambda}. \tag{7}$$

4. Stability

The Jacobian of (1) is

$$J = \begin{pmatrix} J_{11} & -\lambda S + \gamma & rp - aS & -hS \\ \lambda I & \lambda S - hZ - \gamma & 0 & -hI \\ rq - bY & rq - bY & J_{33} & -gY \\ ehZ & Z(eh - j) & egZ & J_{44} \end{pmatrix} \tag{8}$$

with $J_{11} = rp - 2aS - aY - hZ - \lambda I$, $J_{33} = rq - 2bY - b(I + S) - gZ$ and $J_{44} = eh(S + I) + egY - m - jI$.

E_0 is unstable, since its eigenvalues are $0, rp + rq, -m, -\gamma$.

The eigenvalues of the Jacobian evaluated at E_1 are $(ehrpb + egrqa - mab)(ab)^{-1}$, $-r(aq + bp)a^{-1}, (\lambda rp - \gamma a)a^{-1}, -r(aq + bp)b^{-1}$, implying that this point is stable for

$$mab > er(hpb + gqa), \quad \gamma a > \lambda rp. \tag{9}$$

Therefore, the predators and the disease establish themselves in the ecosystem if the disease transmission rate exceeds the threshold

$$\lambda^{\dagger} = \frac{\gamma a}{rp}. \tag{10}$$

On comparing the last conditions in Equations 9 and in 7, we infer that a transcritical bifurcation leading from E_2 to E_3 might arise. The conditional is needed as Equation 7 are only sufficient conditions for the feasibility of the coexistence equilibrium.

Equilibrium E_1, shown in Figure 1, is achieved by the following choice of system's parameter values

$$a = 2.7, \quad b = 2.6, \quad r = 2.42, \quad h = 0.4, \quad g = 0.7, \quad j = 4.5,$$
$$m = 5.5, \quad \lambda = 5.82, \quad \gamma = 3.59, \quad p = 0.4, \quad q = 0.6, \quad e = 0.95. \tag{11}$$

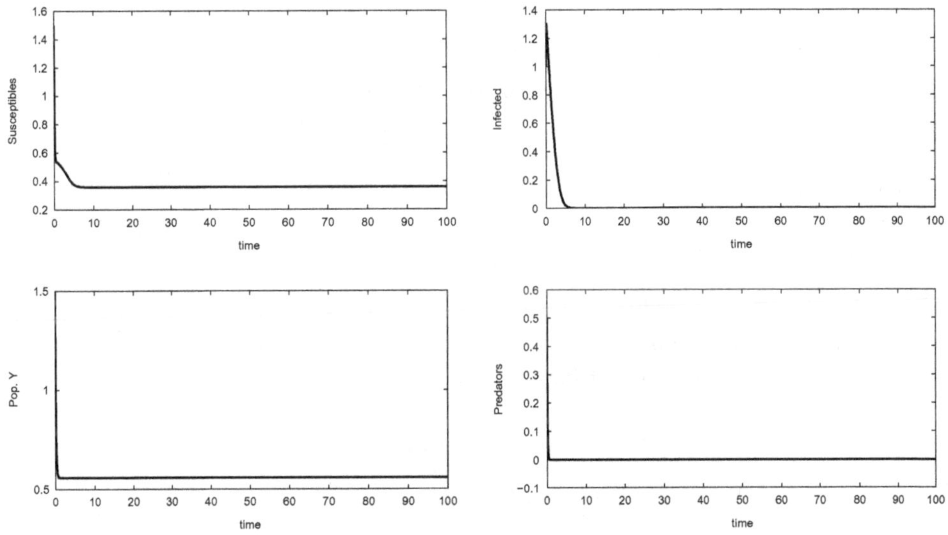

Figure 1. Equilibrium E_1 Parameters listed in Equation 11.

4.1. The disease-free equilibrium E_2

At E_2 one explicit eigenvalue is available: $\lambda S_2 - hZ_2 - \gamma$. It gives the necessary stability condition:

$$\lambda m < \lambda egY_2 + eh(hZ_2 + \gamma). \tag{12}$$

The Routh–Hurwitz conditions on the remaining minor $J^{(2)}$ of order 3 obtained deleting the second row and column of Equation 8 give the following inequalities:

$$\mathrm{tr}(J^{(2)}) <, \quad \det(J^{(2)}) < 0, \quad \mathrm{tr}(J^{(2)})M_2(J^{(2)}) < \det(J^{(2)}),$$

where $M_2(J^{(2)})$ represents the sum of the principal minors of order 2 of the matrix $J^{(2)}$. Now, the condition on the trace gives:

$$rp - 2aS_2 - aY_2 - hZ_2 + rq - 2bY_2 - bS_2 - gZ_2 < 0$$

from which using the value of S_2 we find

$$r - \frac{m}{eh}(2a + b) - \frac{1}{h}Y_2[ag - bh + (g - h)(a + b)] - (g + h)Z_2 < 0.$$

Sufficient conditions for the above inequality are

$$ehr < m(2a + b), \quad ag > bh, \quad g > h. \tag{13}$$

For the determinant, we have

$$eh^2Z_2S_2(rq - bS_2 - 2bY_2 - gZ_2) - eghY_2Z_2(rp - aS_2)$$
$$+ eg^2Y_2Z_2(rp - 2aS_2 - aY_2 - hZ_2) - eghZ_2(rq - bY_2)S_2 < 0$$

which, observing that conditions Equation 13 already imply negativity of the trace and therefore of the above first and third terms, is ensured by

$$aS_2 < rp, \quad bY_2 < rq. \tag{14}$$

The last Routh-Hurwitz condition is then:

$$\begin{aligned}&[(rp - 2aS_2 - aY_2 - hZ_2) + (rq - bS_2 - 2bY_2 - gZ_2)][(rp - 2aS_2 - aY_2\\ &\quad - hZ_2)(rq - bS_2 - 2bY_2 - gZ_2) + eh^2Z_2S_2 + eg^2Y_2Z_2 - (rq - bY_2)(rp - aS_2)]\\ &\quad < eh^2Z_2S_2(rq - bS_2 - 2bY_2 - gZ_2) - eghZ_2(rq - bY_2)S_2\\ &\quad + eg^2Y_2Z_2(rp - 2aS_2 - aY_2 - hZ_2) - eghY_2Z_2(rp - aS_2).\end{aligned}$$

Equilibrium E_2, shown in Figure 2, is indeed achieved by the system, for the following set of parameter values

$$\begin{aligned}&a = 3.1, \quad b = 2.3, \quad r = 6.4, \quad h = 2.1, \quad g = 1.7, \quad j = 3,\\ &m = 3.5, \quad \lambda = 1.8, \quad \gamma = 2.2, \quad p = 0.4, \quad q = 0.6, \quad e = 0.85.\end{aligned} \tag{15}$$

4.2. The coexistence equilibrium E_3

The Jacobian at this point simplifies a bit, in that $J_{22} = J_{44} = 0$.

The Routh–Hurwitz conditions in this case are more involved. The condition on the trace gives here

$$r - (2a + b)\bar{S} - (a + 2b)\bar{Y} - (g + h)\bar{Z} - (\lambda + b)\bar{I} < 0. \tag{16}$$

Denoting by D_{ij} the minor of $J(E_3)$ obtained by deleting the ith row and jth column, the condition on the determinant instead provides: $-\lambda ID_{2,1} - hID_{2,4} > 0$ which is satisfied if we take both minors negative. This holds if we impose the following two conditions:

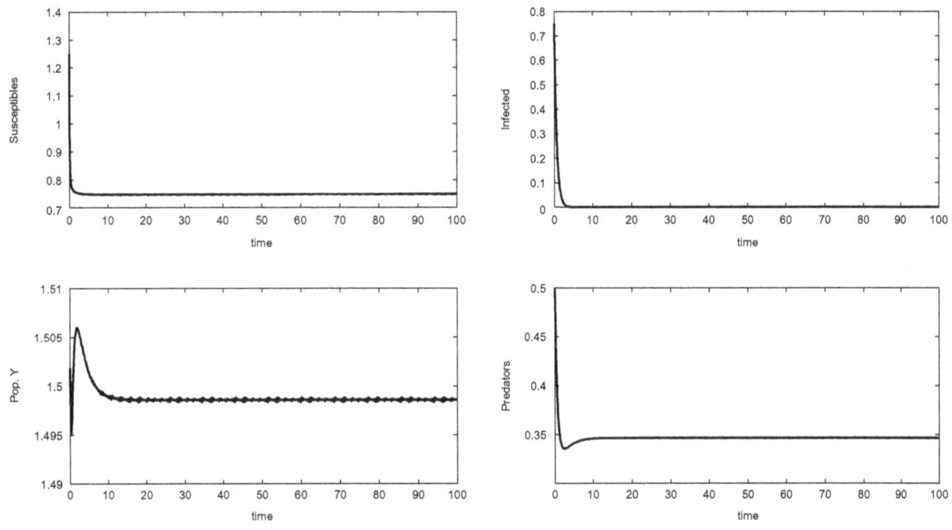

Figure 2. Equilibrium E_2 for the parameters given in Equation 15.

$$h\bar{S}\bar{Z}\left[\bar{J}_{33}(eh-j) - eg(rq-b\bar{Y})\right] + g\bar{Y}\bar{Z}\left[eg(\gamma-\lambda\bar{S}) - (rp-a\bar{S})(eh-j)\right] < 0, \tag{17}$$

$$\begin{aligned} &eg\bar{Z}\bar{J}_{11}(rq-b\bar{Y}) + eh\bar{Z}(\gamma-\lambda\bar{S})\bar{J}_{33} + (rp-a\bar{S})(rq-b\bar{Y})\bar{Z}(eh-j) \\ &\quad - eh\bar{Z}(rp-a\bar{S})(rq-b\bar{Y}) - (eh-j)\bar{Z}\bar{J}_{11}\bar{J}_{33} - eg\bar{Z}(\gamma-\lambda\bar{S})(rq-b\bar{Y}) < 0. \end{aligned} \tag{18}$$

The remaining two conditions are a much more involved. In terms of the coefficients of the monic characteristic equation $\sum_{k=0}^{4} a_k\Lambda^k = 0$, they are

$$a_3a_2 > a_1, \quad a_3a_2a_1 > a_1^2 + a_0a_3^2. \tag{19}$$

These coefficients are, in addition to the trace a_3 and the determinant a_0 of the Jacobian, the sum of its principal minors of order 2 and 3. Explicitly:

$$\begin{aligned} a_2 &= \begin{vmatrix} J_{11} & (\gamma-\lambda S) \\ \lambda I & 0 \end{vmatrix} + \begin{vmatrix} J_{11} & (rp-aS) \\ (rq-bY) & J_{33} \end{vmatrix} + \begin{vmatrix} J_{11} & -hS \\ ehZ & 0 \end{vmatrix} \\ &\quad + \begin{vmatrix} 0 & 0 \\ (rq-bY) & J_{33} \end{vmatrix} + \begin{vmatrix} 0 & -hI \\ Z(eh-j) & 0 \end{vmatrix} + \begin{vmatrix} J_{33} & -gY \\ egZ & 0 \end{vmatrix} \\ &= hSehZ - (\gamma-\lambda S)\lambda I + J_{11}J_{33} - (rp-aS)(rq-bY) + hIZ(eh-j) + gYegZ \end{aligned}$$

and

$$\begin{aligned} a_3 &= -hI\begin{vmatrix} (rq-bY) & J_{33} \\ Z(eh-j) & egZ \end{vmatrix} + \begin{vmatrix} J_{11} & (rp-aS) & -hS \\ (rq-bY) & J_{33} & -gY \\ ehZ & Z(eh-j) & 0 \end{vmatrix} \\ &\quad + \begin{vmatrix} J_{11} & (\gamma-\lambda S) & -hS \\ \lambda I & 0 & -hI \\ ehZ & Z(eh-j) & 0 \end{vmatrix} - \lambda I\begin{vmatrix} (\gamma-\lambda S) & (rp-aS) \\ (rq-bY) & J_{33} \end{vmatrix} \\ &= -hI\left[(rq-bY)egZ - J_{33}Z(eh-j)\right] \\ &\quad - eghYZ(rp-aS) - hSZ(rq-bY)(eh-j) + eh^2SZJ_{33} \\ &\quad + gYZJ_{11}(eh-j)J_{11}hIZ(eh-j) - eh^2IZ(\gamma-\lambda S) \\ &\quad - hS\lambda IZ(eh-j) - \lambda I\left[(\gamma-\lambda S)J_{33} - (rp-aS)(rq-bY)\right]. \end{aligned}$$

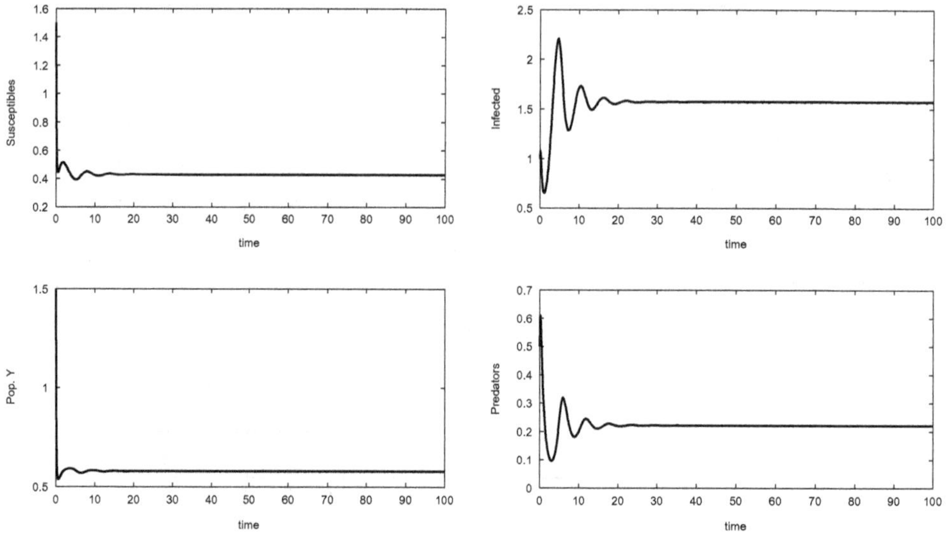

Figure 3. Coexistence equilibrium obtained for the set of parameter values Equation 20.

Also the coexistence equilibrium is empirically shown to be achievable in a stable way in Figure 3 by the following choice of parameters:

$$a = 4.7, \quad b = 3.6, \quad r = 5.42, \quad h = 2.8, \quad g = 1.7, \quad j = 1.5,$$
$$m = 3.9, \quad \lambda = 2.82, \quad \gamma = 0.59, \quad p = 0.6, \quad q = 0.4, \quad e = 0.95. \tag{20}$$

5. Conclusions

The model considered in this paper extends in a new direction the ones that have been introduced in Venturino (2012) and Viberti and Venturino (2014). Namely, we consider the fact that one of the genotypes can be subject to a disease, that however cannot affect the other one. In addition, we assume that the infected genotype has also some toxic effects for a possible specialist predator, that feeds on both genotypes.

Our findings indicate that the ecosystem in these assumptions cannot completely disappear. The model trajectories can settle toward the disease- and predator-free equilibria, toward the predator-free one, or the system can achieve coexistence of all the populations. Thus, to answer one of the questions raised in the Introduction, the disease can be eradicated, and the predators have a fundamental role in it, as the parameters related to their dynamics appear in the necessary stability condition (Equation 12). Both genotypes of the healthy prey can survive in an environment where the predators are absent, as it is found also in Venturino (2012). Thus, the elimination of the predators at first sight does not appear to be tied to the presence of the disease in the prey, elucidating then the statement of the Introduction. It is also interesting to note that in the assumptions of this particular ecosystem, the disease cannot endemically survive among the prey in the absence of the predators, which seems to constitute another fact in support of a negative answer to the second claim stated in the Introduction. However, at a deeper analysis, looking at the stability conditions for the equilibrium in which only the two prey genotypes thrive, (Equation 9), we observe that the first one is essentially a purely demographic condition, which plays the same role of equation (11) in Venturino (2012). The second one however is new and depends on the epidemic parameters, the disease contact rate λ, and its recovery rate γ. Evidently, if in the absence of the disease the two genotypes can thrive without the predators, it is possible that in the presence instead of a particularly virulent, i.e. highly transmissible, disease, combined perhaps with a low recovery rate, or in any case if the contact rate exceeds the threshold $\lambda^{\dagger}$, (Equation 10), the system can be driven away from this equilibrium. In such case certainly the disease is introduced at an endemic level and, since the endemic predator-free equilibrium here does not exist, in addition the invasion of the predators is made possible. Furthermore, if the disease is unrecoverable, i.e. $\gamma = 0$, the second condition in (Equation 9) cannot be satisfied, implying that either the disease or the predators, or both, are always present in the system.

As it in general occurs in ecoepidemic models, the major finding of this investigation is thus the fact that the disease deeply influences the underlying demographics of the system.

Funding
This work has been partially supported by the project "Metodi numerici in teoria delle popolazioni" of the Dipartimento di Matematica "Giu-seppe Peano".

Author details
Ilaria Usville[1]
E-mail: ilaria.usville@edu.unito.it
Cristina Paola Viola[1]
E-mail: cristina.viola@edu.unito.it
Ezio Venturino[1]
E-mail: ezio.venturino@unito.it
ORCID ID: http://orcid.org/0000-0001-7215-5114
[1] Dipartimento di Matematica "Giuseppe Peano", Università di Torino, Torino, Italy.

Citation information
Cite this article as: An ecogenetic disease-affected predator-prey model, Ilaria Usville, Cristina Paola Viola & Ezio Venturino, *Cogent Mathematics* (2016), 3: 1195716.

References

Angulo, O., & López-Marcos, J. C. (1999). Numerical schemes for size-structured population equations. *Mathematical Biosciences, 157*, 169–188.

Auger, P., Mchich, R., Chowdhury, T., Sallet, G., Tchuente, M., & Chattopadhyay, J. (2009). Effects of a disease affecting a predator on the dynamics of a predator--prey system. *Journal of Theoretical Biology, 258*, 344–351.

Bairagi, N., Roy, P. K., & Chattopadhyay, J. (2007). Role of infection on the stability of a predator-prey system with several response functions—A comparative study. *Journal of Theoretical Biology, 248*, 10–25.

Bairagi, N., Sarkar, R. R., & Chattopadhyay, J. (2008). Impacts of incubation delay on the dynamics of an eco-epidemiological system—A theoretical study. *Bulletin of Mathematical Biology, 70*, 2017–2038. doi:10.1007/s11538-008-9337-y

Bairagi, N., Chaudhuri, S., & Chattopadhyay, J. (2009). Harvesting as a disease control measure in an eco-epidemiological system - A theoretical study. *Mathematical Biosciences, 217*, 134–144.

Bate, A. M., & Hilker, F. M. (2013). Complex dynamics in an eco-epidemiological model. *Bulletin of Mathematical Biology, 75*, 2059–2078. doi:10.1007/s11538-013-9880-z

Beltrami, E., & Carroll, T. O. (1994). Modelling the role of viral disease in recurrent phytoplankton blooms. *Journal of Mathematical Biology, 32*, 857–863.

Beretta, E., & Kuang, Y. (1998). Modeling and analysis of a marine bacteriophage infection. *Mathematical Biosciences, 149*, 57–76.

Bhattacharyya, S., & Bhattacharya, D. K. (2006). Pest control through viral disease: Mathematical modeling and analysis. *Journal of Theoretical Biology, 238*, 177–196.

Chattopadhyay, J., & Arino, O. (1999). A predator-prey model with disease in the prey. *Nonlinear Analysis, 36*, 747–766.

Clarke, B. (2003). Heredity - The art of innuendo. *Heredity, 90*, 279–280. doi:10.1038/sj.hdy.6800229

Cushing, J. M. (1998). *An introduction to structured population dynamics*. Philadelphia: SIAM. doi:10.1038/sj.hdy.6800229

Cuthbertson, A. G. S., Blackburn, L. F., & Audsley, N. (2014). Efficacy of commercially available invertebrate predators against *Drosophila suzukii*. *Insects, 5*, 952–960. doi:10.3390/insects5040952

Delgado, M., Molina-Becerra, M., & Suarez, A. (2005). Relating disease and predation: equilibria of an epidemic model. *Mathematical Methods in the Applied Sciences, 28*, 349–362.

Grant, B. S. (1999). Fine tuning the peppered moth paradigm. *Evolution, 53*, 980–984.

Gurtin, M. E., & Levine, D. S. (1979). On predator-prey interactions with predation dependent on age of prey. *Mathematical Biosciences, 47*, 207–219.

Gurtin, M. E., & McCamy, R. C. (1974). Nonlinearly age-dependent population dynamics. *Archive for Rational Mechanics and Analysis, 54*, 281–300.

Gurtin, M. E., & McCamy, R. C. (1979). Some simple models for nonlinear age-dependent population dynamics. *Mathematical Biosciences, 43*, 199–211.

Hadeler, K. P., & Freedman, H. I. (1989). Predator--prey population with parasitic infection. *Journal of Mathematical Biology, 27*, 609–631.

Hooper, J. (2002). *Of moths and men: An evolutionary tale*. New York, NY: W.W. Norton & Company.

Hotopp, I. S., Malchow, H., & Venturino, E. (2010). Switching feeding among sound and infected prey in ecoepidemic systems. *Journal of Biological Systems, 18*, 727–747. doi:10.1142/S0218339010003718

Jaenike, J. (1993). Rapid evolution of host specificity in a parasitic nematode. *Evolutionary Ecology, 7*, 103–108.

Jana, S., & Kar, T. K. (2013). Modeling and analysis of a prey--predator system with disease in the prey. *Chaos, Solitons & Fractals, 47*, 42–53.

Kar, T. K., & Jana, S. (2013). A theoretical study on mathematical modelling of an infectious disease with application of optimal control. *Biosystems, 111*, 37–50.

Kermack, W. O., & McKendrick, A. G. (1927). A contribution to the mathematical theory of epidemics. *Proceedings of the Royal Society of London, 115*, 700–721.

Khan, Q. J. A., Balakrishnan, E., & Wake, G. C. (2004). Analysis of a predator-prey system with predator switching. *Bulletin of Mathematical Biology, 66*, 109–123.

Kooi, B. W., van Voorn, G. A. K. & Das, K. p. (2011). Stabilization and complex dynamics in a predator-prey model with predator suffering from an infectious disease. *Ecological Complexity, 8*, 113–122.

Lively, C. M., & Apanius, V. (1995). Genetic diversity in host-parasite interactions. In: Grenfell, B.T. & Dobson, A.P. (Eds.), *Ecology of infectious diseases* (pp. 419–449). Cambridge: Cambridge University Press.

Lively, C. M., Craddock, C., & Vrijenhoek, R. C. (1990). Red Queen hypothesis supported by parasitism in sexual and clonal fish. *Nature, 344*, 864–866.

Malchow, H., Petrovskii, S., & Venturino, E. (2008). *Spatiotemporal patterns in ecology and epidemiology*. Boca Raton, FL: CRC.

Read, A. F. (1991). Passerine polygyny: A role for parasites? *American Naturalist, 138*, 434–459.

Saenz, R. A., & Hethcote, H. W. (2006). Competing species models with an infectious disease. *Mathematical Biosciences and Engineering, 3*, 219–235.

Scudo, F. M., & Ziegler, J. R. (1976). Vladimir Aleksandrovich Kostitzin and theoretical ecology. *Theoretical Population Biology, 10*, 395–412.

Sieber, M., & Hilker, F. M. (2011). Prey, predators, parasites: Intraguild predation or simpler community modules in disguise? *Journal of Animal Ecology, 80*, 414–421. doi:10.1111/j.1365-2656.2010.01788.x

Singh, B. K., Chattopadhyay, J., & Sinha, S. (2004). The role of virus infection in a simple phytoplankton zooplankton system. *Journal of Theoretical Biology, 231*, 153–166.

Smith, G. D. (1985). *Numerical solution of partial differential equations: Finite difference methods*. Oxford: Oxford University Press.

Venturino, E. (1984). Age-structured predator--prey models. *Mathematical Modelling, 5*, 117–128.

Venturino, E. (1985). A generalization of the classical epidemiology model. *IMACS Transactions on Scientific Computation '85*. Modelling of Biomedical Systems (Vol. 5, pp. 243–248). North-Holland: Amsterdam.

Venturino E. (1987). Non-linearly interacting age-dependent populations. *Computers and Mathematics with Applications, 13*, 901–911.

Venturino, E. (1994). The influence of diseases on Lotka–Volterra systems. *Rocky Mountain Journal of Mathematics, 24*, 381–402.

Venturino, E. (1995). Epidemics in predator-prey models: disease in the prey. In: Arino, O., Axelrod, D., Kimmel, M., & Langlais, M. (Eds.), *Mathematical population dynamics, analysis of heterogeneity* (Vol. 1, pp. 381–393). Winnipeg: Wuerz Publishing.

Venturino, E. (2002). Epidemics in predator–prey models: Disease in the predators. *IMA Journal of Mathematics Applied Medicine and Biology, 19*, 185–205.

Venturino, E. (2012). An ecogenetic model. *Applied Mathematics Letters, 25*, 1230–1233.

Venturino, E. (2016). Ecoepidemiology: A more comprehensive view of population interactions. *Mathematical Modelling of Natural Phenomena, 11*, 49–90.

Viberti, C., & Venturino, E. (2014). An ecosystem with Holling type II response and predators' genetic variability. *Mathematical Modelling and Analysis, 19*, 371–394.

Vrijenhoek, R. C. (1993). The origin and evolution of clones versus the maintenance of sex in *Poeciliopsis*. *Journal of Heredity, 84*, 388–395.

Webb, G. (1985). *Theory of nonlinear age-dependent population dynamics*, Monographs and textbooks in pure & applied mathematics series Vol. 89. New York, NY: Dekker.

Zhen, J., & Haque, M. (2006). Global stability analysis of an eco-epidemiological model of the Salton sea. *Journal of Biological Systems, 14*, 373–385.

14

Assessing the impact of homelessness on HIV/AIDS transmission dynamics

C.P. Bhunu[1,2*]

*Corresponding author: C.P. Bhunu, Department of Mathematics, University of Zimbabwe, Box MP 167 Mount Pleasant, Harare, Zimbabwe; Department of Mathematics and Applied Mathematics, University of Venda, Thohoyandou, South Africa

E-mail: cpbhunu@gmail.com; cp-b@ hotmail.co.uk

Reviewing editor: Ryan Loxton, Curtin University, Australia

Abstract: Care for the people living with HIV/AIDS is more than the provision of antiretroviral therapy. The effects of homelessness on HIV/AIDS transmission are captured through a mathematical model. The mathematical model is rigorously analyzed. The disease-free equilibrium is globally asymptotically stable when the reproduction number is less than unity. Results from the analysis of the reproduction number suggests that homelessness enhances both HIV transmission and progression to the AIDS stage. This is further supported by numerical simulations which show that some elements of homelessness (lack of entertainment) enhances HIV/AIDS transmission.

Subjects: Bioscience; Health and Social Care; Mathematics & Statistics; Medicine; Science

Keywords: mathematical model; HIV/AIDS and homelessness

1. Introduction

Homeless people are amongst the most vulnerable in the society and do not get the help they need to address their health, economic, and social issues. Homelessness and HIV/AIDS are intricately related (National Coalition for the Homeless (NCH), 2009) as homeless worsens HIV and the homeless are doubly affected by HIV. The pressure of daily needs, exposure to violence (including sexual exploitation), alcohol and drug-misuse to cope with stress or mental health issues and other conditions of the homelessness make homeless and unstably housed people extremely vulnerable to HIV infection (Aidala and Sumartjo, 2007). A 1995 survey of homeless adults found that 69% were at risk for HIV infection from unprotected sex with multiple partners, injection drug use (IDU), sex with IDU partners, or exchanging unprotected sex for money or drugs (Adams, 2003). People who

ABOUT THE AUTHOR

C.P. Bhunu (BSc Hons, MSc, DPhil) is a professor in the Department of Mathematics, University of Zimbabwe. He currently serves as an external examiner for Chinhoyi University of Technology, Harare Institute of Technology and Zimbabwe Open University. He was visiting African scientist at Cambridge Infectious Disease Consortium (2010) and visiting professor at the University of Venda (2012), respectively. He is a life member of the Clare Hall College, University of Cambridge. He also serves as an editor and reviewer of several international journals in Applied Mathematics. His research interests lie in the field of mathematical modeling of issues affecting mankind ranging from social issues to biological issues as well as the theoretical analysis of the mathematical models that arise in all these applications.

PUBLIC INTEREST STATEMENT

A model is proposed to explore the impact homelessness has on HIV. Results from model analysis suggest that homeleness enhances HIV transmission and disease progression. Our results suggests that the fight against HIV is not won by the provision of ARVs alone.

are homeless or unstably housed have HIV infection rates as much as 16 times higher than people who have a stable place to live (Robertson et al., 2004). Homeless women and adolescents are particularly at risk (Adams, 2003). Stigma makes it difficult for those infected with HIV to access medical and mental health services (Tomaszewski, 2011). Homeless women have special barriers to health care. Homeless mothers, in particular, have been found to subordinate their own health care needs for the needs of their children (Song, 2003). Single homeless women are more likely to be victims of domestic violence and sexual abuse, both of which have been linked to HIV infection (Song, 2003).

Due to HIV infection, some individuals loose their homes due to the costs of medication and health care. At least half of all people living with HIV/AIDS experience homelessness or housing instability (Aidala, Lee, Abramson, Messeri' & Siegler, 2007). Thus, housing is the greatest unmet need of people living with HIV (Bekele et al., 2013; Shubert & Bernstine, 2007). In the United States of America about one-third to one-half of the HIV infected are either homeless, unable to afford their own housing or at imminent risk of homelessness (NCH, 2009; Song, 2003). The socially and culturally based stigma faced by people living with HIV/AIDS is exacerbated by co-factors of substance misuse, mental illness, and homelessness (Tomaszewski, 2011).

HIV/AIDS disease progression is affected by both medical and social factors which is a double blow for the HIV-infected homeless people. Homeless individuals lack basic needs such as food, clothing, and shelter which are necessary to care for the people living with HIV/AIDS (Tomaszewski, 2011). HIV-infected homeless are less likely to receive and adhere to antiretroviral therapy and are more likely to have higher death rate due to AIDS (Aidala et al., 2007; Kidder, Wolitski Campsmith' & Nakamura, 2007; Leaver, Bargh, Dunn, & Hwanget, 2007). It is against this background that we carry out this study. Mathematical models have been developed to understand the role of social and behavioral processes in HIV transmission (Ajay, Brendan, & David, 2009; Bhunu, Mhlanga, & Mushayabasa, 2014; Pedamallu, Ozdamar, Kropat, & Weber, 2012). However, none have looked into homelessness and HIV from the mathematical point of view. In our past work (Bhunu et al., 2014), we explored the impact of prostitution on HIV and now we model the effects of homelessness on the transmission dynamics of HIV/AIDS.

The paper is structured as follows. The model framework and its analysis are presented in Section 2. Numerical simulations are in Section 3 and the last Section concludes the paper.

2. Model description

The model subdivides the human population based on homelessness and HIV infection. The population is divided into the following sub-groups: non-homeless susceptibles $S_1(t)$, non-homeless HIV infected not yet showing AIDS symptoms $I_1(t)$, non-homeless HIV infected not yet showing AIDS symptoms and on treatment $I_{1_t}(t)$, non-homeless HIV infected displaying AIDS symptoms $A_1(t)$, non-homeless HIV infected displaying AIDS symptoms and on treatment $A_{1_t}(t)$, homeless susceptibles $S_2(t)$, homeless HIV infected not yet showing AIDS symptoms $I_2(t)$, homeless HIV infected showing AIDS symptoms $A_2(t)$. The total sub-populations for the non-homeless and the homeless are given by.

$$N_1(t) = S_1(t) + I_1(t) + I_{1_t}(t) + A_1(t) + A_{1_t}(t) \text{ and } N_2(t) = S_2(t) + I_2(t) + A_2(t), \text{ respectively.} \tag{1}$$

The total population size is given by $N(t) = N_1(t) + N_2(t)$. Individuals in different human sub-groups suffer from natural death at a constant rate μ, which is proportional to the number in each class. We assume that interaction is heterogeneous. The group j members make c_j $(j = 1, 2)$ sexual contacts per unit time and a fraction of the contacts made by a member of group j with a member of group i is p_{ji} $(i = 1, 2)$. Then $p_{11} + p_{12} = p_{22} + p_{21} = 1$. The total number of sexual contacts made in unit time by members of group '2' (homeless people) with members of group '1' (non-homeless people) is $c_2 p_{21} N_2$ and this must be equal to the total number of sexual contacts made by members of group '1' with members of group '2', we have a balance relation

$$\frac{p_{21}c_2}{N_1(t)} = \frac{p_{12}c_1}{N_2(t)} \tag{2}$$

The forces of HIV infection for the non-homeless and the homeless are given by λ_1 and λ_2, with

$$\lambda_1(t) = \frac{p_{11}c_1\beta_1[A_1 + \phi_1 I_1 + \theta(A_{1_t} + \phi_1 I_{1_t})](t)}{N_1(t)} + \frac{p_{12}c_1\beta_2[A_2 + \phi_2 I_2](t)}{N_2(t)}$$

and

$$\lambda_2(t) = \frac{p_{22}c_2\beta_2[A_2 + \phi_2 I_2](t)}{N_2(t)} + \frac{p_{21}c_2\beta_1[A_1 + \phi_1 I_1 + \theta(A_{1_t} + \phi_1 I_{1_t})](t)}{N_1(t)} \tag{3}$$

respectively. In Equation 3, β_i $(i = 1, 2)$ is the probability of one individual being infected by one infectious individual from the 1 or 2 class [$\beta_2 = b_3\beta_1$, $b_3 \geq 1$, as a result of co-infections with other untreated STIs]; c_j $(j = 1, 2)$ is the per capita effective sexual contact rate; $\phi_i \in (0, 1)$ $[i = 1, 2]$ accounts for a reduction in infectiousness for those only infected with HIV not yet displaying AIDS symptoms since the viral load is correlated with infectiousness (WHO, 2005); $\theta \in (0, 1)$ accounts for a reduction in infectiousness for those on treatment when compared to those not yet on treatment. It is important to note that $c_2 = b_1c_1$, $b_1 \geq 1$ as homeless people lack other forms of entertainment, most of them will abuse alcohol/drugs (Didenko & Pankratz, 2007) and have many sexual partners.

Susceptible humans enter the population through sexual maturity at a rate Λ, a proportion π entering the non-homeless susceptibles and the complementary proportion $(1 - \pi)$ entering the homeless susceptibles. The non-homeless susceptibles $S_1(t)$ and homeless susceptibles $S_2(t)$ are infected with HIV at rates $\lambda_1(t)$ and $\lambda_2(t)$ to enter the $I_1(t)$ and $I_2(t)$-classes, respectively. Individuals in $I_1(t)$ and $I_2(t)$-classes progress to the AIDS stage ($A_1(t)$ and $A_2(t)$) at rates ρ_1 and ρ_2, respectively, with $\rho_2 = b_2\rho_1$, $b_2 \geq 1$ as homeless HIV positive individuals are more likely to progress to the AIDS stage of disease progression faster than their counterparts as they are more likely to be doubly infected with other infections and suffer from poor nutrition. Individuals in $I_1(t)$-class are put on antiretroviral therapy at a rate α_i to move into the $I_{1_t}(t)$-class. Individuals in $I_{1_t}(t)$-class progress to the AIDS stage $A_{1_t}(t)$ at a rate ρ_{1_t} with $\rho_{1_t} \leq \rho_1$, as individuals on antiretroviral therapy are likely to progress the AIDS stage at a slower rate than those not yet on treatment. Those in $A_1(t)$-class are put on antiretroviral therapy at α_a to enter $A_{1_t}(t)$-class. AIDS-related deaths are experienced by individuals in the AIDS stage of disease progression at rates ν_1 and ν_2, for the homeless and non-homeless, respectively, with $\nu_2 = b_4\nu_1$, $b_4 \geq 1$ as homelessness experience higher AIDS related than the non-homeless due to failure to access medical care. Individuals who are homeless experience inadequate transportation, lack of comprehensive and/or culturally appropriate services, lack of awareness of services and resources, and poor provider attitudes (Tomaszewski, 2011). For that reason we assume there is no antiretroviral therapy for the HIV-infected homeless people.

We assume any transfer from non-homeless to homeless status or vice versa is negligible. The structure of the model is shown in Figure 1.

Based on these assumptions, the following system of differential equations describes the model.

$$\begin{aligned}
S_1'(t) &= \pi\Lambda - (\lambda_1(t) + \mu)S_1(t),\\
I_1'(t) &= \lambda_1(t)S_1(t) - (\rho_1 + \alpha_i + \mu)I_1(t),\\
I_{1_t}'(t) &= \alpha_i I_1(t) - (\rho_{1_t} + \mu)I_{1_t}(t),\\
A_1'(t) &= \rho_1 I_1(t) - (\alpha_a + \mu + \nu_1)A_1(t),\\
A_{1_t}'(t) &= \rho_{1_t} I_{1_t}(t) + \alpha_a A_1(t) - (\mu + \nu_1)A_{1_t}(t),\\
S_2'(t) &= (1 - \pi)\Lambda - (\lambda_2(t) + \mu)S_2(t),\\
I_2'(t) &= \lambda_2(t)S_2(t) - (\rho_2 + \mu)I_2(t),\\
A_2'(t) &= \rho_2 I_2(t) - (\mu + \nu_2)A_2(t).
\end{aligned} \tag{4}$$

2.1. Invariant region

The model system 4 will be analyzed in a suitable region as follows. We first show that system 4 is dissipative. That is, all solutions are uniformly bounded in a proper subset $\Omega \subset \mathbb{R}_+^8$. Let $(S_1, I_1, I_{1_t}, A_1, A_{1_t}, S_2, I_2, A_2) \in \mathbb{R}_+^8$ be any solution with non-negative initial conditions. Adding all the equations in 4, we have

$$N'(t) = \Lambda - \mu N(t) - v_1(A_1 + A_{1_t} + b_4 A_2)(t). \tag{5}$$

Model system 4 has a varying population size ($N'(t) \neq 0$) and therefore a trivial equilibrium is not feasible. Then,

$$N'(t) \quad \leq \Lambda - \mu N(t). \tag{6}$$

So that (cf. Birkhoff & Rota, 1982)

$$0 \leq N(t) \leq \frac{\Lambda}{\mu} + \left(N(0) - \frac{\Lambda}{\mu} \right) e^{-\mu t}, \tag{7}$$

where $N(0)$ represents the value of 4 evaluated at the initial values of the respective variables. The lower limit comes naturally from the fact that the model variables and parameters are non-negative ($\forall\, t \geq 0$) since they monitor human populations. Thus, as $t \to \infty$, $0 \leq N(t) \leq \frac{\Lambda}{\mu}$. Therefore, all feasible solutions of system 4 enter the region

$$\Omega = \left\{ (S_1, I_1, I_{1_t}, A_1, A_{1_t}, S_2, I_2, A_2) \in \mathbb{R}_+^8 : N \leq \frac{\Lambda}{\mu} \right\} \tag{8}$$

Thus, Ω is positively invariant (it can also be shown that Ω is attracting) and it is sufficient to consider solutions in Ω. Existence, uniqueness, and continuation results for system 4 hold in this region. It can be shown that all solutions of system 4 starting in Ω remain in Ω for all $t \geq 0$. All parameters and state variables for model system 4 are assumed to be non-negative for $t \geq 0$.

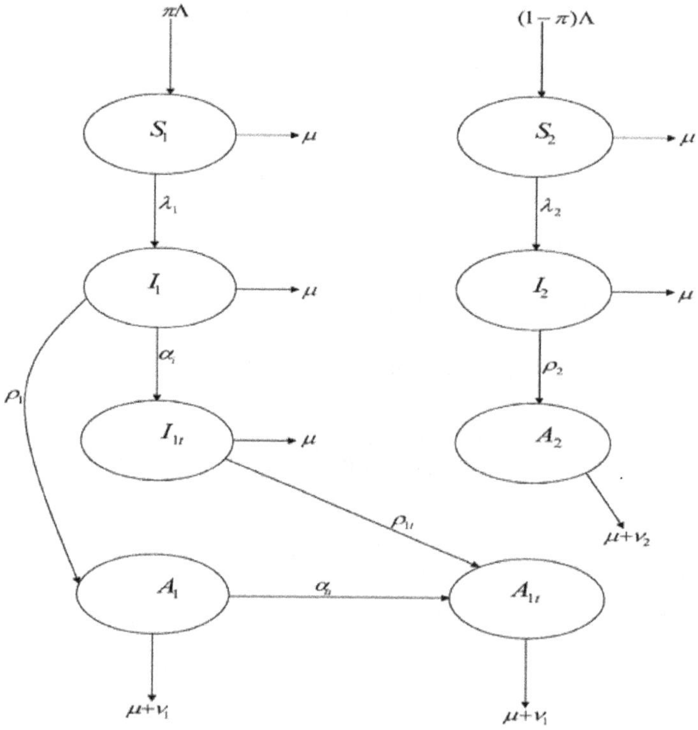

Figure 1. Structure of the model.

2.2. Disease-free equilibrium and stability analysis

The disease-free equilibrium of model system 4 is given by

$$\begin{aligned}\mathcal{E}^0 &= (S_1^0, I_1^0, I_{1_t}^0, A_1^0, A_{1_t}^0, S_2^0, I_1^0, A_2^0)\\ &= \left(\frac{\Lambda\pi}{\mu}, 0, 0, 0, 0, \frac{\Lambda(1-\pi)}{\mu}, 0, 0\right)\end{aligned} \tag{9}$$

Following Van den Driessche and Watmough (2002), we have $\mathcal{R}_{H_H}$ as the reproduction number of the model system 4. $\mathcal{R}_{H_H}$ which is defined as the number of secondary HIV infections produced by one infected individual in the presence of antiretroviral therapy in a completely susceptible population with some homeless people is given by

$$\begin{aligned}\mathcal{R}_{H_H} &= \frac{H_1 + \sqrt{H_1^2 - H_2}}{2g_1g_2g_3g_4g_5g_6}\\ H_1 &= g_5g_6p_{11}c_1\beta_1[\theta g_3\alpha_i(\rho_{1_t} + \phi_1 g_4) + g_2(\theta\alpha_a\rho_1 + (\rho_1 + \phi_1 g_3)g_4)] + g_1g_2g_3g_4p_{22}c_2\beta_2(\rho_2 + \phi_2 g_6),\\ H_2 &= 4g_1g_2g_3g_4g_5g_6c_1c_2\mathcal{P}[\theta g_3\alpha_i(\rho_{1_t} + \phi_1 g_4) + g_2(\theta\alpha_a\rho_1 + (\rho_1 + \phi_1 g_3)g_4)](\rho_2 + \phi_2 g_4)\beta_1\beta_2\end{aligned} \tag{10}$$

with

$$\begin{aligned}g_1 &= \rho_1 + \alpha_i + \mu,\ g_2 = \rho_{1_t} + \mu,\ g_3 = \alpha_a + \mu + v_1,\ g_4 = \mu + v_1,\ g_5 = \rho_2 + \mu,\\ g_6 &= \mu + v_2,\ g_7 = \mu + \rho_1,\ \mathcal{P} = p_{11}p_{22} - p_{12}p_{21} \text{ throughout the manuscript.}\end{aligned} \tag{11}$$

Local stability of the disease-free equilibrium is assured by Theorem 2 (Van den Driessche & Watmough, 2002)

THEOREM 1 *The disease-free equilibrium $\mathcal{E}^0$ for model system 4 is locally asymptotically stable if $\mathcal{R}_{H_H} < 1$ and unstable otherwise.*

Using a theorem from Castillo-Chavez, Feng, and Huang (2002), we show global stability when the reproduction number is less than unity.

THEOREM 2 *The disease-free equilibrium $\mathcal{E}^0$ for model system (4) is globally asymptotically stable provided $\mathcal{R}_{T_P} < 1$*

Proof Following the method by Castillo-Chavez et al. (2002), two conditions should be met to guarantee the global asymptotic stability of the disease-free equilibrium . We write system (4) as

$$\begin{aligned}\frac{dX}{dt} &= F(X(t), Z(t)),\\ \frac{dZ}{dt} &= G(X(t), Z(t)), \quad G(X(t), 0) = 0,\end{aligned} \tag{12}$$

where $X = (S_1, S_2)$ and $Z = (I_1, I_{1_t}, A_1, A_{1_t}, I_2, A_2)$ with $X \in \mathbb{R}_+^2$ representing the number of uninfected individuals and $Z \in \mathbb{R}_+^6$ representing the number of infected individuals. The disease-free equilibrium can now be written as $\mathcal{E}^0 = (\mathbf{X}_0, 0)$ where

$$\mathbf{X}_0 = \left(\frac{\Lambda\pi}{\mu}, \frac{\Lambda(1-\pi)}{\mu}\right) \tag{13}$$

Conditions H_1 and H_2 below must be met to guarantee global asymptotic stability.

$$\begin{aligned}&H_1{:}\quad \text{For } \frac{dX}{dt} = F(X, 0),\ X_0 \text{ is globally asymptotically stable,}\\ &H_2{:}\quad G(X, Z) = AZ - \widehat{G}(X, Z),\ \widehat{G}(X, Z) \geq 0 \text{ for } (X, Z) \in \mathcal{R}_+^6 \subset \Omega,\end{aligned} \tag{14}$$

where $A = D_Z G(X^*, 0)$ is an M-matrix (the off-diagonal elements of A are non-negative) and $\mathbb{R}^6_+$ is the region where the model makes biological sense.

In this case

$$F(X,0) = \begin{bmatrix} \pi\Lambda - \mu S_1 \\ (1-\pi)\Lambda - \mu S_2 \end{bmatrix}, \; G(X,Z) = AZ - \widehat{G}(X,Z),$$

$$A = \begin{bmatrix} -g_1 + p_{11}c_1\beta_1\phi_1 & p_{11}c_1\beta_1\theta\phi_1 & p_{11}c_1\beta_1 & p_{11}c_1\beta_1\theta & \frac{p_{12}c_1\beta_2\phi_2\pi}{1-\pi} & \frac{p_{12}c_1\beta_2\pi}{1-\pi} \\ \alpha_i & -g_2 & 0 & 0 & 0 & 0 \\ \rho_1 & 0 & -g_3 & 0 & 0 & 0 \\ 0 & \rho_{1_t} & \alpha_a & -g_4 & 0 & 0 \\ \frac{p_{21}c_2\beta_1\phi_1(1-\pi)}{\pi} & \frac{p_{21}c_2\beta_1\theta\phi_1(1-\pi)}{\pi} & \frac{p_{21}c_2\beta_1(1-\pi)}{\pi} & \frac{p_{21}c_2\beta_1\theta(1-\pi)}{\pi} & p_{22}c_2\beta_2\phi_2 - g_5 & p_{22}c_2\beta_2 \\ 0 & 0 & 0 & 0 & \rho_2 & -g_4 \end{bmatrix}. \tag{15}$$

It follows that

$$\widehat{G}(X,Z) = \begin{bmatrix} \widehat{G_1}(X,Z) \\ \widehat{G_2}(X,Z) \\ \widehat{G_3}(X,Z) \\ \widehat{G_4}(X,Z) \\ \widehat{G_5}(X,Z) \\ \widehat{G_6}(X,Z) \end{bmatrix} = \begin{bmatrix} p_{11}c_1\beta_1[A_1 + \phi_1 I_1 + \theta(A_{1_t} + \phi_1 I_{1_t})]\left(1 - \frac{S_1}{N_1}\right) + p_{12}c_1\beta_2[A_2 + \phi_2 I_2]\left(\frac{\pi}{1-\pi} - \frac{S_1}{N_2}\right) \\ 0 \\ 0 \\ 0 \\ p_{21}c_2\beta_1[A_1 + \phi_1 I_1 + \theta(A_{1_t} + \phi_1 I_{1_t})]\left(\frac{1-\pi}{\pi} - \frac{S_2}{N_1}\right) + p_{22}c_2\beta_2[A_2 + \phi_2 I_2]\left(1 - \frac{S_2}{N_2}\right) \\ 0 \end{bmatrix} \tag{16}$$

We need to show that $\widehat{G}_i(X,Z) \geq 0, \; (i = 1, 5)$. To do this, we prove by contradiction. Assume statements in 17 and 18 are true.

$$\frac{\pi}{1-\pi} < \frac{S_1}{N_2} \tag{17}$$

and

$$\frac{1-\pi}{\pi} < \frac{S_2}{N_1} \tag{18}$$

From 17 together we have

$$\frac{\pi}{1-\pi} < \frac{S_1}{N_2} \Rightarrow \frac{N_2}{S_1} < \frac{1-\pi}{\pi} \tag{19}$$

From 18 and 19 it follows that

$$\frac{N_2}{S_1} < \frac{1-\pi}{\pi} < \frac{S_2}{N_1} \Rightarrow N_1 N_2 < S_1 S_2 \tag{20}$$

a contradiction as statement 20 is not true. Thus, $N_1 N_2 \geq S_1 S_2$ implying that $\frac{S_1}{N_2} \leq \frac{\pi}{1-\pi}$ and $\frac{S_2}{N_1} \leq \frac{1-\pi}{\pi}$. Thus, $\widehat{G}(X,Z) \geq 0$. Therefore, the disease-free equilibrium is globally asymptotically stable. □

2.2.1. Analysis of the reproduction number $\mathcal{R}_{H_H}$

In the case that the like only have sexual contacts with the like, $p_{11} = p_{22} = 1,\ p_{12} = p_{21} = 0$, then $\mathcal{R}_{H_H} = \max\{\mathcal{R}_{H_1}, \mathcal{R}_{0_{H_2}}\}$ where

$$\mathcal{R}_{H_1} = \frac{\beta_1 c_1[\theta g_3 \alpha_i(\rho_{1_t} + \phi_1 g_4) + (\theta \alpha_a \rho_1 + (\rho_1 + \phi_1 g_3)g_4)g_2]}{g_1 g_2 g_3 g_4}$$
$$\mathcal{R}_{0_{H_1}} = \frac{\beta_1 c_1(\rho_1 + \phi_1 g_4)}{g_4 g_7},\ \mathcal{R}_{0_{H_2}} = \frac{\beta_2 c_2(\rho_2 + \phi_2 g_6)}{g_5 g_6} \tag{21}$$

These are: (i) the antiretroviral-induced reproduction number for HIV transmission when non-homeless people have sexual contacts only with the non-homeless ($\mathcal{R}_{H_1}$) and (ii) the basic reproduction number for HIV transmission when the homeless only have sexual contacts with the homeless ($\mathcal{R}_{0_{H_2}}$). In the absence of any intervention strategy $\mathcal{R}_{H_1}$ becomes $\mathcal{R}_{0_{H_1}} = \frac{\beta_1 c_1(\rho_1 + \phi_1 g_4)}{g_4 g_7}$, $\mathcal{R}_{0_{H_2}} = \frac{\beta_2 c_2(\rho_2 + \phi_2 g_6)}{g_5 g_6}$. This allows us to compare the various components of the two basic reproduction numbers $\mathcal{R}_{0_{H_2}}$ and $\mathcal{R}_{0_{H_1}}$ for different scenarios such as lack of entertainment, poor nutrition, co-infections with other STIs, and reduced socio-economic status.

In Table 1, the various components of homelessness are singly assessed: Case 1 suggests that homeless people are at a comparative disadvantage when it comes to entertainment, as lack of entertainment leaves sexual intercourse as the only form of entertainment, making the homeless people more prone to HIV infections than their non-homeless counterparts. Case 2 attempts to describe the effect of poor nutrition by capturing the increased probability of progressing to the AIDS stage among the homeless than among the non-homeless. Poor nutrition tends to compromise one's immunity, thus contributing to an increase in the progression to the AIDS stage of disease progression for the HIV infected. Due to the lack of entertainment, proper medical advice and treatment, co-infections with other STIs are common among the homeless, making them more prone to HIV infections than the non-homeless as noted in Case 3. Generally homeless people in the AIDS stage of disease progression experience higher AIDS-induced death rates than their counterparts. Results from Table 1 suggest that homeless people are at an increased disadvantage when it comes to HIV infection. All the signifiers of homelessness serve to exasperate the risk of HIV transmission between the homeless and the non-homeless. These results suggest that fighting homelessness should be addressed to alleviate the plight of the homeless HIV-infected people who find it impossible to access medical care. This is in total agreement with Wolitski et al. (2010) and Buchanan, Kee, Sadowski, and Garcia (2009) who showed a positive linkage between housing assistance for low-income people living with HIV/AIDS and better access to health care services. Results from Table 1 are further illustrated in Figure 2.

Table 1. Effects of lack of entertainment, poor nutrition, co-infections with other STIs, and reduced socio-economic status on HIV/ AIDS transmission dynamics. In each case, all parameters are equalized between the homeless and non-homeless, except those under the heading "Conditions"

Case	Description	Conditions	$\mathcal{R}_{0_{H_2}} - \mathcal{R}_{0_{H_1}}$
1	Lack of entertainment	$c_2 = b_1 c_1,\ b_1 > 1$	$\frac{(b_1 - 1)c_1\beta_1(\rho_1 + \phi_1 g_4)}{g_4 g_7} > 0$
2	Poor nutrition	$\rho_2 = b_2\rho_1, b_2 > 1$	$\frac{c_1\beta_1(b_2 - 1)(\mu - \phi_1 g_4)}{g_4 g_5 g_7} > 0,\ 0 < \phi_1 < \frac{\mu}{g_4}$
3	Co-infections with other STIs	$\beta_2 = b_3\beta_1, b_3 \geq 1$	$\frac{(b_3 - 1)c_1\beta_1(\rho_1 + \phi_1 g_4)}{g_4 g_7} > 0$
4	Reduced socio-economic status	$v_2 = b_4 v_1, b_4 \geq 1$	$\frac{(b_4 - 1)c_1\beta_1 v_1(\phi_1 v_1 - \rho_1)}{g_4 g_6 g_7} > 0,\ \frac{\rho_1}{v_1} \leq \phi_1 < 1$

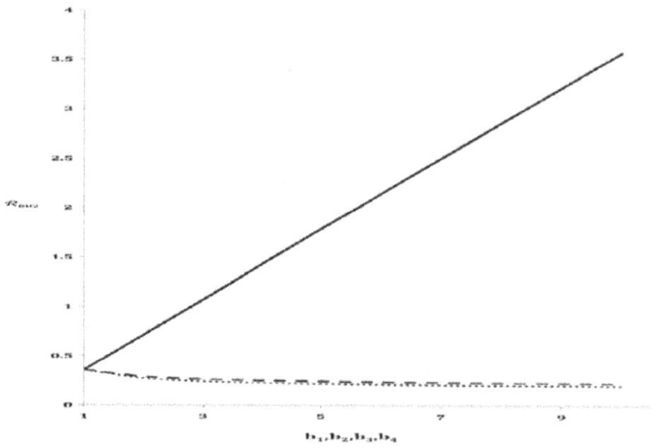

Figure 2. Effects of lack of entertainment or co-infections with other STIs (solid line), poor nutrition (dashed line), and reduced socio-economic status (dotted line) noting that $c_2 = b_1 c_1$, $\rho_2 = b_2 \rho_1$, $\beta_2 = b_3 \beta_1$, and $v_2 = b_4 v_1$. Parameter values are as in Table 2.

Figure 2 shows the effects of homelessness on HIV acquistion. It shows that lack of entertainment or co-infection with other STIs enhance the growth of the homeless-induced reproduction number. This suggests that lack of entertainment and/or co-infection with other STIs enhance the transmission of HIV.

2.3. Endemic equilibria

Model system 4 has three possible equilibria states: the homeless-only, the non-homeless-only, and the co-existence equilibrium. It is worth mentioning that the homeless and non-homeless-only equilibria states are simple HIV/AIDS endemic equilibria not worth discussing as so many researchers have analyzed them (see Bhunu, Garira, & Magombedze 2009 and references cited there in). The co-existence equilibrium occurs when there is sexual interaction between the homeless and non-homeless and is given by $\mathcal{E}^*$ where

$$\mathcal{E}^* = \left(S_1^*, I_1^*, I_{1_t}^*, A_1^*, A_{1_t}^*, S_2^*, I_2^*, A_2^*\right)$$

where

$$S_1^* = \frac{\pi\Lambda}{\lambda_1^* + \mu}, \; I_1^* = \frac{\pi\Lambda\lambda_1^*}{g_1(\lambda_1^* + \mu)}, \; I_{1_t}^* = \frac{\pi\alpha_i\Lambda\lambda_1^*}{g_2 g_1(\lambda_1^* + \mu)}, \; A_1^* = \frac{\pi\rho_1\Lambda\lambda_1^*}{g_3 g_1(\lambda_1^* + \mu)}, \tag{22}$$
$$A_{1_t}^* = \frac{\pi\Lambda\lambda_1^*[\rho_{1_t}\alpha_i g_3 + \rho_1\alpha_a g_2]}{g_1 g_2 g_3 g_4(\lambda_1^* + \mu)}, \; S_2^* = \frac{(1-\pi)\Lambda}{\lambda_2^* + \mu}, \; I_2^* = \frac{(1-\pi)\Lambda\lambda_2^*}{g_5(\lambda_2^* + \mu)}, \; A_2^* = \frac{(1-\pi)\rho_2\Lambda\lambda_2^*}{g_6 g_5(\lambda_2^* + \mu)}.$$

Substituting equation 22 into the equation for the forces of infection λ_1^* in equation 3 we obtain

$$\lambda_1^* = \frac{p_{11}c_1\beta_1\lambda_1^* D_1}{D_5 + \lambda_1^* D_2} + \frac{p_{12}c_1\beta_2 D_3\lambda_2^*}{D_6 + \lambda_2^* D_4},$$
$$D_1 = \theta g_3\alpha_i(\rho_{1_t} + \phi_1 g_4) + g_2(\theta\alpha_a\rho_1 + (\rho_1 + \phi_1 g_3)g_4), \; D_3 = \rho_2 + \phi_2 g_6, \; D_4 = g_6 + \rho_2 \tag{23}$$
$$D_2 = g_2 g_3 g_4 + \alpha_i g_3 g_4 + \rho_1 g_2 g_4 + \rho_{1_t}\alpha_i g_3 + \rho_1\alpha_a g_2, \; D_5 = g_1 g_2 g_3 g_4, \; A_6 = g_5 g_6$$

Substituting equation 22 into the equation for the forces of infection λ_1^* in equation 3 we obtain

$$\lambda_2^* = \frac{p_{22}c_2\beta_2 D_3\lambda_2^*}{D_6 + \lambda_2^* D_4} + \frac{p_{21}c_2\beta_1\lambda_1^* D_1}{D_5 + \lambda_1^* D_2} \tag{24}$$

Expressing λ_1^* as the subject of the formula in equations 23 and 24 and equating the two forms of λ_1^* to obtain an equation in λ_2^* which upon being solved we obtain $\lambda_2^* = 0$ corresponding to the disease-free equilibrium, $\lambda_2^* = \dfrac{B_1(\mathcal{R}_{H_H} - 1)}{B_2}$, $[B_1,\ B_2$ are positive and in terms of D_k's, p_{ij}, c_i, i, j=(1,2), k=(1, 2, 3, 4, 5, 6)] being the endemic equilibrium which clearly exists when $\mathcal{R}_{H_H} > 1$ and the other two complex roots which are going to be discarded since we are dealing with real populations. This result is summarized in Lemma 1.

LEMMA 1 *The endemic equilibrium $\mathcal{E}^*$ exists whenever* $\mathcal{R}_{H_H} > 1$

3. Numerical simulations

Unless otherwise stated, values used in the analysis and simulations are given in Table 2

The fourth-order Runge–Kutta numerical scheme coded in C + + programming language is used to graphically depict disease progression over time. Numerical simulations using a set of reasonable parameter values in Table 2 are carried out for illustrative purpose and to support the analytical results.

Figure 3 shows antiretroviral therapy for the non-homeless also has a beneficial effect on the homeless as noted by a decline in the number of new HIV cases among the homeless whenever levels of antiretroviral therapy are increased. This result suggests effective control of HIV lie in antiretroviral therapy for the non-homeless. This result further suggests control of HIV require strategies that remove people from the streets (homeless) into decent homes where they can be tracked and put on therapy. This result further suggests that care and control of HIV goes beyond provision of drugs. This is all in support of Wolitski et al. (2010) and Buchanan et al. (2009) who found positive linkage between provision of housing for the people living with HIV/AIDS and improved access to health care services.

Figure 4 is a graphical representation showing the effects of lack of entertainment with regard to cumulative new HIV cases in the homeless and non-homeless communities, respectively. It shows that lack of entertainment increases the rate of acquiring HIV as noted by increase in new HIV cases

Table 2. Default (baseline) model parameters used in the analysis and simulations. Here, ZIMSTAT stands for Zimbabwe National Statistics Agency (2012)

Definition	Symbol	Value (Range)	Source
Recruitment rate	Λ	0.029 yr^{-1}×4.2 × 10^6	ZIMSTAT
Natural mortality rate	μ	0.02 yr^{-1}	ZIMSTAT
Modification parameters	ϕ_1, ϕ_2, θ	0.125 (0.1–1)	Bhunu and Mushayabasa (2012)
Rate of being put on treatment	α_i, α_a	0.33(0.01–1) yr^{-1}	Bhunu et al., (2009)
Product of effective contact rate			
for HIV transmission and probability	$\beta_i c_j\ (i, j = 1, 2)$	0.125(0.011–0.95) yr^{-1}	Hyman, Li, and Stanley (1999)
of HIV transmission per sexual contact			
Rate of progression to AIDS	ρ_1, ρ_{1_t}	0.1(0.075–0.95) yr^{-1}	Bhunu et al. (2009)
AIDS-related death (non-homeless)	ν_1	0.333 (0.3–0.75) yr^{-1}	Bhunu et al. (2009)
Homegeneous mixing	p_{11}, p_{22}	0.67	Assume
Heterogeneous mixing	p_{12}, p_{21}	0.33	Assume
Modification parameter	$b_k\ (k = 1, \cdots, 4)$	1.125(≥1)	Assume

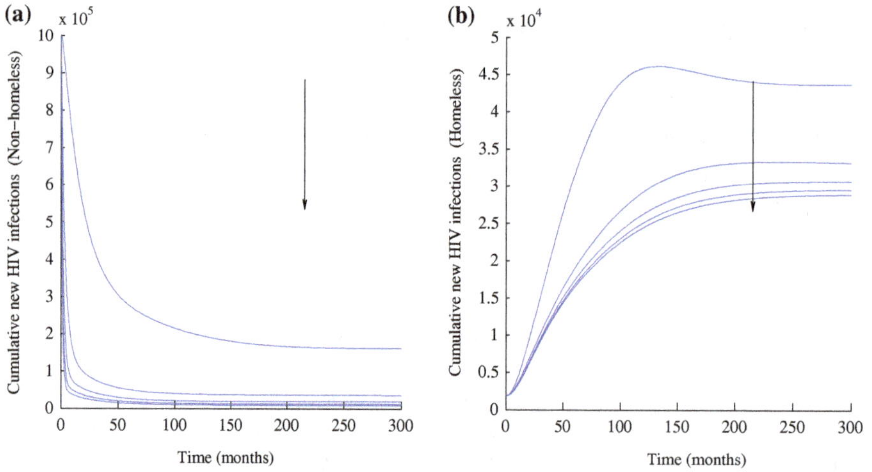

Figure 3. Simulations showing the possible effects of antiretroviral therapy using parameter values in Table 2. The direction of the arrow shows an increase in the levels of treatment.

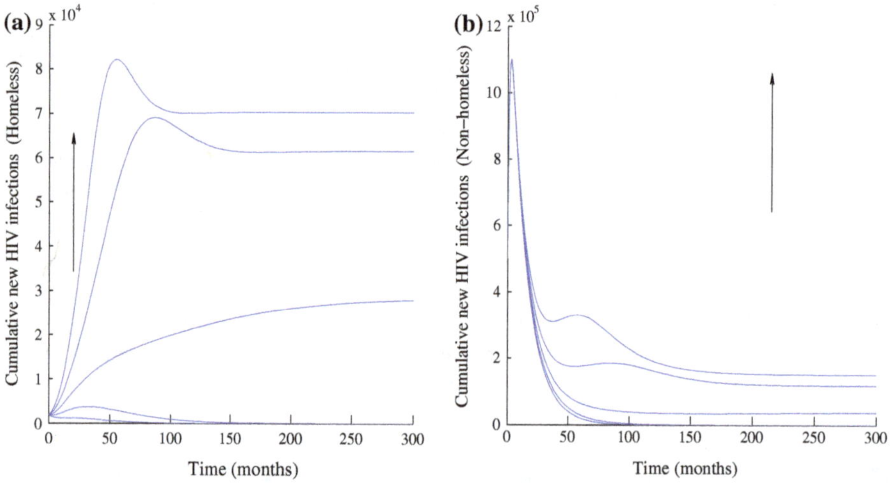

Figure 4. Simulations showing the possible effects of lacking entertainment using parameter values in Table 2 in the presence of antiretroviral therapy. The direction of the arrow shows a decrease in the levels of entertainment.

with decrease in entertainment levels. This greatly affects the homeless more than the non-homeless as noted in Figure 4(a). This suggest that provision of entertainment facilities which cuts across the homeless and non-homeless boundaries will play a crucial role in reducing the spread of HIV as some people resort to sexual intercourse due to the lack of entertainment.

4. Discussion

Homelessness expose an individual to a number of social, economic, and health risks and challenges. A mathematical model to explore the potential effects of homelessness on HIV/AIDS transmission dynamics is presented as a system of non-linear differential equations. The reproduction number of the model is computed and analyzed. The disease-free equilibrium is shown to be globally asymptotically stable whenever the reproduction number is less than unity. Numerical simulations show that antiretroviral therapy for the non-homeless also has a beneficial effect on the homeless as noted by a decline in the number of cumulative new HIV cases among the homeless with increase in antiretroviral therapy among the non-homeless. This is in total support of Wolitski et al. (2010) and Buchanan et al. (2009) who showed a positive linkage between provision of housing for those living with HIV/ AIDS and improved access to health care services which in turn improves the quality of their lives (living longer healthy productive lives). This result suggests the effect of creation of homes for the homeless where antiretroviral therapy and monitoring can be easily administered will have a more beneficial effect on the general population. Furthermore, numerical simulations show that the lack of entertainment also play a significant role in the spread of HIV and mostly so for the homeless. Thus, there is a need for policy-makers to provide recreational facilities which cuts across the homeless and non-homeless boundaries as some people resort to sex due to lack of other forms of entertainment. This on its own creates more social, economic, and health problems: unwanted pregnancies and increased risk of contracting HIV among others. Results from this theoretical study (comparison of homeless- and non-homeless-induced reproduction numbers) show that lack of entertainment, poor nutrition, and co-infections with other STIs worsen HIV/AIDS (HIV transmission and AIDS-related deaths are higher among the homeless than their counterparts). There are a number of limitations to our study, which should be acknowledged. We assumed that "homeless" versus "non-homeless" is a state assigned at birth, with no possibility of transfer later in life. While true to some degree in large parts of the world, this is obviously not always the case.

Acknowledgements
The author thanks the handling editor and reviewers for their insighful comments which improved the manuscript.

Funding
The author did not receive any funds from any organization to conduct this work.

Cover image
Source: Author.

Author details
C.P. Bhunu[1,2]
E-mail: cpbhunu@gmail.com; cp-b@hotmail.co.uk
[1] Department of Mathematics, University of Zimbabwe, Box MP 167 Mount Pleasant, Harare, Zimbabwe.
[2] Department of Mathematics and Applied Mathematics, University of Venda, Thohoyandou, South Africa.

Citation information
Cite this article as: Assessing the impact of homelessness on HIV/AIDS transmission dynamics, C.P. Bhunu, *Cogent Mathematics* (2015), 2: 1021602.

References

Adams, M. (2003). HIV and Homeless shelters: Policy and practice, American civil liberties union AIDS project. Retrieved June 20, 2003, from archive.aclu.org/issues/gay/hiv_homeless.html

Aidala, A. A., Lee, G., Abramson, D. M., Messeri, P., & Siegler, A. (2007). Housing need, housing assistance, and connection to HIV medical care. *AIDS & Behavior, 11*, 101–115.

Aidala, A., & Sumartojo, E. (2007). Why housing? *AIDS & Behavior, 11*(6, Suppl. 2), S1–S6.

Ajay, M., Brendan, O., & David, E. B. (2009). *Needle sharing and HIV transmission: A model with markets and purposive behavior.* NBER Working Papers. Retrieved from http://ideas.repec.org/p/nbr/nberwo/14823.html

Bekele, T., Rourke, S. B., Tucker, R., Greene, S., Sobota, M., Koornstra, J., ... Guenter, D. (2013). Direct and indirect effects of perceived social support on health-related quality of life in persons living with HIV/AIDS. *AIDS Care, 25*, 337–346.

Bhunu, C. P., Garira, W., & Magombedze, G. (2009). Mathematical analysis of a two strain HIV/AIDS model with antiretroviral treatment. *Acta Biotheoretica, 57*, 361–381.

Bhunu, C. P., & Mushayabasa, S. (2012). Assessing the impact of using antiretroviral drugs as pre-exposure vaccines. *HIV & AIDS Review, 11*, 42–48.

Bhunu, C. P., Mhlanga, A. N., & Mushayabasa, S. (2014). Exploring the impact of prostitution on HIV/ AIDS transmission. *International Scholarly Research Network, 2014*, 10.

Buchanan, D., Kee, R., Sadowski, L. S., & Garcia, D. (2009). The health impact of supportive housing for HIV-Positive homeless patients: A randomized controlled trial. *American Journal of Public Health, 99* (Suppl. 3), S675–S680.

Birkhoff, G., & Rota, G. C. (1982). *Ordinary differential equations.* Boston, MA: Ginn.

Castillo-Chavez, C., Feng, Z., & Huang, W. (2002). On the computation of R_0 and its role on global stability. In C. Castillo Chavez, S. Blower, P. Driessche, D. Kirschner, & A. Yakubu (Eds.), *Mathematical approaches for emerging and reemerging infectious diseases: Models, methods, and theory.* IMA, (Vol. 126, pp. 215–230) New York, NY: Springer.

Didenko, E., & Pankratz, N. (2007). Substance use: Pathways to homelessness? Or a way of adapting to street life? *Visions: BC's Mental Health and Addictions Journal, 4*, 9–10.

Hyman, J. M., Li, J., & Stanley, E. A. (1999). The differential infectivity and staged progression models for the transmission of HIV. *Mathematical Biosciences, 155*, 77–109.

Kidder, D., Wolitski, R., Campsmith, M., & Nakamura, G. (2007). Health status, health care use, medication use, and medication adherence in homeless and housed people living with HIV/AIDS. *American Journal of Public Health, 97*, 2238–2245.

Leaver, C. A., Bargh, G., Dunn, J. R., & Hwang, S. W. (2007). The effects of housing status on health-related outcomes in people living with HIV: A systematic review of the literature. *AIDS & Behavior, 11* (Suppl. 2), S85–S100.

National Coalition for the Homeless. (2009). HIV/AIDS and homelessness. Retrieved from www.nationalhomeless.org/factsheets/hiv.html

Robertson, M. J., Clark, R. A., Charlebois, E. D., Tulsky, J., Long, H. L., Bangsberg, D. R., & Moss, A. R. (2004). HIV seroprevalence among homeless and marginally housed adults in San Francisco. *American Journal of Public Health., 94*, 1207–1217.

Pedamallu, C. S., Ozdamar, L., Kropat, E., & Weber, G. W. (2012). A system dynamics model for intentional transmission of HIV/AIDS using cross impact analysis. *CEJOR, 20*, 319–336.

Shubert, V., & Bernstine, N. (2007). Moving from fact to policy: Housing is HIV prevention and health care. *AIDS & Behavior, 11* (Suppl. 2), S167–S171.

Song, J. (2003). AIDS housing of Washington. *AIDS housing Survey* (p. 1). Washington, DC: Springer.

Tomaszewski, E. P. (2011). HIV/ AIDS and homeless. National Association of Social Workers. Retrieved from www.socialworkers.org/practice/hiv_aids/spectrum

Van den Driesche, P., & Watmough, J. (2002). Reproduction numbers and sub-threshold endemic equilibria for the compartmental models of disease transmission. *Mathematical Biosciences, 180*, 29–48.

Wolitski, R. J., Kidder, D. P., Pals, S. L., Royal, S., Aidala, A., Stall, R., & Courtenay-Quirk, C. (2010). Randomized trial of the effects of housing assistance on the health and risk behaviors of homeless and unstably housed people living with HIV. *AIDS & Behavior, 14*, 493–503.

WHO. (2005). Guidelines for HIV diagnosis and monitoring of antiretroviral therapy. Geneva: WHO

Permissions

All chapters in this book were first published in CM, by Cogent OA; hereby published with permission under the Creative Commons Attribution License or equivalent. Every chapter published in this book has been scrutinized by our experts. Their significance has been extensively debated. The topics covered herein carry significant findings which will fuel the growth of the discipline. They may even be implemented as practical applications or may be referred to as a beginning point for another development.

The contributors of this book come from diverse backgrounds, making this book a truly international effort. This book will bring forth new frontiers with its revolutionizing research information and detailed analysis of the nascent developments around the world.

We would like to thank all the contributing authors for lending their expertise to make the book truly unique. They have played a crucial role in the development of this book. Without their invaluable contributions this book wouldn't have been possible. They have made vital efforts to compile up to date information on the varied aspects of this subject to make this book a valuable addition to the collection of many professionals and students.

This book was conceptualized with the vision of imparting up-to-date information and advanced data in this field. To ensure the same, a matchless editorial board was set up. Every individual on the board went through rigorous rounds of assessment to prove their worth. After which they invested a large part of their time researching and compiling the most relevant data for our readers.

The editorial board has been involved in producing this book since its inception. They have spent rigorous hours researching and exploring the diverse topics which have resulted in the successful publishing of this book. They have passed on their knowledge of decades through this book. To expedite this challenging task, the publisher supported the team at every step. A small team of assistant editors was also appointed to further simplify the editing procedure and attain best results for the readers.

Apart from the editorial board, the designing team has also invested a significant amount of their time in understanding the subject and creating the most relevant covers. They scrutinized every image to scout for the most suitable representation of the subject and create an appropriate cover for the book.

The publishing team has been an ardent support to the editorial, designing and production team. Their endless efforts to recruit the best for this project, has resulted in the accomplishment of this book. They are a veteran in the field of academics and their pool of knowledge is as vast as their experience in printing. Their expertise and guidance has proved useful at every step. Their uncompromising quality standards have made this book an exceptional effort. Their encouragement from time to time has been an inspiration for everyone.

The publisher and the editorial board hope that this book will prove to be a valuable piece of knowledge for researchers, students, practitioners and scholars across the globe.

List of Contributors

Godfrey Chagwiza, Brian C. Jones, Senelani D. Hove-Musekwa
Department of Applied Mathematics, National University of Science & Technology, P.O. Box AC939, Ascot, Bulawayo, Zimbabwe

Sobona Mtisi
Overseas Development Institute, London, SE1 7JD, UK

Uğur Yucel
Faculty of Arts and Sciences, Department of Mathematics, Pamukkale University, Denizli 20070, Turkey

Bapurao C. Dhage and Shyam B. Dhage
Kasubai, Gurukul Colony, Ahmedpur, 413 515 Maharashtra, India

Jürgen Geiser
The Institute of Theoretical Electrical Engineering, Ruhr University of Bochum, Universitätsstrasse 150, D-44801 Bochum, Germany

Ioannis K. Argyros
Department of Mathematical Sciences, Cameron University, Lawton, OK 73505, USA

Santhosh George
Department of Mathematical and Computational Sciences, National Institute of Technology Karnataka, Mangaluru, Karnataka 757 025, India

Masato Shinjo, Masashi Iwasaki, Akiko Fukuda, Emiko Ishiwata, Yusaku Yamamoto and Yoshimasa Nakamura
Graduate School of Informatics, Kyoto University, Kyoto, 606-8501, Japan

Hiroshi Shiraishi
Department of Mathematics, Keio University, Yokohama, Kanagawa, Japan

Elias Munapo
Graduate School of Business and Leadership, University of KwaZulu-Natal, Westville Campus, Durban, South Africa

Santosh Kumar
Department of Mathematics and Statistics, University of Melbourne, Parkville, Australia

Xu-Yang Guo and Bo-Yan Xi
College of Mathematics, Inner Mongolia University for Nationalities, Tongliao City 028043, China

Feng Qi
Department of Mathematics, College of Science, Tianjin Polytechnic University, Tianjin City 300387, China

Xin Chen, Zhu Zhang
Department of Industrial Engineering, Southern Illinois University Edwardsville, Edwardsville, IL 62026-1805, USA

Ryan Fries
Department of Civil Engineering, Southern Illinois University Edwardsville, Edwardsville, IL 62026-1800, USA

Muhammad Hanif
Department of Mathematics and Statistics, PMAS-Arid Agriculture University, Rawalpindi, Pakistan

Waseem Ghazi Alshanti
Department of Mathematics, University of Hail, Hail, Kingdom of Saudi Arabia

Ilaria Usville, Cristina Paola Viola and Ezio Venturino
Dipartimento di Matematica "Giuseppe Peano", Università diTorino, Torino, Italy

C.P. Bhunu
Department of Mathematics, University of Zimbabwe, BoxMP 167 Mount Pleasant, Harare, Zimbabwe
Department of Mathematics and Applied Mathematics,University of Venda, Thohoyandou, South Africa

Index

www.ingramcontent.com/pod-product-compliance
Lightning Source LLC
Chambersburg PA
CBHW082014190326
41458CB00010B/3186
* 9 7 8 1 6 3 2 3 8 5 5 8 1 *